Human Life Before Birth

Second Edition

Human Life Before Birth

Second Edition

Frank J. Dye, PhD
Western Connecticut State University
Danbury, Connecticut

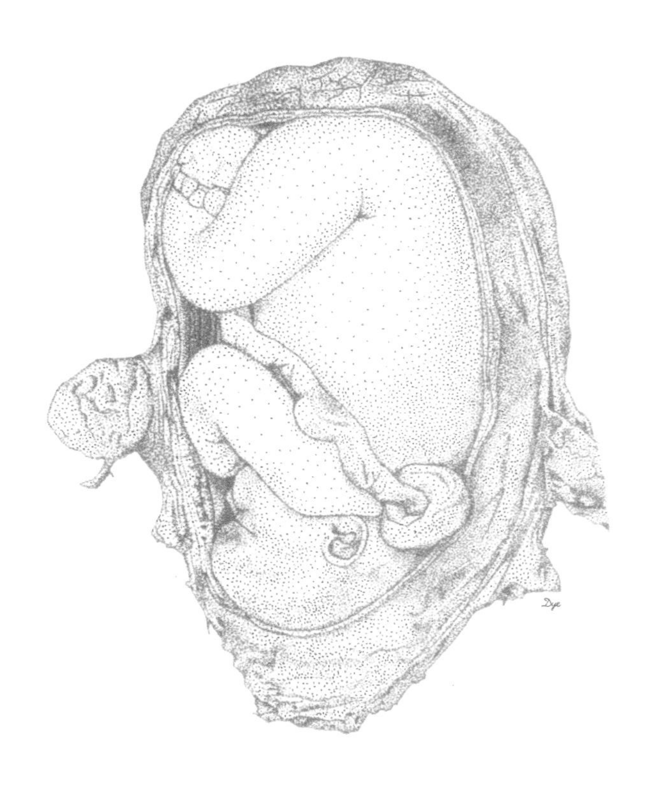

CRC Press is an imprint of the
Taylor & Francis Group, an **informa** business

Front cover graphic: By John Dye. Figure 9-8. Week 28, third trimester.

CRC Press
Taylor & Francis Group
6000 Broken Sound Parkway NW, Suite 300
Boca Raton, FL 33487-2742

Printed on acid-free paper

International Standard Book Number-13: 978-0-8153-5524-3 (Paperback)
978-0-367-13630-7 (Hardback)

Library of Congress Cataloging-in-Publication Data

Names: Dye, Frank J. (Frank John), 1942- author.
Title: Human life before birth / Frank J. Dye.
Description: Second edition. | Boca Raton : Taylor & Francis, 2019. |
Includes bibliographical references and index.
Identifiers: LCCN 2018044512| ISBN 9780815355243 (paperback : alk. paper) |
ISBN 9780367136307 (hardback : alk. paper) | ISBN 9781351130264 (epub) |
ISBN 9781351130257 (mobi/kindle) | ISBN 9781351130288 (general) | ISBN
9781351130271 (pdf)
Subjects: LCSH: Embryology, Human. | Fetus--Growth. | Fetus--Physiology.
Classification: LCC QM601 .D94 2019 | DDC 612.6/4--dc23
LC record available at https://lccn.loc.gov/2018044512

Visit the Taylor & Francis Web site at
http://www.taylorandfrancis.com

and the CRC Press Web site at
http://www.crcpress.com

To the memory of my mother Lucia Concetta, who always said, 'be happy,' and to the memory of my father John Lester, who was a self-educated naturalist on the streets of New York City.

Contents

PART II
Some details of human development

PART III
Society and human development

Preface

For me, biology is the most fascinating undertaking to pursue, and human development is the most fascinating part of biology. As a child, I had the good fortune to receive a gift of a microscope and to have access to the shores of Putnam Lake, New York. The shores of the lake provided endless encounters with living creatures—often in the early stages of their development. The wonder I encountered there has stayed with me all the days of my life.

As an undergraduate at Western Connecticut State University (then Danbury State College), I chose to study biology and chemistry. It is a choice I have never regretted. By the time I applied to graduate school, I wanted to learn more about cells and embryos. My professors at Fordham University and a National Institutes of Health predoctoral fellowship provided me with the opportunity to do just that. Dr. James Forbes introduced me to the wonderful details of descriptive embryology and Dr. Alexander Wolsky introduced me to the insights of comparative and experimental embryology. Later, as a National Institute of Dental Research postdoctoral fellow, I came to learn about the formation of organs in the laboratory of Dr. Ed Kollar at the University of Connecticut Health Center, and, as a visiting fellow in the laboratory of Dr. Clement Markert at Yale University, I was able to learn how to manipulate early mammalian embryos.

Nothing gives me more pleasure than helping someone discover the fascination of development. Teaching courses in embryology and development has provided me with the means to do this. As I reflect on my own study of development, I am confident that the best approach to understanding development is to first understand some of the details of *normal* developmental biology. In my teaching and in my writing, I have tried to keep the horse before the cart; I have tried to remember where the student beginning a study of development is and to remember that normal development is at least as fascinating as is the experimental manipulation of it.

Therefore, I have written this book for readers with diverse backgrounds: intelligent laypersons, college undergraduates not majoring in science but interested in human prenatal development, biology majors who want to learn more about human development than is found in their general biology textbooks, and nursing students. Biology majors who plan to take an upper-level undergraduate course in embryology or developmental biology will appreciate this book's focus on human development rather than animal development. Nursing students will find that the treatment of human development has sufficient detail to provide a good foundation for their studies in obstetrical nursing. Also, students heading for medical school, who have never studied human development, will likely find this book a smooth introduction to the subject, which should make medical school embryology much more fascinating.

Acknowledgments

During the twenty years between the first and second editions of this book, great strides have been made in biology and medicine, which have given us deeper insights into understanding development, in general, and human development, in particular. This second edition incorporates these recent insights. My interactions with students and colleagues have motivated me to write this second edition. Western Connecticut State University, my undergraduate alma mater, provided the 'implantation site' for the development of my career as a teaching biologist, and nurtured the early work on the first edition of this book with a sabbatical leave. My son John, who created some of the illustrations herein, often provided much needed comic relief. Some of the physicians at Danbury Hospital graciously provided illustrations. The editors and staff members with whom I have worked at Taylor & Francis, S. R. Crumly, Jennifer Blaise, Laurie Oknowsky, Marsha Hecht, and Pam Tagg, provided just the right mix of encouragement and daunting tasks to keep the momentum going. Without all these people, this book would not exist. To the extent that this book makes a contribution to understanding human development is due in large measure to their efforts. Any errors are my responsibility, and I hope that readers will call them to my attention (dye@snet.net).

Author

Frank J. Dye is a native New Yorker who lives with his wife in Danbury, Connecticut. He is Emeritus Professor of Biology at Western Connecticut State University. He has been published in cell and developmental biology journals, is the author of two dictionaries of developmental biology, and, together with his son, an illustrator, has published two field guides to the ferns and wildflowers of the Westside Nature Preserve at WCSU.

PART I

An overview of human development

Biology has no story to tell that is more fascinating than that of human development. A tiny fragment of matter, in the appropriate environmental context, is able to develop into a human baby. Birth, like the reality of stars in the heavens, is so common that we almost take it for granted. Yet for this event to occur, 266 days of cellular activity must pass with a precision that would impress the most critical of engineers.

To read and understand this text properly, the knowledge of basic terms, concepts, and anatomic landmarks is needed (Chapter 1). And to understand development, it is necessary to know something about cells, the interacting entities that result in development (Chapter 2) and genetics (Chapter 4). Two types of cell division (Chapter 3) are necessary for reproduction (Chapter 5) and development to occur. In addition to cell division, cells must also differentiate and become specialized members of the cellular society that is the human body. All of cell differentiation is fascinating, but none more so than gametogenesis (Chapter 6), the process by which seemingly ordinary cells give rise to the tiny motile sperm and the large expecting egg.

An enormous amount of human activity is concerned with fertilization—how to encourage it, how to prevent it, and humankind's preoccupation with sexual behavior, originally designed to culminate in fertilization (Chapter 7). Fertilization launches human development.

In addition to continuing cell division and cell differentiation during embryogenesis (Chapter 8) and development of the fetus (Chapter 9), we see a third dramatic component of development, morphogenesis—the origin of form. Human development does not involve the growth of preformed parts but, rather, the gradual emergence of eyes and ears and arms and legs.

Development would not progress beyond the first week if it were not for the placenta, the most unique of human organs, and the umbilical cord (Chapter 10), through which the developing human communicates with the placenta. As the embryo and fetus grow, they need this life-support system. When studying the development of the baby, one must be mindful of the context in which this development occurs—the pregnant woman (Chapter 11). Then comes the day we celebrate annually for the rest of our lives, our "birth" day. We have no direct recollection of this event, but it is life-altering not only for us, but especially for our parents. Moreover, as dramatic as birth is, humankind, at least until the advent of fertility drugs, has been particularly enthralled by multiple births.

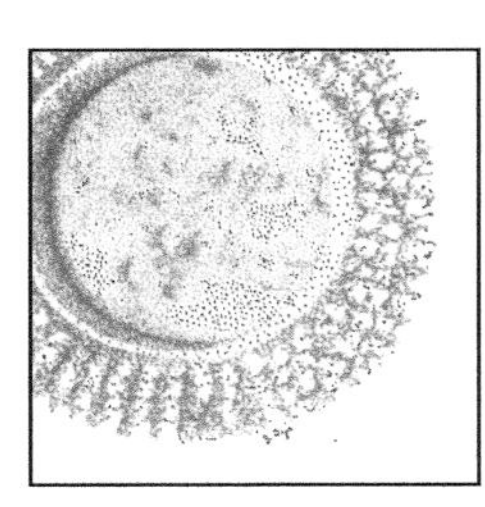

CHAPTER ONE

Before you begin

CHAPTER OBJECTIVES

After studying this chapter, you should be able to:

1. Correctly use and interpret the following anatomic landmarks and terms: anatomic position, superior, inferior, cephalic, caudal, rostral, medial, lateral, proximal, distal, dorsum, dorsal, venter, ventral.
2. Understand symmetry and explain the difference between radial symmetry and bilateral symmetry.
3. Distinguish between planes and sections, and explain how the sagittal, frontal, and transverse planes divide the human body.
4. Explain the differences between the developmental timelines used by obstetricians and embryologists.

Before beginning our discussion of human development in the ensuing chapters, we need to learn some basic concepts that will guide the more detailed descriptions that follow later in the book. Our discussion will take place in this order.

- First, we will take up the "geographic" descriptions essential to any study of anatomy—adult or developmental.
- We will move to the fourth dimension—time—and briefly discuss the terminology used to discuss the progression of pregnancy.
- Finally, we will talk about **comparative embryology**. Although this book is primarily concerned with human development, many aspects of human development are particularly interesting when compared with that of other animals.

Anatomic descriptions

A large part of the content of human developmental biology is **developmental anatomy**. Anatomy is never static, but it is especially dynamic before birth. When you study the development of the human face during the second half of the embryonic period (the fourth through eighth weeks), you will be especially impressed by **dynamic anatomy (morphogenesis)**.

To communicate about anatomy, it is necessary to have a common vocabulary shared by teacher and student; otherwise, the information transferred is not precise. This is especially true when the anatomy described is constantly changing. The terms explained here are frequently used in the descriptions in this text.

Anatomic landmarks

The **anatomic position** for a human is standing straight and upright, with arms at the sides and palms facing forward (Figure 1.1). The head

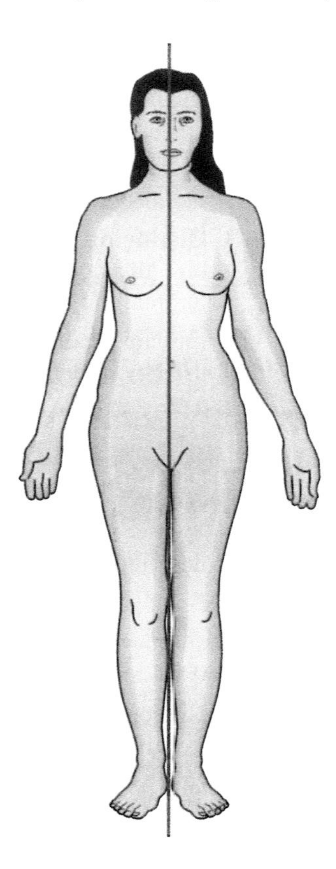

FIGURE 1.1 Anatomic position and bilateral symmetry. The body is erect, the arms are at the sides of the body, and the head, palms, and feet face forward. The line is in the median plane of the body, which is also the plane of bilateral symmetry.

is at the **superior** (top) end of the body, and other parts of the body are relatively **inferior** (below) to it. Because a dog walks on four legs, its head is anterior (front), and its tail is posterior (back). To avoid the confusion caused by four-legged and two-legged animals, we will use the term **cephalic** to refer to the head and the term **caudal** to refer to the tail of both. If we use these terms as adverbs rather than adjectives, they become **cephalad** (toward the head end) and **caudad** (toward the tail end). We may even refer to something on the head that is closer to the very end of the head than to some other reference point as **rostral**.

Let's imagine a line running down the middle of a person in the anatomic position, that is, down the middle of the nose, chin, "belly button," and so on. This line would pass down the midline of the person and would be **medial**. The part of the body on either side of the midline is **lateral** to the midline. If we move toward the midline, we move **mediad**; if we move to either side of it, we move **laterad**.

Two other terms of relative position are important. Compared with the wrist, the elbow is **proximal** (closer) to the shoulder, whereas the wrist is **distal** to (farther from) the shoulder (Figure 1.2).

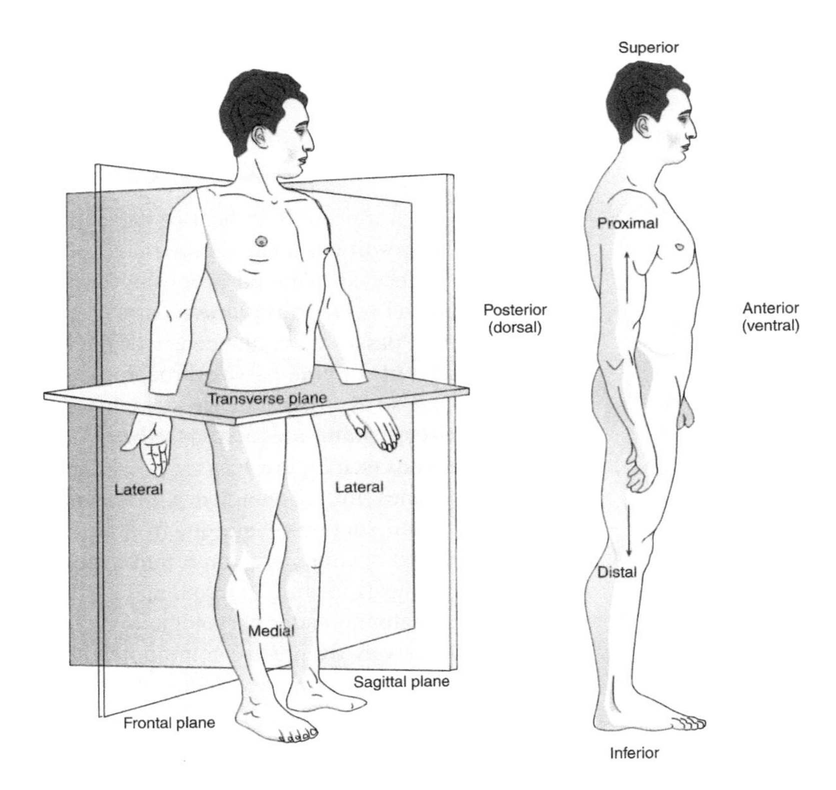

FIGURE 1.2 Relative positions and planes of the body. Relative positions are reference points. Note that the elbow is closer (proximal) to the shoulder than is the wrist (distal to the shoulder). Three important planes–transverse, sagittal, and frontal–divide the body into anatomically important parts (see text).

When you pet a cat or dog on its back, you are stroking its **dorsum**, almost always referred to as the **dorsal** side. The opposite side of the animal is its **venter** (belly), or **ventral** side (see Figure 1.2).

Symmetry

An important feature of an animal's anatomy is its **symmetry**. A starfish with arms radiating from its center is said to have **radial symmetry** and obviously has a body shape unlike ours. We humans have **bilateral symmetry**, which means that only one plane divides us into two mirror images—the plane that passes through the medial line referred to above (see Figure 1.1). Not all organisms exhibit bilateral symmetry, but we exhibit it. For example, some organisms have a spherical shape and therefore can be divided into two mirror-image halves by passing any plane through them that includes their diameter. This is like cutting an apple into two halves. Only one plane will accomplish this for organisms with bilateral symmetry.

Imagine a plane passing through your body; it will divide you into two mirror-image halves only if it passes down the center of your body (**sagittal plane**), separating your left half from your right half. Don't worry that some of your internal organs (e.g., heart, liver, spleen) will not be divided into mirror-image halves; it is the concept that is important.

Planes and sections

An example of a plane is a sheet of paper. With no thickness to speak of, it has a two-dimensional geometric shape. There are three kinds of planes, and they are defined according to how they pass through the human body: (1) a **sagittal plane** separates the left and right sides of the body when it passes down the center of the body, including the medial line; (2) a **frontal plane** passes through the body at right angles to the sagittal plane so as to separate the dorsal from the ventral surfaces; and (3) a **transverse plane** goes across the body, separating the cephalic from the caudal ends (see Figure 1.2).

When planes are not imaginary, but actually cut through the body, they are called **sections**. Sagittal, frontal, and transverse (also called cross-sections) sections are important sections that allow anatomists and embryologists to study internal and surface anatomy. In this text as well as in anatomic and embryologic texts in general, the terminology is used extensively for precise communication. Nevertheless, "everyday language" is also used here to allow the student to make the transition from "plain English" to technical jargon.

The developmental timeline

In addition to anatomic descriptions used to describe the developing human form, some terminology is necessary to describe the progression from fertilization to birth. Most of you are familiar with the term "trimester," which divides the 9-month pregnancy into three equal periods. Most of you also know that a developing baby is referred to as a fetus. However, the different terminologies used by embryologists (scientists who study development) and obstetricians (physicians who care for pregnant women) may be unfamiliar. Embryologists, studying development from fertilization onward, place the beginning of development at what seems to be a very logical point in time—fertilization.

Obstetricians use a different set of terminology that is equally logical, given the role that obstetricians play in women's lives. Obstetricians use the first day of a woman's last menstrual cycle as the reference point because most women know the date of this event, but may not know the date on which conception (fertilization) took place.

The result of the difference in terminology is that the embryologist considers human development to take 38 weeks, whereas the obstetrician considers it to last 40 weeks. Where appropriate, this book uses the embryologist's terminology to give the student an idea of how long developmental processes take to occur. However, the obstetrician's terminology is also important, especially the familiar division into trimesters. And the first trimester is a very critical period of life, encompassing the entire embryonic period and the early fetal period.

One prominent embryologist uses the term "preembryo." Grobstein, in his book, *Science and the Unborn*, uses the term "preembryo" to refer to the conceptus during the first 2 weeks after fertilization and considers an embryo to begin its existence with the appearance of the primitive streak. Such usage of terminology notwithstanding and with the realization that a significant portion of the early conceptus does go into the formation of extraembryonic structures, we will continue to use the term "embryo" for the entire 8-week period *preceding* the appearance of the fetus. Incidentally, the **conceptus** is the product of conception (fertilization), and initially consists of the embryo and extraembryonic membranes, which then become the fetus and extraembryonic membranes.

Comparative embryology

Although this book is concerned with exploring human development with the college undergraduate, it is nonetheless instructive to consider from time to time the field of comparative embryology, which compares the developmental patterns of different animals, including humans. For example, this approach explains the consequence of the human egg's small amount of yolk—the mother provides the embryo with necessary nutrients. Comparative embryology also looks at the way in which the human egg divides, explaining why humans begin their development differently from the way other animals do.

Study questions

True or *false*. If the answer is false, supply the term that makes the statement true:

1. **T F** The head is at the superior end of the body.
2. **T F** Caudal refers to the head end of the body.
3. **T F** When we move from the midline to the side of the body, we move mediad.
4. **T F** Compared with the wrist, the elbow is proximal to the shoulder.
5. **T F** Humans exhibit bilateral symmetry.
6. **T F** The frontal plane separates the left and right sides of the body.
7. **T F** The sagittal plane separates the dorsal from the ventral surface of the body.

Critical thinking

1. Why did obstetricians and embryologists adopt different methods of timing human development?
2. Distinguish among preembryo, embryo, and fetus.

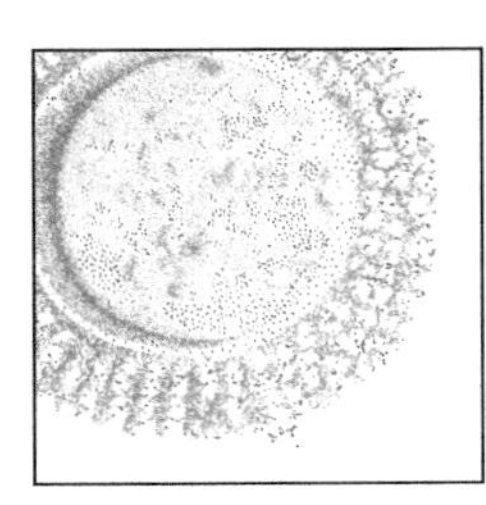

CHAPTER TWO

Cells

CHAPTER OBJECTIVES

After studying this chapter, you should be able to:

1. Explain the significance of the cell in development.
2. Draw a simple diagram of a cell showing the nucleus, cytoplasm, and plasma membrane, and understand the general role of each.
3. Describe each of the following parts of the nucleus and give their functions in cells: chromosomes, chromatin, genes, nuclear membrane, nuclear pores.
4. Distinguish among these parts of the cytoplasm: organelles, cytoskeleton, and cytoplasmic matrix.
5. Give the *general* cellular roles of the following cytoplasmic organelles and a *specific* function of each in human development: mitochondria, Golgi bodies, lysosomes, and centrioles.

General background

Early in the 20th century, E. B. Wilson pointed out that the key to every biological problem must be sought in the cell: every organism begins (or remains) as a single cell. More recently, Alberts and others hypothesized that every organism had its origin in a single cell that lived billions of years ago!

The concept of the cell is one of the organizing concepts of modern biology. Most of us realize that we are each composed of cells, but this idea is not really so old. Credit for this concept goes to the biologists, Schwann and Schleiden, who made their major contributions during the early years of the 19th century.

Any consideration of human reproduction, genetics, and development should begin with the cell. The initial objective of human reproduction is the bringing together of two cells—the spermatozoon and the egg. Human heredity depends on the passage of genetic information through these cells to the fertilized egg (**zygote**), which is the first cell of the next generation. The expression of this information results in the interaction of the many cells produced during development and finally results in a new human being.

The interactions of cells during development include the formation of junctions between cells. These junctions allow the cells to directly communicate with each other and to separate the organism's body from its surroundings. Later, the cells develop the capacity to communicate at longer distances through hormones and nerves, as well as through these direct contacts. For the development of a multicellular organism, cells must interact over long distances. Otherwise, how would the brain, for example, develop control over distant parts of the body?

Anatomy of the cell

We first consider the three general parts of the cell: (1) the nucleus, (2) the cytoplasm, and (3) the plasma membrane (Figure 2.1). After we develop a fundamental understanding of the cell, we will consider some of its more specific parts.

Nucleus

The **nucleus** is found within the cytoplasm of the cell and generally has a spherical shape (see Figure 2.1). Within the nucleus of the cell, we find the **chromosomes**. They carry the genes, which are made up of deoxyribonucleic acid (DNA) (for an explanation of DNA, see Chapter 4). The chromosomes are best observed during cell division. When the cell is not dividing, the chromosomes are too dispersed within the nucleus to be visible as separate entities. In this dispersed state, the chromosomes are collectively referred to as **chromatin**. Although the chromosomes are less visible when the cell is not dividing, it is during this time that the genes are most active.

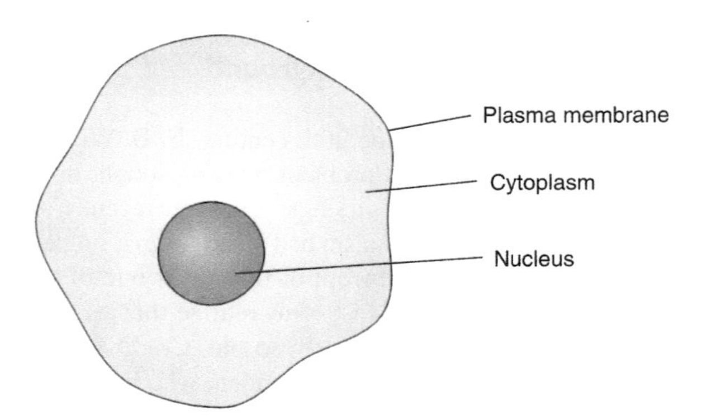

FIGURE 2.1 General parts of the cell. Cells of the body come in different sizes and shapes. Generally, cells have three basic parts: a plasma membrane, cytoplasm, and a nucleus.

At its periphery, the nucleus has a **nuclear membrane**, which separates the nuclear contents from the cytoplasm. If we examined the nuclear membrane with an electron microscope, we would see a two-layered "envelope" consisting of two membranes separated by a fluid-filled space. We would also see a large number of **nuclear pores** perforating the nuclear membrane.

The nucleus is central to the process of fertilization. In fact, fertilization is complete when the nucleus of a spermatozoon (the male sex cell) fuses with the nucleus of an egg.

Cytoplasm

Physically, the **cytoplasm** is the material that fills the space between the nuclear membrane and the plasma membrane. Initially, we may consider the cytoplasm to have three general components: (1) various **organelles**, which are subcellular components that carry out specific functions; (2) the **cytoskeleton**, made up of threadlike components (microtubules, microfilaments, and intermediate filaments), which control many aspects of cell shape and movement; and (3) the **cytoplasmic matrix**, an ill-defined component pervading the cytoplasm, in which the organelles and cytoskeleton are found.

Plasma membrane

At the edge of the cytoplasm, we find the **plasma membrane**, which is the boundary membrane of the cell. Composed of protein, lipid, and carbohydrate, its responsibility is to control the passage of materials into and out of the cell. Consequently, the plasma membrane is responsible for the interior of the cell having a composition different from that of its surroundings. This is a vital function because a cell that has the same composition as its environment is a dead cell!

Cytoplasmic organelles

Having discussed some of the important functions of the nucleus and the plasma membrane, let us single out a few organelles found in the cytoplasm. We will examine the roles of four organelles in human reproduction, heredity, and development: mitochondria, Golgi bodies, lysosomes, and centrioles.

Mitochondria

Individual cells contain numerous **mitochondria**, which provide energy for the cell in the form of the "high-energy" chemical adenosine triphosphate (ATP) (Figure 2.2). These mitochondria are barely visible at the light microscope level. During the formation of spermatozoa, the mitochondria of the cell become incorporated into a region just behind the head of the spermatozoon; they provide ATP for movement of the sperm. Although the male mitochondria enter the egg during fertilization, they

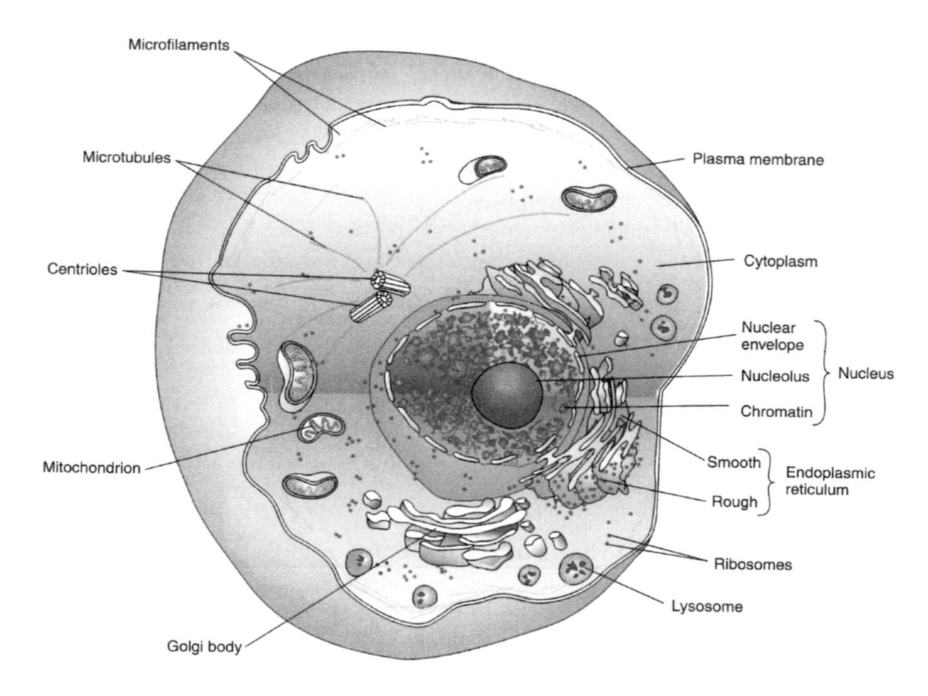

FIGURE 2.2 Detailed parts of the cell. The cytoplasm of the cell contains numerous structures, including the organelles. The organelles that play important roles in our story are mitochondria, Golgi bodies, centrioles, and lysosomes.

apparently disintegrate and play no further role in development. The egg accumulates a large number of its own mitochondria, which *do* function in development after fertilization. We inherit our mitochondria from our mothers, not our fathers. Although most of the genes of the cell are in its nucleus, mitochondria contain DNA with genes; mitochondrial DNA has been closely scrutinized in recent years to gain insight into human evolution.

Some organelles have ordinary "housekeeping" functions that are the same for a great variety of cell types, and some have specialized functions in specific cell types. For example, in a wide range of cells, mitochondria provide ATP as a source of energy. On the other hand, mitochondria in brown fat cells are designed to produce heat rather than ATP.

Golgi bodies

A human cell may contain one or more **Golgi bodies** (see Figure 2.2). The *general* functions of Golgi bodies are to chemically modify glycoproteins (chemical complexes of carbohydrate and protein) produced by the cell, act as a "traffic director" for macromolecules produced by the cell, and package cell products (e.g., enzymes) to be secreted from the cell. In addition, Golgi bodies play a role in the formation of organelles called lysosomes, which we will consider in the next section. In the formation of spermatozoa and eggs, Golgi bodies have the *specialized* functions of producing acrosomes (organelles found in sperm) and cortical granules (organelles found in eggs), respectively. These two highly specialized organelles, which are peculiar to gametes (sex cells), play important roles in fertilization.

Lysosomes

Human cells generally contain a number of **lysosomes**, which may be thought of as tiny "bags" of digestive enzymes. Enzymes are protein molecules that function as catalysts: they dramatically increase the rates of chemical reactions that allow for the existence of cells. Digestive enzymes thus catalyze (increase the rate of) digestion or breakdown of various molecules in cells. The membrane (or "bag") at the periphery of each lysosome keeps the digestive enzymes from destroying the cell that contains these enzymes. Lysosomes play important roles in development—let's consider three examples.

Cell death It may come as a surprise, but cell death is a normal part of development. In fact, cell death occurs normally even before we are born. A newborn female infant has fewer eggs (germ cells) than she had as a fetus. Similarly, all of us had fewer nerve cells as newborn infants than we had as fetuses. A particularly interesting example of cell death in action has to do with the formation of the hand and its fingers. Limb buds appear during the 4th week of development and subsequently give rise to arms and legs, hands and feet, and fingers and toes. When the precursor of the hand first appears during arm development, it has the form of a flattened paddle. The fingers must be "sculpted" from this paddle by cell death (Figure 2.3).

Cell death is not a haphazard process, it is genetically programmed (called **apoptosis**)—the cells die "on cue." Lysosomal enzymes bring about the demise of the cells that are to die, both in human limb buds and in the tadpole's disappearing tail, for example.

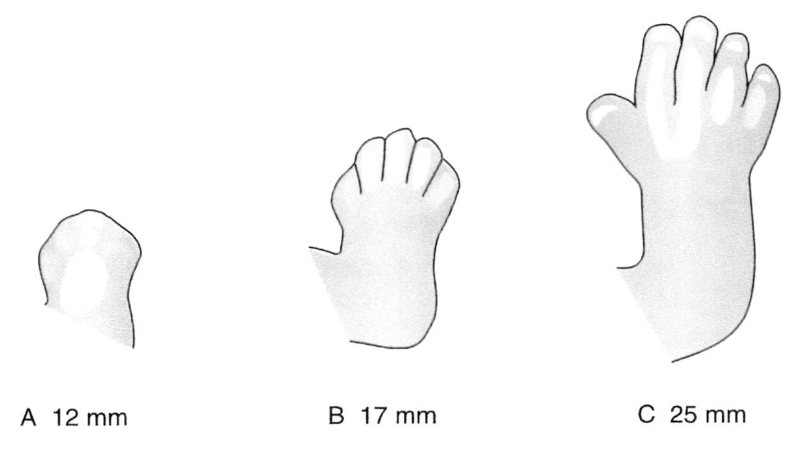

FIGURE 2.3 Development of the human hand. When the prospective hand appears, it has the shape of a flattened paddle. The fingers are "sculpted" out of this paddle by the death of columns of cells.

Fertilization Another developmentally important example of the special roles of lysosomes occurs during fertilization. As the spermatozoon approaches the egg, it faces two barriers that must be breached before fertilization can occur. The spermatozoon must first cross the corona radiata (a layer of cells) and then the zona pellucida (a noncellular layer). We will consider the passage of the spermatozoon through these layers in detail during our consideration of fertilization (see Chapter 5). At this point, however, it is important to know that the spermatozoon can pass through these barriers by means of its acrosome, which is actually a modified lysosome. Thus, the digestive enzymes of the acrosome play a key role in fertilization.

Abnormal roles Lysosomes play other important roles during human development, including abnormal roles leading to disease. If any of the enzymes normally found in lysosomes are missing, certain molecules cannot be broken down. These undigested molecules accumulate in the lysosomes and lead to **lysosomal storage diseases** (accumulation of cell debris caused by missing or defective lysosomal enzymes), to which nerve cells are particularly sensitive. Table 2.1 lists some of these diseases, the missing enzymes, the molecules not digested, and the tragic human consequences.

Table 2.1 The glycosphingolipidoses

Disorder	Enzyme deficiency	Storage product(s)	Effect on health
Fabry's disease	α-galactosidase	Trihexosylceramide	Kidney, heart, and central nervous system (CNS) disorders; death in 50s
Farber's disease	Ceramidase	Ceramide	Growth retardation; CNS disorders; death by age 3
Niemann–Pick disease, types A, B, C, D, E	Sphingomyelinase	Sphingomyelin	Visceral organ damage; severe psychomotor degeneration
Sandhoff's disease	Hexosaminidase B	GM_2 ganglioside	Tay–Sachs features
Tay–Sachs disease	Hexosaminidase A	GM_2 ganglioside	Psychomotor deterioration from 5 months; blindness, deafness, seizures; death by age 5

Source: Adapted from Stine, GJ. *The New Human Genetics*. Dubuque, IA: WC Brown Publishers, 1989.

Centrioles

Under an electron microscope, the **centriole** appears as a hollow, cylindrical body that is made up of microtubules. Centrioles occur in pairs, with the two centrioles oriented at right angles to each other (see Figure 2.2). The centriole has many roles. Its housekeeping roles include assisting in cell division and, as a component of the cell center, organizing at least part of the cytoskeleton in the nondividing cell. Here, however, we are most interested in the centriole's specialized role in development.

During the formation of the spermatozoon (see Figure 6.4), in a process called **spermiogenesis**, the centriolar pair assumes a position just behind the forming head of the spermatozoon. Since the two centrioles are oriented at right angles, one is in a position with its long axis parallel to that of the forming spermatozoon and organizes the microtubules that make up the core of the forming tail. The activity of these microtubules allows the spermatozoon to reach the surface of the egg. The other centriole persists as such and after fertilization plays the pivotal role in organizing the spindle (an organelle that separates chromosomes during cell division; the reader won't learn about this until the next chapter) required for the first division of the fertilized egg.

A little biochemistry

Cells possess a family of molecular motors made of proteins called dynein. The microtubules (made up of tubulin protein) that make up the core (axoneme) of the sperm tail (flagellum) have dynein protein motors associated with them. These dynein molecules are able to breakdown ATP to release energy. This released energy in the sperm flagellum causes microtubules to *tend* to slide past each other. However, other protein molecules associated with the microtubules prevent the microtubules from sliding and instead cause them to bend. It is the bending of the microtubules of the flagellum that causes the flagellum to "beat" and the sperm to swim.

Study questions

1. Name the three general parts of the cell.
2. Where in the cell are chromosomes found?
3. Explain the relationship between chromosomes and chromatin.
4. What two parts of the cell are separated by the nuclear membrane?
5. Where in the cell is the cytoplasm found?
6. What part of the cell controls the passage of materials into and out of the cell?
7. What is the energetically important product of mitochondria?
8. What is the role of mitochondria in brown fat cells?
9. Give three general functions of Golgi bodies.
10. Lysosomes may be thought of as tiny "bags" of what? Give two of their important roles in development.
11. Undigested materials accumulating in lysosomes give rise to what category of diseases?

12. Give two "housekeeping" roles of centrosomes.
13. During spermiogenesis, what does a centriole organize?
14. Dynein protein motors, associated with the axoneme of spermatozoa, break down ATP, releasing energy required for the sperm tail (flagellum) to do what?

Critical thinking

1. In what sense would the key to every developmental problem be found in the cell?
2. Multicellular organisms such as humans must have cells that do what exquisitely well? Explain.

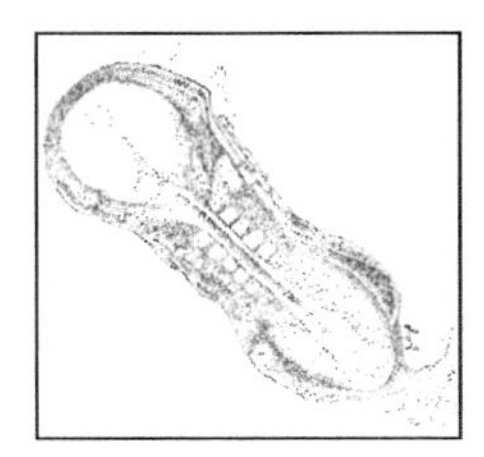

CHAPTER THREE

Cell division

> **CHAPTER OBJECTIVES**
>
> After studying this chapter, you should be able to:
>
> 1. Explain why cell division occurs and why there are two types—mitosis and meiosis.
> 2. Describe the cell cycle and explain the role of mitosis in the cell cycle.
> 3. Describe the five phases of mitosis, explaining what happens to the chromosomes and the role of the spindle.
> 4. Discuss the significant events in meiosis: reduction of chromosome number, random segregation of chromosomes, and crossing over.
> 5. Discuss the role of genes and how meiosis affects natural selection.

Role and types of cell division

An understanding of cell division is central to the study of development. Many genetic and birth defects are due to abnormalities in cell division, and most people studying human development are interested in these defects. Cell division is one of three fundamental processes in development, along with cell differentiation (also called cytodifferentiation) and morphogenesis (discussed in Chapter 8). If cell division is abnormal, especially during certain critical periods of development, so too will be the development it accompanies.

Two types of cell division occur in humans: mitotic cell division and meiotic cell division. **Mitosis** (or karyokinesis) refers specifically to division of the cell nucleus. **Meiosis** refers to two consecutive nuclear divisions. In addition, these two types of cell division are fundamentally different in their effect on chromosome number. Mitosis is conservative in that the two resulting daughter cells each have the same number of chromosomes as in the original (mother) cell. Meiosis results in four daughter cells, with each having half the number of chromosomes as in the mother cell. The term **cytokinesis** is reserved for division of the cytoplasm of the cell.

Mitosis

The purpose of mitosis is to increase cell numbers in a conservative fashion. That is, when a cell divides, the two resulting daughter cells are genetically identical. Each daughter cell must receive an exact copy of the mother cell's DNA, the chemical that makes up the genes. To understand how this is accomplished, you need to realize that mitosis is part of the cell cycle (the life cycle of the cell) (Figure 3.1).

The cell cycle

In a population of dividing (called cycling) cells, each cell passes through four consecutive stages: $\mathbf{G_1}$, **S**, $\mathbf{G_2}$, and **M** (gap 1, DNA synthesis, gap 2, and mitosis). During G_1, preparations are made for **DNA synthesis**; this is the stage that usually occupies the longest portion of the cell cycle. Also, differences in the duration of cell cycles between different kinds of cells can usually be attributed to differences in the duration

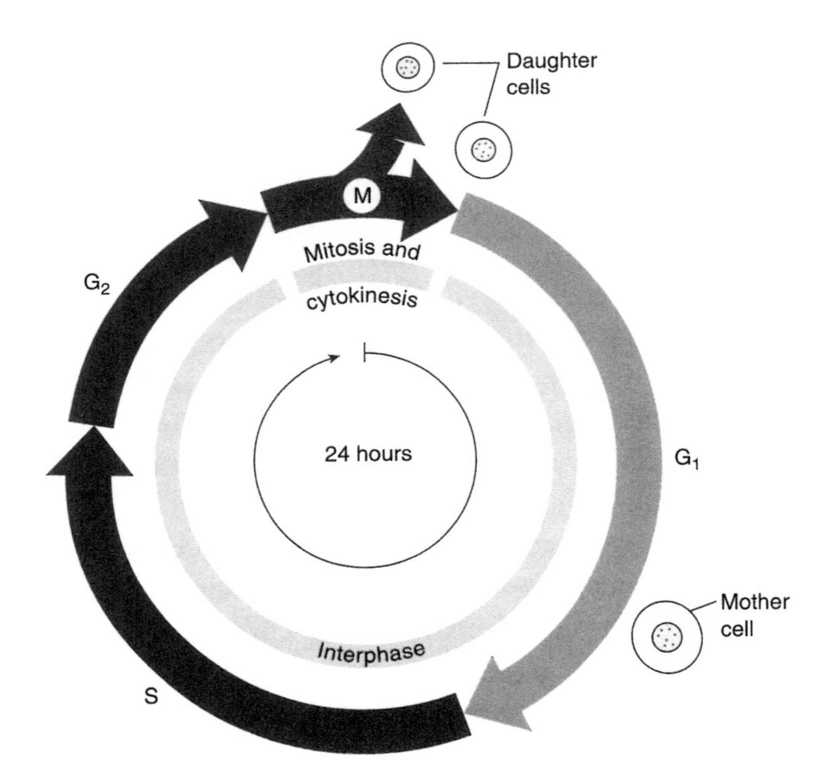

FIGURE 3.1 The cell cycle. Between the birth of a cell by cell division and its own division to form two daughter cells, the cell passes through the four stages that make up the cell cycle: G_1, S, G_2, and M.

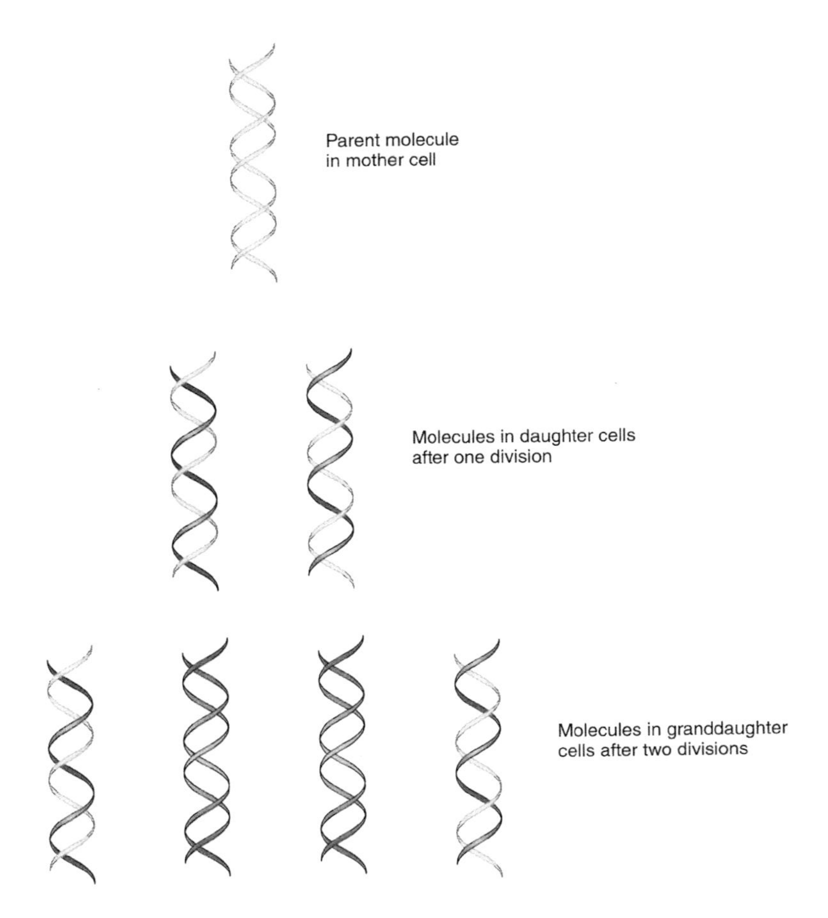

FIGURE 3.2 Semiconservative replication of DNA. When the double-stranded DNA molecule replicates, its two strands (light gray) separate and each strand acts as a template, directing the formation of a new strand (dark gray). As a result of this mode of replication, each of the two identical new DNA molecules has one strand of the original molecule.

of G_1. DNA undergoes a process called **semiconservative replication** during the S stage. The two complementary strands of the mother cell's DNA separate, and each one acts as a **template** (a guide; see details in Chapter 4) for a new complementary strand. Barring mutations, two complete identical copies of the original DNA are formed (Figure 3.2).

G_2 is the stage in which preparations are made for mitosis. Collectively, G_1, S, and G_2 are referred to as **interphase**—that period of time when the cell is *not* dividing. Although the cell is not dividing during interphase, a necessary prerequisite for mitosis occurs: the replication (duplication) of the DNA just described.

After interphase is complete, the function of mitosis is to distribute the replicated DNA (genes) to the two daughter cells that will result from cell division. For this distribution to happen, two kinds of characters must appear on the stage of mitosis: chromosomes and the spindle.

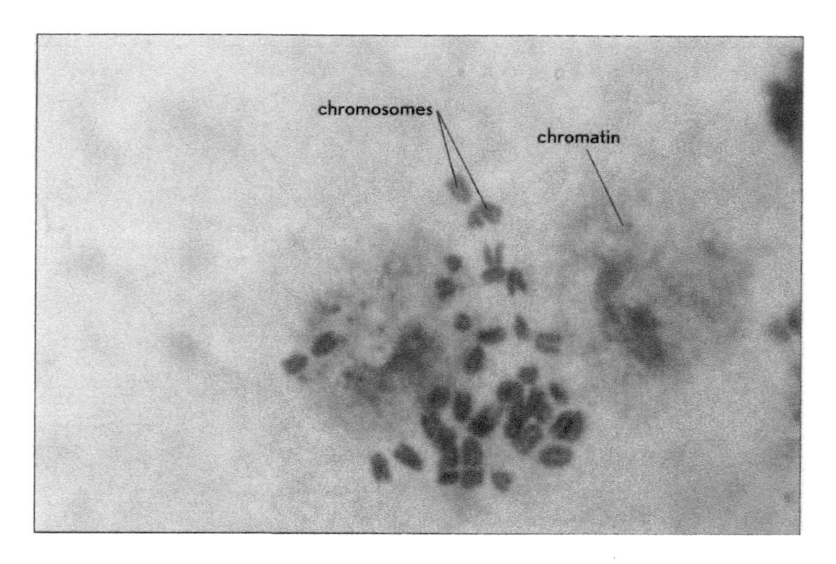

FIGURE 3.3 Chromosomes and chromatin. Chromosomes (colored bodies) are visible when cells divide because they are then condensed entities. When cells are not dividing, during interphase, the same chromosomes are in a dispersed state called chromatin. (Photo by the author.)

Chromosomes

Chromosomes exist throughout the cell cycle, but they are so dispersed during interphase that they are not visible through the ordinary light microscope (Figure 3.3). In this highly dispersed (**chromatin**) state, it is virtually impossible for the replicated chromosomes to precisely distribute their DNA and genes to the daughter cells. Consequently, before the chromosomes are distributed to the daughter cells, they undergo a condensation process. When condensed by proteins called **histones**, the chromosomes are visible through the microscope. In fact, the word "chromosome" means "colored body," referring to the fact that chromosomes are easily stained with various dyes.

We can think of each chromosome as being composed of two identical lengthwise halves called **chromatids** (Figure 3.4). The chromatids are held together at a constricted region called the **centromere**. On either side of the centromere, there protrudes an arm. If the centromere is located midway between the two ends of the chromosome, it has arms of equal length and is called a **metacentric chromosome** (if the centromere is not exactly at the midpoint, the resulting chromosome will be **submetacentric**). Conversely, if the centromere is closer to one end of the chromosome than the other, it is an **acrocentric chromosome**. The short and long arms of an acrocentric chromosome are referred to by the letters p (petite) and q (the next letter in the alphabet), respectively. We humans have metacentric, submetacentric, and acrocentric chromosomes (Figure 3.5).

Many human birth defects are due to abnormalities of chromosome number, chromosome structure, or a combination of the two. (We will return to this topic in greater detail later in our discussion of birth defects in Chapter 19.)

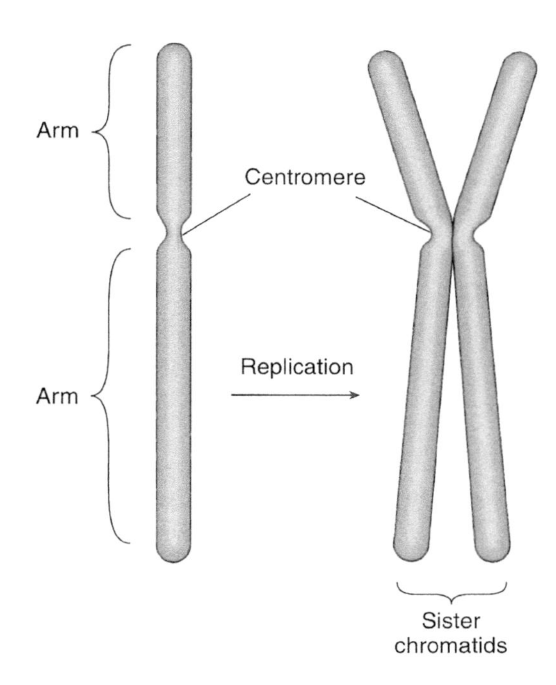

FIGURE 3.4 Chromosome structure. At the level of the light microscope, chromosomes have a constriction along their length called the centromere (primary constriction). The part of the chromosome on each side of the centromere is called the arm of the chromosome. All normal human chromosomes have two arms.

Before S stage, each chromosome is composed of a single chromatid, but after S stage, each chromosome is composed of two chromatids. Chromosomes consist of more than just DNA; they also contain proteins. When DNA is replicated during S stage, so too are the chromatids that contain it.

During mitosis (M stage), the sister chromatids of each chromosome are separated from each other, carrying their identical cargo of DNA (genes) with them. Although mitosis is a continuous process, we study it by dividing it into five phases: **prophase, prometaphase, metaphase, anaphase**, and **telophase** (Figure 3.6). If you observe a film of mitosis, you will see why it is referred to as the "dance of the chromosomes." For the choreography of this "dance" to be orderly and accomplish its purpose, the cell needs a spindle (see later in this chapter).

Our interest in chromosomes goes beyond studying their shapes and tracing their movements through the five phases of mitosis. We are primarily interested in chromosomes because they bear our genetic heritage—the genes that we inherited from our parents and that we may pass on to our children—and so a discussion of genes follows.

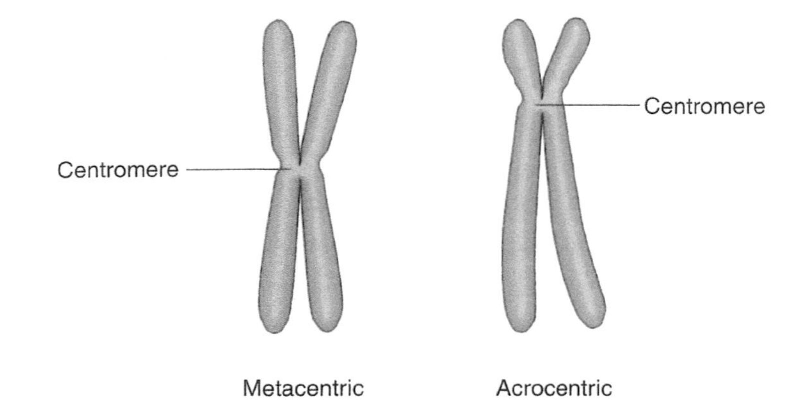

FIGURE 3.5 Types of chromosomes. Normal human chromosomes are of three types—metacentric, submetacentric, and acrocentric—depending on the position of the centromere. If the centromere is midway between the two ends of the chromosome, it is a metacentric chromosome with arms of equal length. If the centromere is close to one end of the chromosome, it is an acrocentric chromosome with arms of unequal length. Submetacentric chromosomes have centromeres that are nearly midway between the two ends of the chromosome.

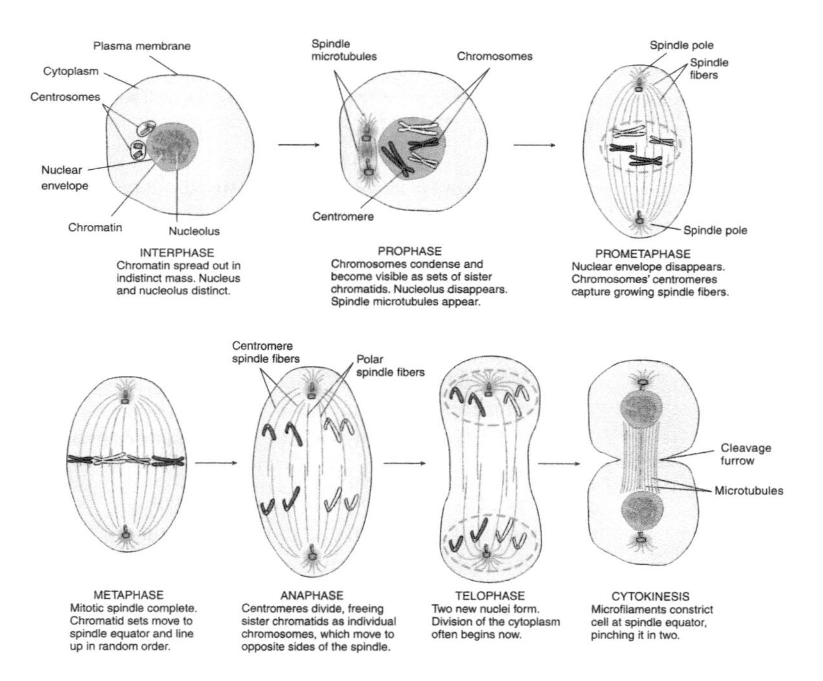

FIGURE 3.6 The five phases of mitosis. Interphase is not part of mitosis but represents a nondividing cell. The criteria for distinguishing among these five phases have to do with the appearance of the chromosomes through the light microscope. However, also note the changes in the centrosomes and the spindle.

Genes For each kind of gene, there is a position—a **locus**—where it is found along the length of a chromosome. The locus for most genes is found on two chromosomes—one inherited from our mother and one from our father. That is, we have two copies (**alleles**) of most genes. These copies may be identical or different. If they are different, they may both be expressed, or one copy (**dominant**) may mask the other (**recessive**).

In addition, the chromosomes we receive from our mother and father are not equivalent—they may contain different copies of a given gene. That is, we may have received an Rh+ allele from our mother and an Rh– allele from our father, but this is not what is referred to here. In our discussion of formation of the embryo (embryogenesis), we consider three kinds of information coming to the zygote: genes, maternally derived substances, and differential chemical modification of parental genes. We are here referring to the latter (**genomic imprinting**). Moreover, the first kind of information is genetic information, whereas the latter two kinds are referred to as **epigenetic information**—beyond genetic information. The maternally derived epigenetic substances are deposited in the cytoplasm of the developing egg during **oogenesis** (formation of the egg), leading to **maternal inheritance**, and the chemical modification of the genes involves a modification of the already formed DNA. Evidence indicates that this chemical modification is in the form of **methylation** (addition of what chemists call a methyl group) of the respective DNA of the chromosomes, with chromosomes from a female parent having a different pattern of methyl groups from chromosomes coming from the male parent.

Spindle

Like the chromosomes, the **spindle** plays an important role in mitosis. Unlike the chromosomes, the spindle is a short-lived cellular structure that comes and goes according to the cell's needs. Two microtubule-organizing centers (MTOCs; classically called **centrosomes**) make the spindle during the earliest stage of mitosis (**prophase**). The spindle is composed of a protein called **tubulin** that has been organized into **microtubules**. As the chromosomes condense during **prophase**, the centrosome divides, and the two resulting daughter centrosomes move to opposite ends of the nucleus, organizing the spindle as they go (see Figure 3.6). When the nuclear membrane disappears at the end of prophase, initiating **prometaphase**, the condensed chromosomes are captured by **spindle fibers** (bundles of microtubules).

The spindle derives its name from its long narrow shape: its two ends are its **poles**, and the region midway between the two poles is referred to as its **equator**. Chromosomes attach to the spindle fibers by means of their centromeres, with the arms of the chromosomes flailing in different directions. Upon alignment of the chromosomes on the spindle equator, prometaphase is over and the cell enters **metaphase** of mitosis (Figure 3.7).

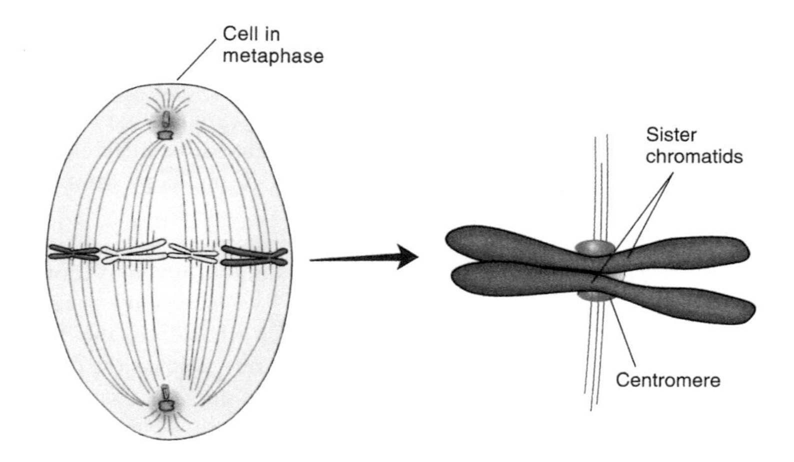

FIGURE 3.7 Chromosomes and chromatids. As cells progress through the cell cycle, the number of chromatids in each of their chromosomes changes. From anaphase to G_1, each chromosome consists of a single chromatid. During S stage of interphase, the chromosomes are in transition, replicating from one chromatid to two chromatids. From G_2 to metaphase, each chromosome consists of two chromatids.

Metaphase ends and **anaphase** begins when the sister chromatids of each chromosome begin their poleward movement. As soon as the sister chromatids separate, they are referred to as chromosomes. During anaphase, chromosomes each consist of only a single chromatid. Note that chromosomes consist of two chromatids from the beginning of G_2 until the end of metaphase, and they consist of one chromatid from the beginning of anaphase to the end of G_1 of the next cell cycle. The chromosomes make the transition from one to two chromatids during S stage (Figure 3.8).

When the chromosomes reach the poles of the spindle and stop moving, anaphase is over and **telophase** begins. During telophase, the spindle disappears and a new nuclear membrane forms around each of the two groups of daughter chromosomes. Gradually, the chromosomes disperse into a chromatin condition, and the process of **cytokinesis** divides the cytoplasm of the cell into two approximately equal halves.

As a result of mitosis, we have two **daughter cells**, each about half the size of the mother cell. Barring mutations, each of these smaller cells is genetically identical to the mother cell. Because each of the three cells (mother cell and two daughter cells) has the same number of chromosomes, mitosis is said to be "conservative" as far as chromosome number is concerned.

Meiosis

Meiosis is the mechanism that reduces the number of chromosomes to offset the doubling of the chromosome number that occurs at fertilization.

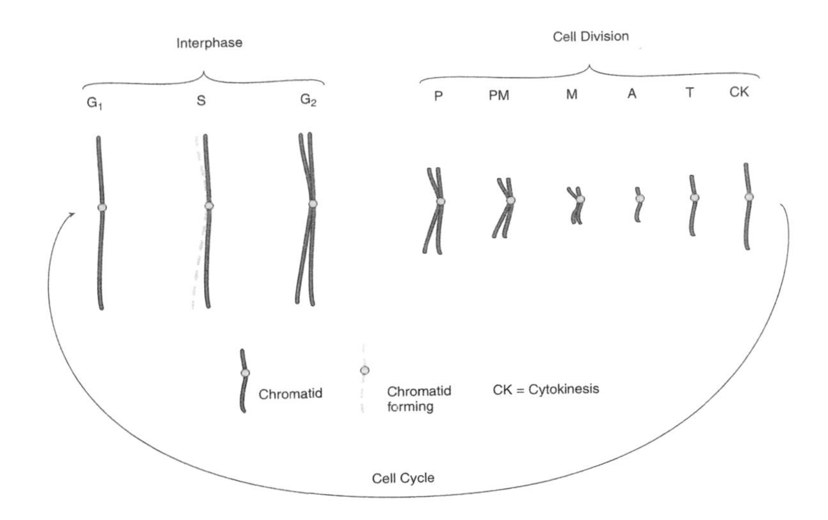

FIGURE 3.8 What the chromosome consists of depends on the stage of the cell cycle. From G_2 through M (metaphase), it consists of two chromatids; from A (anaphase) through G_1, it consists of one chromatid. During S (DNA synthesis), the chromosome replicates from one to two chromatids.

It is a necessary part of sexual reproduction. A good understanding of mitosis provides the foundation for understanding meiosis, but meiosis is a more complicated process. Unlike mitosis, meiosis is *not* conservative as to chromosome number; rather, during meiosis, the number of chromosomes is reduced to half the number in the original mother cell. Because meiosis entails two consecutive nuclear divisions, the process results in four daughter cells rather than two.

The process of meiosis is divided into **meiosis I** and **meiosis II**, each further divided into phases (prophase I, metaphase I, prophase II, metaphase II, and so on) with names familiar from mitosis (Figure 3.9). Unlike mitosis, which is widespread in both time and space during the human life cycle, meiosis is restricted in time and very restricted in space; only the **gonads** (ovaries and testes) contain cells undergoing meiosis (Figure 3.10). More will be said about this in Chapter 6.

We can think of meiosis as having three significant aspects: (1) reduction of chromosome number, (2) random segregation of chromosomes, and (3) crossing over.

Reduction of the number of chromosomes

Each species has a specific number of chromosomes called the **diploid number** and expressed as $2n$. We humans have a $2n$ of 46 chromosomes. When a human cell divides by mitosis, each of the two daughter cells receives a set of 46 chromosomes carrying identical genes. However, when a human cell divides by meiosis, each of the four daughter cells receives a set of 23 chromosomes, and the sets do not carry identical genes. Twenty-three is the **haploid number** (n) of chromosomes for humans.

At the time of fertilization, two cells (an egg and a spermatozoon, each with the haploid number of chromosomes) fuse, resulting in the

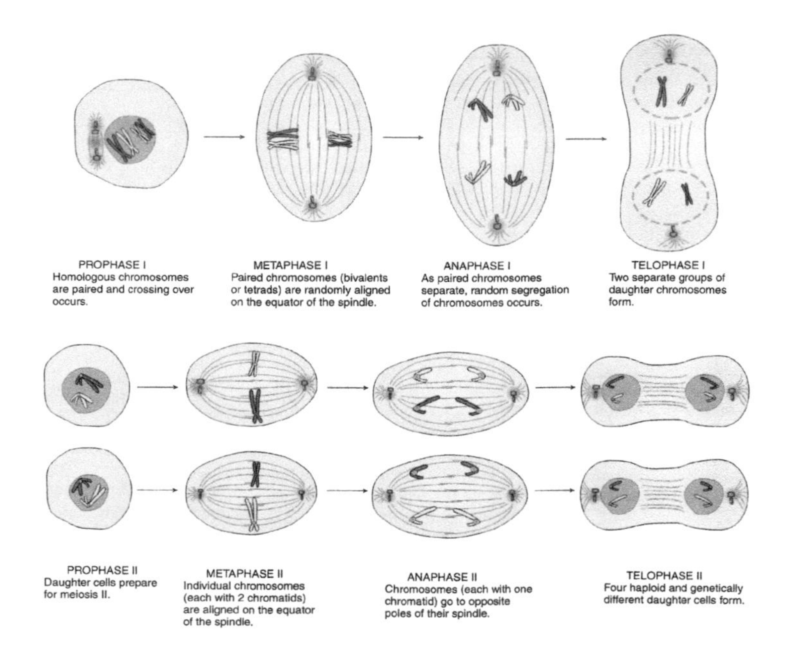

FIGURE 3.9 Phases of meiosis (in order): prophase I, metaphase I, anaphase I, telophase I, prophase II, metaphase II, anaphase II, and telophase II.

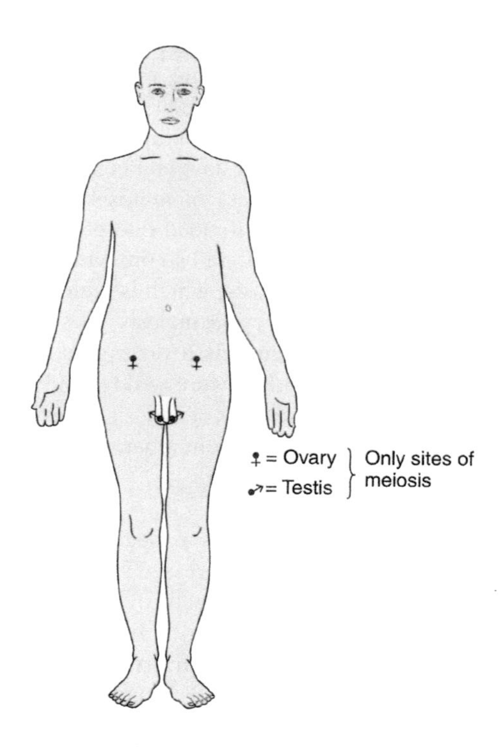

FIGURE 3.10 Meiosis and the gonads. Meiosis in humans occurs only in the gonads—ovaries in women and testes in men.

summation of their chromosome number. Because the spermatozoon and egg each contain only a haploid number of chromosomes, chromosome number does not double with each generation. The reason that the spermatozoon and egg contain only a haploid number of chromosomes is that during meiosis, the chromosome number is halved. Thus, meiosis is an integral part of the formation of eggs and sperm (see Figures 6.2 and 6.6).

During meiosis, the chromosome number is reduced from $2n$ to n; then, during fertilization, the $2n$ is restored. As a result, the species-specific chromosome number remains constant from generation to generation.

Random segregation of chromosomes

Our 46 chromosomes are composed of two sets—one set of 23 from each parent. With the exception of sex chromosomes, each chromosome in one set has a **homologous chromosome** (one that it structurally resembles) in the other set. During mitosis, chromosomes in one set do not associate with their homologue in the other set. However, during prophase I of meiosis, homologous chromosomes **synapse** (pair together) to yield 23 pairs of homologous chromosomes. The resulting pair of homologous chromosomes may be called either a **bivalent** (because it has two chromosomes) or a **tetrad** (because it has four chromatids).

When the pairs of chromosomes—rather than single chromosomes, as in mitosis—go to the equator of the spindle during metaphase I, the arrangement of the chromosome pairs is random. For example, all the chromosomes from mom may face one pole, and all those from dad may face the other pole of the spindle. On the other hand, any combination of chromosomes from mom and dad may face either pole. In other words, the alignment of the chromosome pairs on the equator of the spindle is random (Figure 3.11).

When the chromosomes of each pair move toward the opposite poles of the spindle at anaphase I, maternal and paternal chromosomes **segregate** (separate) randomly: 23 chromosomes (each composed of two chromatids) go to each pole of the spindle. Thus, it is during anaphase I that the actual reduction of chromosome number occurs. Since humans have 23 pairs of homologous chromosomes, more than 8 million genetically different kinds of either spermatozoa or eggs may result! Because each of us had our origin from such an egg and sperm, we can see here the basis of our individuality. Therefore, the likelihood that any other human being living in the past, today, or in the future was, is, or will be genetically identical to either you or me is essentially nil. Is it any wonder that a given medication may have vastly different side effects in different people?

Crossing over

When homologous chromosomes undergo synapsis (pairing), two chromosomes (thus, four chromatids) are involved. While in this paired condition, **nonsister chromatids** (chromatids from different chromosomes) may break and rejoin, but in the process exchange the broken pieces (cross over) (Figure 3.12). Remember that one of the chromosomes in the pair came from mom and one came from dad, so they are not genetically identical.

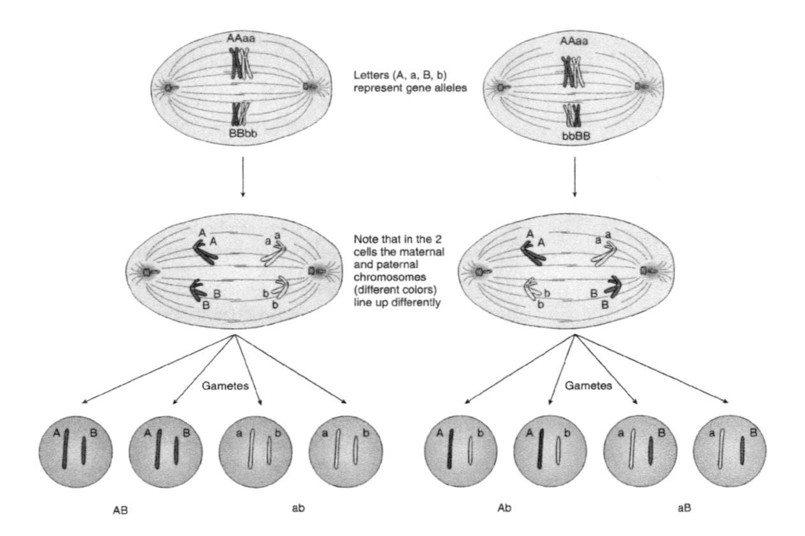

FIGURE 3.11 Random segregation of chromosomes. Note that in the two cells at the top of the figure, the alignment of the pairs of chromosomes on the spindles at metaphase I is different. Consequently, the separation (segregation) of the chromosomes at anaphase I is also different in the two cells. As a result, the chromosomes in the two sets of four daughter cells are different (different letters represent different genes).

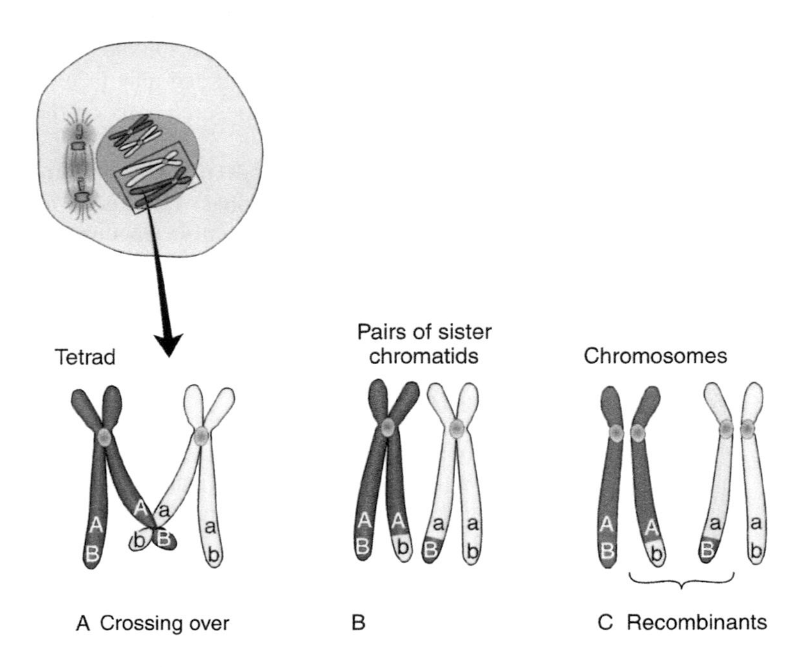

FIGURE 3.12 Crossing over. Each of the two chromosomes in (A) has identical chromatids. (B) During crossing over, there is a reciprocal exchange of homologous segments of homologous chromosomes. As a result, each of the two chromosomes in (C) has nonidentical chromatids (different letters represent different genes).

When **crossing** over occurs, new combinations of genes are created (**recombination**). For example, let us designate the four chromatids in a single pair as M, M, D, and D, two coming from each parent. Before crossing over, the two M sister chromatids and the two D sister chromatids in each chromosome pair are identical; recall chromatid replication during S stage. After crossing over, we are likely to have M, M/D, D/M, and D. Whereby we initially had *two* kinds of chromatids genetically, we now have *four* kinds of chromatids genetically. In other words, **genetic diversity** has been increased. If the meiosis was involved in egg production, now four kinds are possible rather than the two kinds possible before crossing over. The same is true for sperm production.

Consider this: if we add the genetic diversity generated by crossing over to that generated by random segregation of chromosomes, we truly have a great deal of genetic diversity resulting from meiosis.

Significance of meiosis

Because meiosis reduces the number of chromosomes from the diploid number to the haploid number, the fusion of cells during sexual reproduction is manageable. If, on the other hand, the chromosome number increased with each generation, developmental problems would result. We know this because occasionally humans are conceived with a **tetraploid number** ($4n$) of chromosomes. This genetic makeup is not compatible with human life, and the situation leads to spontaneous abortions (miscarriages) or stillbirths. As a species, we do not tolerate chromosomal abnormalities very well, and almost invariably such developmental chromosomal abnormalities carry with them intellectual disability. Perhaps chromosomal abnormalities almost always cause intellectual disability because our brains are so complex that all of our chromosomes carry genes that are vital for normal brain development. Therefore, if any chromosome is abnormal, some gene involved in brain development is damaged.

The genetic diversity generated by random segregation of chromosomes and by crossing over during meiosis generates our individuality. But the broader significance of genetic diversity—other than making each of us unique—is this. This diversity, expressed on the population level rather than on the individual level, is the substratum on which **natural selection** acts to create evolution. If it were not for this genetic diversity within populations of organisms, whole populations would become extinct. Fortunately, this does not happen.

A toxicologist who reports the effect of a harmful drug states the LD_{50} (median lethal dose) of the drug; that is, the dose that is lethal to 50% of the population tested. Why do some members of the population die and others live? Because of genetic diversity within the population.

Environmental changes, such as an increase in ultraviolet light, climate change, and so on, will affect some members of a population more than others. In other words, some will survive and others will not. Those who have genes that allow them to adapt will persist, whereas the others will not live to reproduce and their genes will be eliminated. This change in the gene frequency of a population is what we call **evolution**.

We will turn to the context in which meiosis occurs in humans in the chapter on gametogenesis (see Chapter 6).

Epigenesis—histone proteins and deoxyribonucleic acid methylation Genes are made up of DNA, and DNA molecules are polymers, strings of subunits (monomers) called nucleotides. The string of nucleotides (of four different kinds, abbreviated A, T, G, C) encodes the informational content of the DNA (gene); more specifically, it is the order of the nucleotides along the DNA molecule that encodes the genetic message. If genes are to have an effect on development, they need to be expressed; all the cells of the body (with a few exceptions, e.g., certain cells of the immune system) carry the same genes, but different subsets of these genes are expressed in different cells; that is, kidney cells and liver cells have the same genes, but different subsets of these genes are expressed in the two types of cells. If a genetic mutation occurs, that is, an alteration in the sequence of nucleotides, then the message the gene expresses is changed, which may be detrimental or even lethal to the developing embryo or fetus. However, the expressed genetic message may be altered without an alteration of the sequence of nucleotides; in this case, we have an epigenetic change. As one example, certain nucleotides may undergo a slight chemical alteration by the addition or removal of a small chemical entity called a methyl group, causing methylation or demethylation of parts of the gene, respectively, resulting in altered expression of the gene without altering the sequence of nucleotides. Another example, histone protein molecules, associated with the DNA in chromosomes, may have small chemical entities like methyl or acetyl groups added or removed from them, and this may alter the expression of the gene, again without altering the nucleotide sequence.

Stem cells Generally, cell division is symmetric; these symmetric cell divisions give rise to daughter cells with the same fates. In times of growth or regeneration, stem cells can also divide symmetrically to produce two identical copies of the original cell. Notably, *stem cells may also* divide asymmetrically to give rise to two distinct daughter cells: one copy of the original stem cell, as well as a second daughter programmed to differentiate into a non–stem cell fate. Stem cells leave the pool of mitotically dividing cells to begin a process of cell differentiation. Stem cells are, in effect, an embryonic population of cells, continually producing cells that can undergo further development within an adult organism. The path of differentiation that a stem cell descendant enters depends on the molecular environment (niche) in which it resides; for example, erythrocytes, granulocytes, neutrophils, platelets, and lymphocytes shared a common precursor cell, the pluripotential hematopoietic stem cell. See Chapter 6, "Gametogenesis," for further information on stem cells.

Study questions

1. What two types of cell division occur in humans?
2. In what sense is mitosis conservative regarding chromosome number?

3. List in order the four stages of the cell cycle and describe what happens during each stage.
4. Which stages of the cell cycle collectively make up interphase?
5. During interphase, chromosomes exist in what state?
6. Distinguish between a metacentric chromosome and an acrocentric chromosome.
7. What are the five phases of mitosis (in order)?
8. How many chromatids make up a chromosome in the following stages: G_1, S, G_2, metaphase, anaphase?
9. What are the chromosomes doing during each of the five phases of mitosis?
10. Which cell organelle is responsible for organizing the spindle?
11. What are the three regions of the spindle?
12. What is the role of meiosis in sexual reproduction?
13. How many nuclear divisions make up meiosis? What are they called?
14. What is meant by the following terms: diploid (number of chromosomes), haploid, homologous chromosomes, bivalent, tetrad?
15. What is crossing over and what does it increase?
16. Is methylation of DNA a genetic change or an epigenetic change? Explain.
17. Distinguish between symmetric and asymmetric cell division.

Critical thinking

1. The diameter of a fertilized human egg is 100 μm. You throw a piece of chalk against a chalkboard, smashing it into many tiny particles, one of which is 100 μm in diameter. What is the basis of the difference in potential between the zygote and the piece of chalk?
2. Three fundamental aspects of development are cell proliferation, cell differentiation, and morphogenesis. In your own words, what does each contribute to the development of the new individual?
3. What two aspects of meiosis constitute the basis of our individuality? Explain.

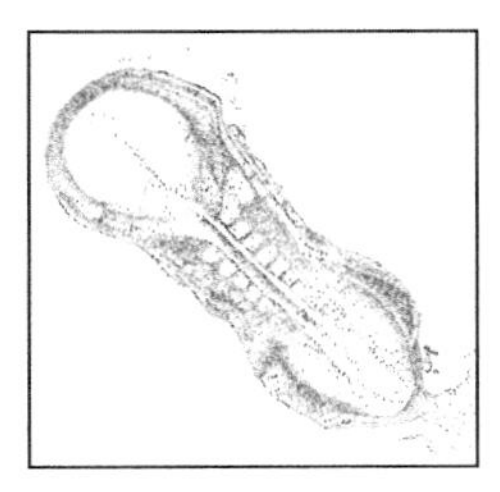

CHAPTER FOUR

Genetics

> **CHAPTER OBJECTIVES**
>
> After studying this chapter, you should be able to:
>
> 1. Understand the origin of genetics.
> 2. Describe simple Mendelian genetics.
> 3. Explain the importance of cytogenetics and chromosomes in human development.
> 4. Explain the roles of DNA replication, transcription, and translation in human development.

Over an 8-year period (1856–1864), Gregor Mendel carried out a series of experiments with pea plants in the garden of the Austrian monastery where he was a monk. He presented the results of his work at a meeting of a scientific society in 1865, and the work was published in 1866. As a result of his work, Mendel is known as the "father of genetics." Mendel's "offspring," the science of genetics, had a very long gestation period of 34 years. His work had no effect on biological thought until 1900. The "birth" of genetics occurred in 1900, when Mendel's work was rediscovered by three European researchers working independently: Correns, DeVries, and von Tschermak.

Genetics and Mendel

Why is Mendel considered the father of genetics? Mendel made two contributions: (1) he proposed that hereditary material had a particulate nature, and (2) he proposed two "laws" for the distribution of these particles—the Law of Segregation and the Law of Independent Assortment. It is also important to know what Mendel did not do. He did not comment on the chemical nature of the hereditary material (Friedrich Miescher, a Swiss contemporary of Mendel's, was just discovering nucleic acids), and he did not comment on the intracellular localization of the hereditary material (at the time, not much was known about chromosomes, and the details of mitosis and meiosis were worked out between 1875 and 1900).

Why did Mendel's work have *no* influence for 34 years? A number of reasons have been proposed. For example, it has been proposed that biology was unprepared to assimilate Mendel's work because his work was quantitative and biology at that time was not very quantitative. Another reason proposed is that biology was preoccupied with Darwin's theory of evolution. Darwin's work was published in 1859 and exploded on the biological scene.

Mendel's **Law of Segregation** resulted from experiments involving crossing tall pea plants with short pea plants (the parental generation). The resulting pea plants (F1 generation) were all tall. When the F1 tall plants were bred with each other, the resulting offspring plants (F2 generation) were mostly tall, but some were short. When Mendel observed his plants, he saw their **phenotypes** (the expression of their genes, i.e., their outward appearance). Mendel's genius was in deducing what was going on with the invisible hereditary material (**genotypes**). He proposed that each unit characteristic, such as the height of the plants, is controlled by a pair of factors—what we now call **alleles**—and when the plants produce gametes (eggs and sperm), the factors separate (segregate). For example, the tall plants of the P generation would have two alleles for tall, TT. The P-generation short plants would have two alleles for short, tt. The gametes of these plants would possess either a "T" or a "t," respectively. All the F1 plants had tall phenotypes, so tall was dominant over short, which was recessive; or $T > t$; genotypically, each of the F1 plants was Tt.

When two tall plants (Tt) of the F1 generation produced gametes, half of them would carry the "T" allele and half of them would carry the "t" allele. When the gametes came together in fertilization, the resulting zygotes would have the genotypes TT, Tt, and tt, in the ratio of 1:2:1; indeed, Mendel observed phenotypic ratios of tall to short plants of 3:1 (remember that since $T > t$, both TT and Tt would give tall plants). Because the TT and tt plants had identical alleles in the zygotes, such plants were referred to as **homozygous** (same in the zygote), whereas Tt plants were referred to as **heterozygous**. To recapitulate, Mendel's Law of Segregation states that in the formation of gametes, the members of the pair of alleles (he called them *formbildungelementen*) for each unit character separate. That is, the gametes each get a single allele; when fertilization occurs, a pair of alleles is reconstituted in the zygote for each unit characteristic (Figure 4.1).

Mendel's **Law of Independent Assortment** resulted from experiments in which two unit characters were followed—not just one character such as height. He crossed pea plants that produced yellow-smooth peas with plants that produced green-wrinkled peas—the two characters here are pea color and pea texture. The F1 plants from this cross were all yellow-smooth–pea producing plants. That is, yellow was dominant over green, and smooth was dominant over wrinkled.

Mendel was interested in heredity, so it is not surprising that he would cross F1 plants with each other. The Law of Segregation would dictate that each gamete produced would carry one allele for color and one allele for texture. If we let "Y" stand for yellow, "y" for green, "S" for

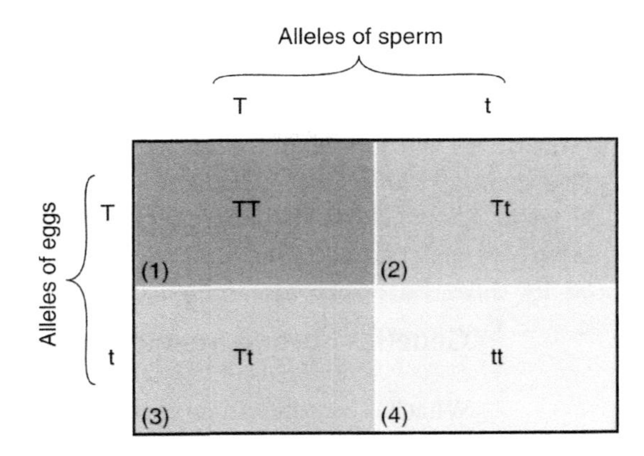

FIGURE 4.1 Segregation. The genotype of the gametes produced by the F1 plants and, in the boxes, the genotypes of the zygotes of the F2 plants are shown. Boxes 1 and 4 have homozygous zygotes, whereas boxes 2 and 3 have heterozygous zygotes. See text.

smooth, and "s" for wrinkled, it should be obvious that the **genotypes** of all the F1 plants would be YySs. What would have been the genotypes of the gametes produced by the parental generation plants? Therefore, when Mendel crossed the F1 plants, the genotypes involved would have been YySs × YySs. Each plant could produce four kinds of gametes; YS, Ys, yS, and ys (Figure 4.2). If the way in which the alleles for color (Yy) segregated had no effect on the way in which the alleles for texture (Ss) segregated; that is, if independent assortment occurred, we would expect the four kinds of gametes to be produced in equal numbers. If this were true, we would expect four kinds of phenotypes to be produced,

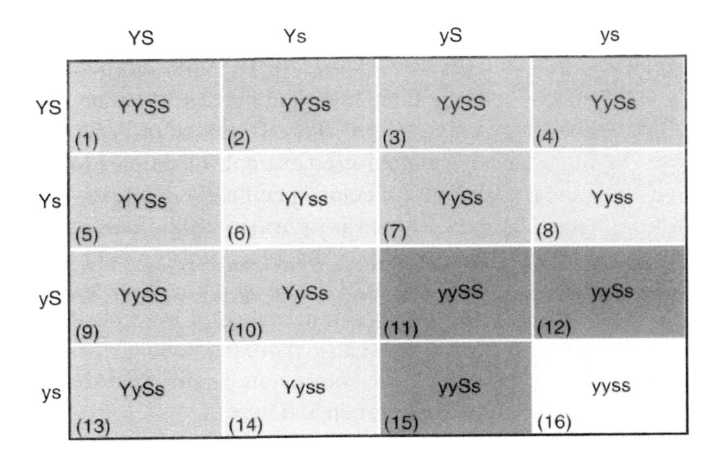

FIGURE 4.2 Independent assortment. The genotypes of the gametes produced by the F1 plants and the genotypes of the F2 plants are shown. Boxes 1–4, 5, 7, 9, 10, and 13 have zygote genotypes that will produce yellow-smooth pea-producing plants. Boxes 6, 8, and 14 produce yellow-wrinkled plants. Boxes 11, 12, and 15 produce green-smooth plants. Box 16 produces green-wrinkled pea-producing plants.

yellow-smooth, yellow-wrinkled, green-smooth, and green-wrinkled, in a ratio of 9:3:3:1. This is what Mendel found (see Figure 4.2).

As it turns out, independent assortment frequently does not occur because organisms, including peas and humans, have many more genes than chromosomes. As a generalization, one can say that alleles on different chromosomes display independent assortment, but here we are getting ahead of the story.

Genetics beyond Mendel

What has been described so far is simple Mendelian inheritance. However, inheritance is not always so simple. Take the examples of intermediate inheritance, multiple allelism, gene interaction, polygenic traits, and multifactorial inheritance. In the previous examples of inheritance in pea plants, we considered what is called **dominant-recessive inheritance**, that is, tall is dominant over short. But this is not always the case. In some instances, the two alleles for a character are *both* expressed; this is called **intermediate inheritance**.

It was also pointed out that there is a pair of alleles for each unit character, and this is generally true in a given *individual*. But a *population* of individuals may have more than one pair of alleles for a given character. This is called **multiple allelism**. The human ABO blood typing system exhibits both multiple allelism and intermediate inheritance. There are three alleles for ABO in the human population, designated I^A, I^B, and i, even though a given individual will have only two. The alleles I^A and I^B exhibit intermediate inheritance (or codominance). Neither one is dominant or recessive to the other and both influence the phenotype. On the other hand, both I^A and I^B are dominant over i, which is recessive. Because the three different alleles may be combined in six different pairs—I^AI^A, I^Ai, I^BI^B, I^Bi, I^AI^B, and ii—there are six different genotypes in the human population that generate four different phenotypes: Type A (I^AI^A or I^Ai), Type B (I^BI^B, I^Bi), Type AB (I^AI^B), and Type O (ii) blood.

An often-used example of simple Mendelian inheritance in humans is that of eye color, specifically regarding the inheritance of blue or brown eyes. Brown is considered dominant over blue, with "B" standing for brown and "b" for blue. So, three genotypes exist, BB, Bb, and bb, in a ratio of 1:2:1 and generating a phenotypic ratio of 3:1, brown to blue. However, the inheritance of eye color in humans is more complicated than this because more than one gene is involved. This is not surprising because eye colors come in different shades of blue and brown as well as shades of green and hazel.

Other human characteristics, such as intelligence and height, involve a number of genes. They are **polygenic** characters; when environmental factors also cause variation in the character, we have **multifactorial inheritance**. Most **congenital malformations**, or birth defects, are considered to be multifactorial disorders. It is important to keep in mind that alleles are alternative forms of a given gene found at the same locus (place) along a chromosome. When we talk about multiple genes, we are talking about multiple genes located at different loci along chromosomes. But when we talk about multiple alleles, we are talking about multiple alternative gene forms found at a given locus.

Cytogenetics

By the time Mendel's work was rediscovered in 1900, it was known that gametes contain one set of chromosomes and zygotes contain two sets. In 1902, Sutton and Boveri pointed out parallels in the *observed* distribution patterns of chromosomes and in the *inferred* distributions of alleles. This resulted in the intimate relationship between Mendelian genetics and cytology called **cytogenetics**.

Beginning in 1910, T. H. Morgan and his colleagues initiated a decade of genetics research with fruit flies, resulting in observations leading to the **Chromosomal Theory of Heredity**—that the alleles (genes) are carried on the chromosomes. The observations included that: (1) each gene appears to have a fixed and unique location (a locus) along the length of a specific chromosome, (2) genes located on the same chromosome are said to be linked and tend to be transmitted together, and (3) linked genes may recombine by reciprocal exchange of homologous segments of homologous chromosomes (crossing over; see Chapter 3).

To accurately determine the number of chromosomes for a given species, it is necessary to get the chromosomes condensed, spread out, and stained. In the first half of the twentieth century, the techniques did not exist to do this with human chromosomes. Around the middle of the century, it was incorrectly believed that the diploid number of chromosomes for humans was 48. By 1956, techniques had improved to such an extent that geneticists Tjio and Levan were able to demonstrate that the correct diploid number for human chromosomes was 46. By 1959, Down's syndrome, known since the nineteenth century, was the first human birth defect shown to be caused by a chromosomal abnormality—an extra chromosome (trisomy) belonging to the G group. By 1970, improved techniques, called **banding techniques**, allowed each human chromosome to be unequivocally identified. For example, Down's syndrome was shown to be caused by trisomy 21 (the human chromosome 21 belongs to the G group of human chromosomes).

Banding of chromosomes allows subtle genetic damage to be picked up. For example, a chromosome deletion (missing part) of the short arm of chromosome 5 gives rise to the birth defect *cri du chat*. In addition, many kinds of cancers have been found to be associated with specific chromosome (genetic) damage, as in the case of chronic myelogenous leukemia, involving a chromosome 9 and 22 translocation (exchange), producing the so-called Philadelphia chromosome. Another example is Burkitt's lymphoma, involving a translocation between chromosome 8 and chromosome 2, 22, or 14 and resulting in inappropriate expression of the c-*myc* proto-oncogene. When a **proto-oncogene** (a normal gene often involved in the control of cell division) becomes inappropriately expressed, it becomes a cancer gene or **oncogene**.

Genes

Again, we are primarily interested in chromosomes because they bear our genetic heritage—the genes that we inherited from our parents and that we may pass on to our children (Figure 4.3; review Chapter 3).

Deoxyribonucleic acid

The twentieth century was a century pregnant with scientific discoveries. Possibly, history will record that the most significant discovery of this century was the discovery of the structure of DNA. The DNA molecule is double-stranded, and each of the two strands is a polymer (a large molecule made up of many monomers of similar repeating parts). Each DNA polymer is made up of numerous building blocks—generally called **monomers,** and more specifically called **deoxyribonucleotides**. The sequence (linear order) of these deoxyribonucleotides encodes information, genetic information, which, among other things, provides instructions for the development of the embryo and fetus.

Identical DNA is contained in almost all the cells of the body. Notable exceptions are mature red blood cells, which possess no DNA; cells of the immune system, in which different cells change part of their DNA in response to their exposure to antigens; and cells that have acquired somatic mutations.

Deoxyribonucleic acid replication

So that all the cells of the body can have identical DNA, when cells divide (beginning with the fertilized egg), their DNA must be replicated in a highly precise manner (Figure 4.4A). This process of replication, which involves the separation of the two strands of DNA and the use of each strand to act as a template for the formation of a new strand, is under the control of a team of enzymes. Enzymes are protein molecules.

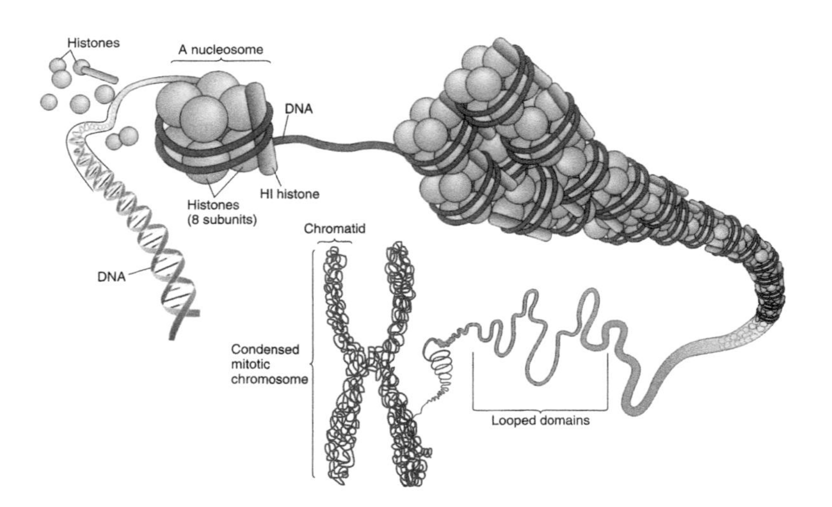

FIGURE 4.3 The organization of DNA in chromosomes. DNA is highly condensed in visible chromosomes, along with small proteins called histones. Histones and DNA make up the subunits of chromosomes called nucleosomes.

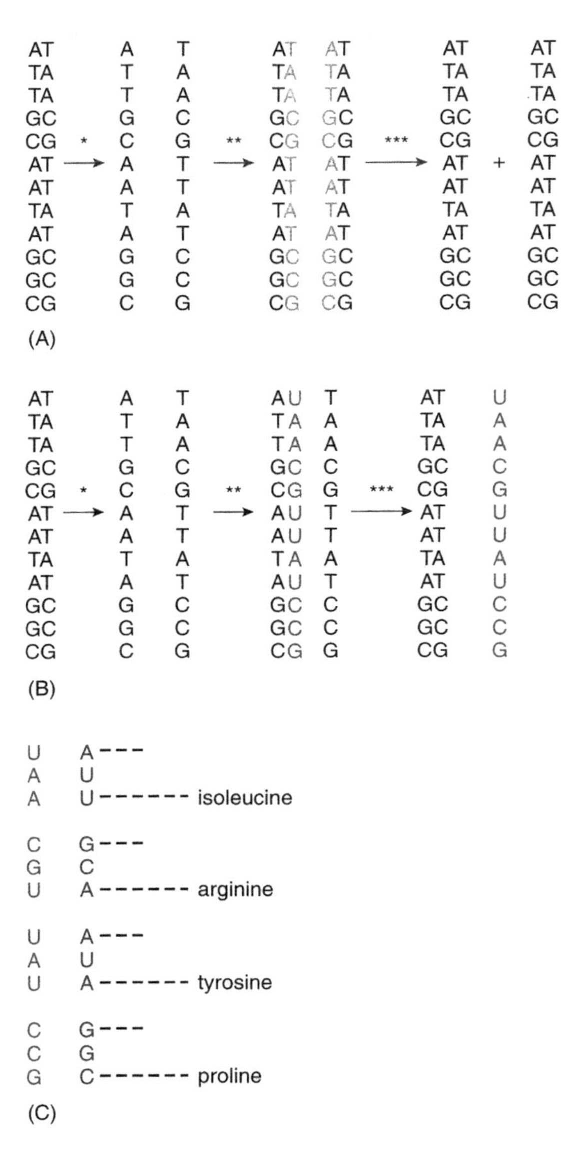

FIGURE 4.4 Replication of DNA and expression of genetic information. (A) DNA replication. The two strands of the DNA molecule are separated (*). Each strand acts as a template according to the base-pairing rules (**, new strand deoxyribonucleotides are light purple), and two identical DNA molecules result (***). (B) Transcription. The two strands of the DNA molecule are separated (*). One strand acts as a template for RNA according to the base-pairing rules (**, ribonucleotides in medium purple) and genetic information encoded in the DNA (gene). (C) Translation. Triplets of ribonucleotides (codons) in messenger RNA (medium purple) call forth specific triplets of ribonucleotides (anticodons) in transfer RNA (deep purple) according to the base-pairing rules. As a result, a specific sequence of amino acids, such as isoleucine, arginine, tyrosine, proline, and so on, is created (black). A specific sequence of amino acids makes up a protein.

The job of enzymes is to make chemical reactions in the cell happen fast enough that our biochemistry can sustain our existence. Different enzymes are highly specific in their jobs. Often, a given enzyme will speed up only one specific type of chemical reaction.

First, the enzymes of DNA replication will cause the two strands of a DNA molecule to separate—the strands are held together by weak chemical bonds, called **hydrogen bonds**, so the replication enzyme team can do this readily at body temperature. Then, each strand will be used as a template for the formation of a complementary strand; this is done according to the base-pairing rules (explanation follows).

What is meant by "being used as a template"? The deoxyribonucleotides that make up each strand of DNA are of four different kinds, depending on whether they contain adenine, guanine, cytosine, or thymine as part of their structure. Such deoxyribonucleotides are designated A, G, C, or T. The **base-pairing rules** state that in DNA, hydrogen bonds form only between A and T or G and C. For example, if one strand of DNA has the deoxyribonucleotide sequence ATTGCAATAGGC, the other strand will have the complementary sequence TAACGTTATCCG. Although the deoxyribonucleotides in a given strand are held together by strong (covalent) chemical bonds, the strands are held together by hydrogen bonds according to the base-pairing rules.

Suppose that the enzyme team has separated two DNA strands with the following sequences: strand X has the sequence ATTGCAATAGGC, and strand Y has the sequence TAACGTTATCCG. The enzyme team will use strand X as a template by building on it, according to the base-pairing rules, a new strand with the sequence TAACGTTATCCG; simultaneously, an enzyme team will use strand Y similarly to build a new strand with the sequence ATTGCAATAGGC. In other words, each new DNA molecule is identical to the original DNA molecule.

Our consideration of DNA replication has used a simple example—having to do with scale, among other things. When two deoxyribonucleotides are hydrogen-bonded together in the DNA molecule (e.g., A-T), they are referred to as a **base pair**. The DNA in the nucleus of a single human cell contains 3 billion base pairs. Indeed, it was the goal of the multibillion-dollar international scientific effort called the **Human Genome Project** to determine the sequence of all 3 billion base pairs. Why? Because the sequence is information, and one use of information is improved medical care. Note that each **diploid** (having two sets of chromosomes) human cell has 46 chromosomes, and each chromosome contains a single DNA molecule, and therefore each chromosome has, on average, millions of base pairs (our example of DNA replication above, with 12 base pairs, was obviously a simplification of scale).

Transcription

To make an embryo, it is necessary to use, or express, the genetic information encoded in the DNA molecule (see Figure 4.4B). To reiterate, this information is in a sequence of DNA deoxyribonucleotides (a gene). When the information is expressed, a sequence in DNA is used to dictate a sequence of ribonucleotides in a molecule of RNA. At this point, a team of enzymes separates the two strands of a part of a DNA molecule. Next, one of the two separated strands is used as a template for the formation of an RNA molecule, according to base-pairing rules. With RNA molecules, we have a slight change in these rules because these molecules do not contain ribonucleotides with thymine; uracil is found in place of thymine. The RNA rules are that U hydrogen-bonds with A, and G with C.

Let's take an example from the X strand of DNA from earlier in the text. The transcription enzyme team will use the DNA sequence ATTGCAATAGGC as a template to form an RNA molecule with the sequence UAACGUUAUCCG. Again, we are using a simplification of scale, since single-stranded RNA molecules have many more than 12 ribonucleotides. RNA molecules of three general types—**messenger RNA (mRNA)**, **transfer RNA (tRNA)**, and **ribosomal RNA (rRNA)**—are relevant here. All three are made, as described, according to the information encoded in sequences of DNA (genes). All three types of RNA molecules play a role in the further expression of genetic information—a process called **translation**.

Translation

This process involves the conversion of the information in a sequence of ribonucleotides in a mRNA molecule to information in a sequence of amino acids in a protein molecule (see Figure 4.4C).

During translation, the ribonucleotides in mRNA are used in triplets called **codons**. If we use the RNA molecule we transcribed previously—UAACGUUAUCCG—we find that we have four codons: UAA, CGU, UAU, and CCG. These are used by tRNA molecules to bring certain amino acids, the building blocks of proteins, into position. This is done, again, according to base-pairing rules.

Transfer RNA molecules have specific triplets of ribonucleotides, called **anticodons**, along their lengths, and a specific amino acid carried on another part of the tRNA molecule. For example, if the tRNA molecule has the anticodon AUU, it will carry the amino acid isoleucine. In addition, GCA, AUA, and GGC go with the amino acids arginine, tyrosine, and proline, respectively. The relationship between the codons in mRNA and the corresponding amino acids is called the **genetic code**.

The genetic information encoded in a specific sequence of deoxyribonucleotides (gene) of DNA has been transcribed into a specific sequence of ribonucleotides (and therefore a specific sequence of codons) in mRNA, which in turn has called forth a specific sequence of anticodons (of tRNA) and translated the genetic information into a specific sequence of amino acids, which makes up a specific protein. Again, we are dealing with a simplification of scale, since an average protein has 400–500 amino acids and our example has four.

What of the role of rRNA? Proteins in the cell are made only on ribosomes; that is, translation occurs only on ribosomes found in the cytoplasm of the cell. The structure of ribosomes consists of many proteins and several kinds of rRNA molecules. The ribosome is a complicated, protein-making machine.

The expression of genetic material

Most genes are expressed as protein; some are expressed as tRNA and rRNA molecules. But most of the DNA in the nucleus of the cell is expressed as neither protein molecules nor RNA molecules. In fact, it is not clear what most of the DNA in the cell nucleus does. The Human Genome Project is helping to solve this mystery of the purpose of the "unused" DNA. For example, a whole "universe" of RNA molecules, beyond the three types considered here, has recently been discovered.

Let's concentrate on the proteins made by genes. Almost every function carried out by cells (and, therefore, organisms) depends on the activities of proteins. For instance, **digestive proteins** eliminate waste in cells, **motor proteins** carry cargo from one part of a cell to another, **cytoskeletal proteins** determine the shapes and movements of cells, **receptor proteins** are positioned in the plasma membranes of cells so that they can respond to the instructions of the body. The list goes on and on.

Many specific proteins are specific enzymes, of which there are many different kinds in each cell. Since, generally, a specific enzyme speeds up a specific chemical reaction, such enzymes control the numerous individual chemical reactions that make up the biochemistry and metabolism of the cell (and, therefore, the embryo).

If you have followed this discussion, you should have a grasp of a profound realization of the twentieth century—*that the specific sequences of deoxyribonucleotides (genes) in the DNA of the nucleus of the cell control the metabolism of the cell* (Figure 4.5).

Why may a number of different genes have an effect on a given phenotypic characteristic? As seen in Figure 4.5, a sequence of deoxyribonucleotides in the DNA of a gene programs the information for a given enzyme that controls a single chemical reaction in a biochemical pathway made up of a number of chemical reactions. Let us suppose that the end product of the biochemical pathway shown in the figure is responsible for a characteristic such as eye color. If the end product is missing, the eye color is different. Each arrow in the biochemical pathway requires an enzyme, which in turn depends on a specific gene (DNA sequence). If a mutation (a heritable, genetic change) occurs in a given gene, the corresponding enzyme may be ineffective and no end product is produced.

In our example of a biochemical pathway, three arrows are shown corresponding to three different enzymes controlled by three different genes. If any one of the three genes mutates, the nonfunctional enzyme will result in the end product not being produced and the normal phenotype not being expressed.

Genes and development

As a consequence of sexual reproduction, a new individual receives two sets of genes from his or her parents. Explaining how genes are

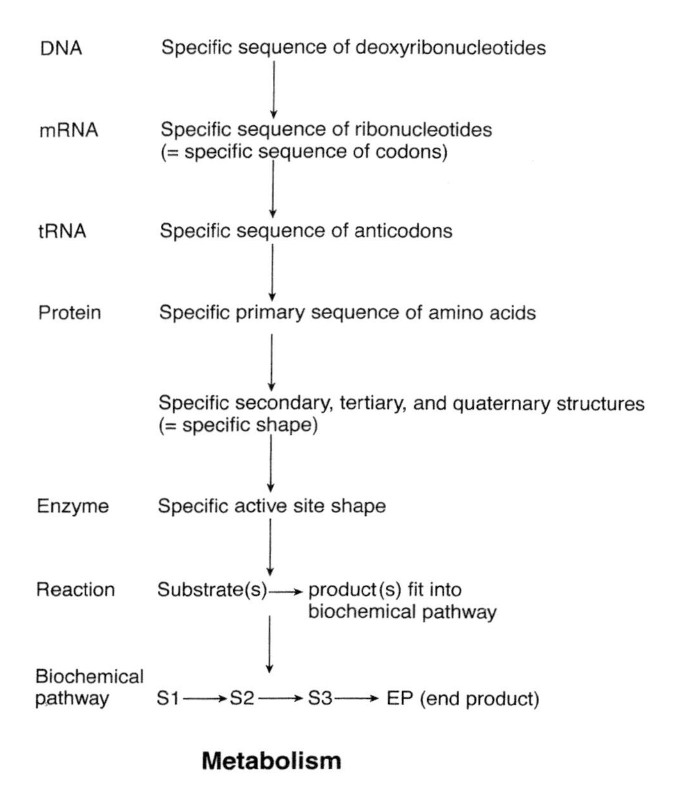

FIGURE 4.5 The connections between information encoded in the genes and the metabolism of the cell (or organism). See text. mRNA = messenger ribonucleic acid; tRNA = transfer RNA.

transmitted from parents to offspring is the task of the **geneticist**. Some of these genes direct the individual's development. Explaining how these genes control development is the task of the **developmental geneticist**.

The offspring's genes take over control of development from the maternal genes at different times in different types of organisms. In sea urchins and frogs, this occurs at the time of gastrulation (the process in which three germ layers are produced). In mammals, the transfer of control takes place early in cleavage (the mitotic divisions of early development).

For genes to control development, they must be expressed. Late in the nineteenth century, it was proposed that during development, different types of cells arose because they lost different genes. This is not generally true; different types of cells all have the same genes. What makes cells different is that different types of cells express different sets of genes. For example, liver cells and kidney cells express different genes, even though they have identical sets of genes.

What determines which cells express which specific genes? That is, what determines differential gene expression? At first, it seems illogical that cells identical in every way would express different sets of genes. Therefore, as a first approximation, we will assume that the cells resulting from the cleavage of the fertilized egg are not identical. What is the basis of this nonidentity? Two alternatives are possible: (1) the fertilized egg has regional differences in its cytoplasm that are incorporated into different cells with cleavage, and (2) the environment of the fertilized egg imposes regional differences on the different parts of the zygote.

Nature provides examples of the possible alternatives. In a simple marine organism, the sea squirt, *Styela partita*, the fertilized egg has a nonuniform distribution of pigments. One pigment is organized into the shape of a crescent—the yellow crescent. Cells that incorporate this pigmented cytoplasm, as they are formed by cleavage, are destined to give rise to the tail muscles of the larval stage of this organism. Obviously, these cells are not identical with those cells that did not incorporate the yellow cytoplasm, giving rise to other, nonmuscle, tissues. *Fucus* is a genus of brown seaweed that includes a number of species common to the intertidal zones around the world. The fertilized eggs of this organism are spherical. Yet at the first division of the egg cell, differentiation is evident. One daughter cell gives rise to a rootlike structure (rhizoid), and the other gives rise to the beginning of the rest of the plant. Here, environment plays a role in setting up an axis through the spherical egg. An environmental factor that can set up the axis through the egg is light. If the zygote is exposed to a light gradient so that one pole of the egg is in the light and the opposite pole of the egg is in the dark, the dark pole of the egg will give rise to the rhizoid.

Returning to human development, the first observable instance of cell differentiation is when the blastocyst (an early form of the human embryo) forms and is observed to consist of two different kinds of cells: inner cell mass cells and trophoblast cells. What is the basis of this differential gene expression?

The inner cell mass cells and the trophoblast cells have different types of cell junctions among themselves. The inner cell mass cells are joined together by gap junctions, and the trophoblast cells are joined together by tight junctions. What is the basis of the differential gene expression leading to the formation of the different kinds of proteins that make up the two different kinds of cell junctions?

Apparently, as indicated by experiments with mouse embryos, whether a cell becomes an inner cell mass cell or a trophoblast cell depends on the position of the cell in the compaction stage that precedes the blastocyst stage; cells on the outside of the aggregate of cells become trophoblast cells, whereas those on the inside of the aggregate become inner cell mass cells. So here, the determining factor is environment.

Another early differential event in the development of mammals is the segregation of the "germplasm" from the "somatoplasm." By 21 days of human development, a distinct population of cells may be discerned on a membrane called the yolk sac, the primordial germ cells (cells that will become germ cells, e.g., eggs and sperm). These cells may be distinguished from other cells by their appearance, which is enhanced by staining the cells for the presence of an enzyme, alkaline phosphatase. But what is it that selectively causes these primordial germ cells to express the gene for the production of this proteinaceous enzyme? Not all the questions in developmental biology have been answered. The most interesting questions are still unanswered.

Epigenesis and epigenetics

Two important concepts in development and genetics are epigenesis and epigenetics. Epigenesis is a school of thought that maintains that the developing organism gradually comes into existence. The alternative school of thought is preformationism, which maintains that the organism is preformed and development involves the growth of the organism. Preformationists historically have been referred to as spermists (or animalculists) on the one hand and ovists on the other hand, depending upon whether they believed the preformed organism is initially found in the sperm or egg, respectively. Preformationism, in its historical sense, is not accepted by modern developmental biologists, whereas epigenesis is the accepted school of thought.

A different concept with a similar-sounding name is epigenetics. Traditionally, genes, which are composed of DNA, were considered to control development by the sequence of nucleotides (building blocks) in the DNA of the gene. More recently it has been realized that development is partially controlled by chemical modifications of the nucleotides without alteration of the sequence of nucleotides, thus by epigenetics. One example of epigenetics is the addition of small chemical entities, called methyl groups, to some of the nucleotides of genes. The addition of these methyl groups, methylation, can alter the expression of the affected genes. Here, it is important to remember that essentially all the cells of the body have the same genes with the same nucleotide sequences, and cells become different from each other, differentiate, because different subsets of cells in the embryo *express* different subsets of the common pool of genes. Consequently, during development, we have the appearance of different kinds of cells: liver cell, skin cells, brain cells, and so on.

Additionally, epigenetic modification may be inherited, that is, passed from one generation to another. For example, paternal epigenetic effects are epigenetic changes in the male germline that can be passed from one generation to the next, for example, as caused by environmental agents, such as the endocrine disruptors, bisphenol A (BPA), or vinclozolin. The endocrine disruptor vinclozolin, a fungicide, has been reported to induce epigenetic transgenerational adult-onset disease. Another specific example is the estrogen mimic diethylstilbestrol, once prescribed for women to prevent miscarriage and subsequently found to enhance breast cancer risk in exposed women and birth-related adverse outcomes in their daughters.

Gene editing

A system, CRISPR-Cas9, initially discovered in prokaryotic cells (e.g., bacteria) has revolutionized gene editing in eukaryotic cells (e.g., mammalian and human cells). In prokaryotic cells, this system acts as a form of acquired immunity, whereby bacteria may protect themselves against bacteriophages ("phages"), viruses that attack the bacteria. The bacteria do this by incorporating some of the phage genetic material (DNA) into their own genetic material (DNA). When a given phage again attacks the bacteria, the bacteria recognizes the phage DNA and makes an RNA molecule complementary to the phage DNA that has been stored in the bacteria's CRISPR system. This "guide" RNA (gRNA) attaches to an enzyme called Cas9 (for "CRISPER-associated system"), which has the ability to cut the phage DNA, thereby destroying the phage's ability to harm the bacteria.

Over approximately the past 10 years, researchers have learned how to utilize this system to edit the mammalian genome, including that of embryos. A synthetic gRNA, complementary to the DNA of a target gene, is attached to a Cas9 protein and introduced into the mammalian cell. Like a guided missile, the gRNA carries the Cas9 enzyme to the target gene, which is then cut by Cas9. The mammalian cell "perceives" this damage and mobilizes mechanisms to repair the damage. One such mechanism, trying to fix the damage, tends to alter the target gene and by so doing renders it nonfunctional, an example of gene deletion. Another type of repair mechanism may be co-opted to replace the original targeted gene, which may be the cause of a disease, with a copy of the normal, that is, non–disease-causing gene. In this way, researchers are able to edit the mammalian, including human, embryonic genomes.

In research published during 2017, researchers claimed to have used CRISPR gene-editing technology to repair a genetic mutation causing heart disease in human embryos. This claim will have to be independently confirmed to dispel questions about whether what is claimed has actually been accomplished. And, of course, there are many ethical issues that need to be addressed regarding research carried out on human embryos.

Study questions

1. Why is Gregor Mendel regarded as "the father of genetics"? When, where, and with which organism did he work?
2. What does Mendel's Law of Segregation state?
3. What are polygenic characters? What is multifactorial inheritance?
4. What is cytogenetics?
5. What is the Chromosomal Theory of Inheritance?
6. What was the first human birth defect shown to be caused by a chromosomal abnormality? When was this discovered?
7. What is perhaps the most significant scientific discovery of the twentieth century?
8. How many base pairs are in the DNA in the nucleus of a single human cell? What was the goal of the Human Genome Project?
9. What is the role of transcription in the making of a human embryo? What is the role of translation in the expression of genetic information?

10. What three kinds of RNA are involved in translation, and what is the role of each?
11. What do enzymes control?
12. For genes to control development, what must they be? What makes different kinds of cells different?
13. Distinguish between epigenesis and preformation.
14. Distinguish between epigenesis and epigenetics.
15. How does CRISPR-Cas9 allow for the editing of genes?

Critical thinking

1. What are practical applications of knowing the sequence of the entire human genome?
2. All the cells of the embryo possess the same genes. If genes determine what cells are, how do you explain that we are made up of over 200 different kinds of cells?

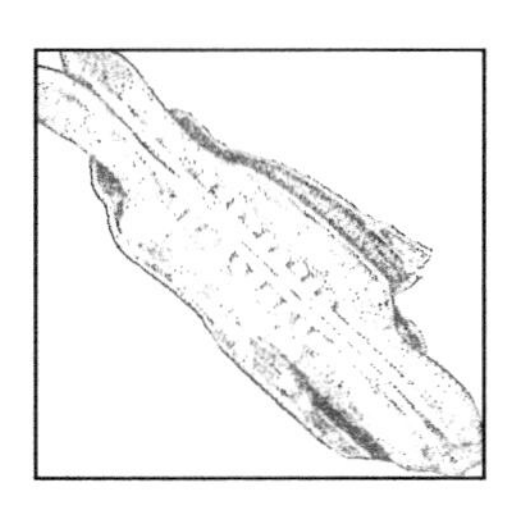

CHAPTER FIVE

Reproduction

CHAPTER OBJECTIVES

After studying this chapter, you should be able to:

1. Explain the relationship between human reproduction and human development and describe the functions of the male and female reproductive systems.
2. Trace the pathway of spermatozoa from their origin in the seminiferous tubules to the urogenital opening on the glans penis.
3. Trace the pathway of the egg from its origin in the ovary to the vaginal opening.
4. Describe the menstrual cycle.

Reproduction is the process by which organisms produce offspring to continue the species.

In both sexes, the reproductive system consists of gonads, reproductive ducts, glands, and external genitalia. These structures are designed by natural selection to produce gametes (sex cells), deliver them to the site of fertilization in a viable and functional condition, and, in the case of females, provide a site and support system for development of offspring.

Male reproductive system

The functions of the male reproductive system are to produce the male gametes (spermatozoa)—in a nurturing and protective fluid—and deliver them to the vagina of the female. It is then the responsibility of the spermatozoa to reach the egg and participate in the process of fertilization.

External genitalia

The external genitalia of the male are the penis and scrotum (Figure 5.1). The scrotum provides a home for the testes, maintaining temperatures lower than that of the abdominal cavity. Normal body temperature is damaging to human sperm production, and testes that do not descend

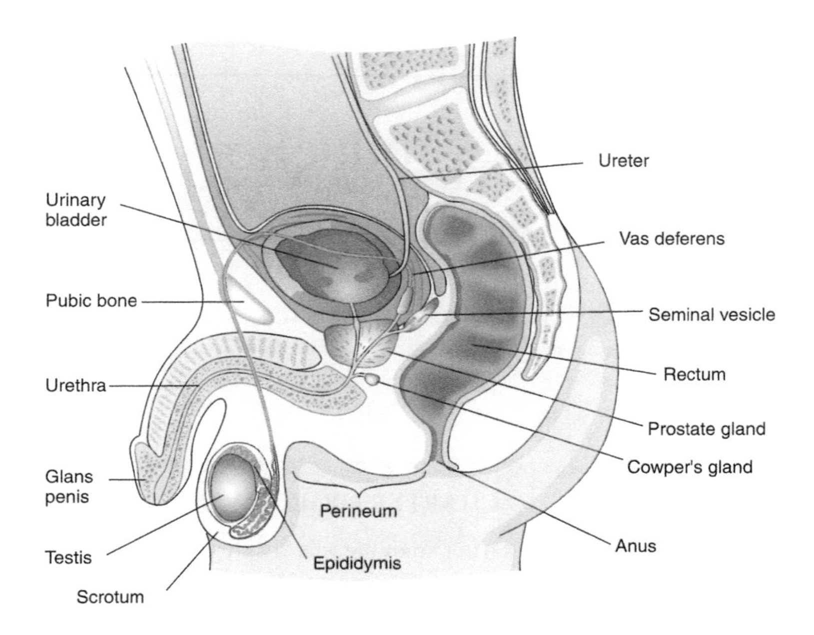

FIGURE 5.1 The male reproductive tract. A sagittal section through the male pelvis shows the external genitalia and internal reproductive organs.

into the cooler environment of the scrotum during development will not produce sperm. The reproductive function of the penis is penetration into the female vagina during intercourse and delivery of ejaculate to the posterior fornix of the vagina.

Gonads

The male gonads are the paired **testes** located outside of the abdominal cavity in the **scrotum**. Each testis has a surface covering, the **tunica albuginea**. Connective tissue partitions arise from this covering and penetrate into the substance of the testis, dividing it into **lobules**. Within the lobules are the **seminiferous tubules**, which are the sites of sperm production where **spermatogenesis** takes place, and the beginnings of the male reproductive duct system (Figure 5.2). The seminiferous tubules contain two general kinds of cells: **germ cells** (spermatogonia and other cells) and **somatic cells** (Sertoli cells) (Figure 5.3). These cells play roles in spermatogenesis, as we will see later.

Within the tunica albuginea and between the seminiferous tubules is the **interstitial tissue**, which includes "housekeeping" structures (blood vessels, lymphatic vessels, and connective tissue) and endocrine cells called the **interstitial cells of Leydig**. The latter interstitial cells produce the steroid hormone **testosterone**, which is necessary for male secondary sex characteristics and spermatogenesis.

Reproductive ducts

Beginning with the seminiferous tubules in the testes, there is a continuous passageway for the sperm to the urogenital opening on the tip of the penis (see Figure 5.2). The passageway begins where the seminiferous tubules of a testis open into a network of tubules within the testis, the **rete testis**.

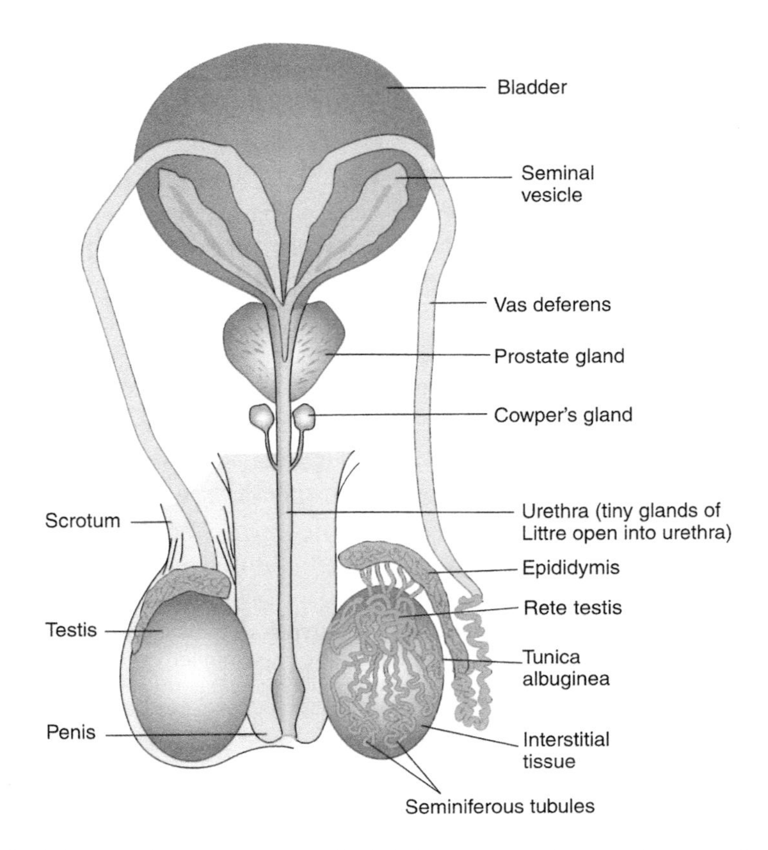

FIGURE 5.2 The male gonads. Each testis is covered by a tough connective tissue coat, the tunica albuginea, from which connective tissue partitions arise and divide the interior of each testis into lobules. The lobules contain the seminiferous tubules, which are the sites of spermatogenesis.

Epididymis From the rete testis, a number of small tubules leave the testis as the **efferent ductules of the epididymis**. The **epididymis** consists of the efferent ductules and the **duct of the epididymis**. Note that the term "epididymis" means "upon the testis," and the epididymis is located on one border of the testis.

The duct of the epididymis requires further explanation. It is a highly contorted tube with three parts: the head (**caput epididymis**), body (**corpus epididymis**), and tail (**cauda epididymis**) (Figure 5.4). The cauda epididymis is the major storehouse of sperm in the male. While in transit through the epididymis, the sperm appear to undergo some sort of "physiological ripening" without which they are not capable of fertilizing an egg.

Vas deferens and urethra The sperm then enter the **vas deferens**, which arises at the end of the epididymis. The vas deferens is the sperm duct proper that will pump the sperm out of the epididymal "well" at the time of ejaculation. Note that the vasa deferentia are the ducts cut during the contraceptive **vasectomy** procedure (Figure 5.5). The vasa deferentia,

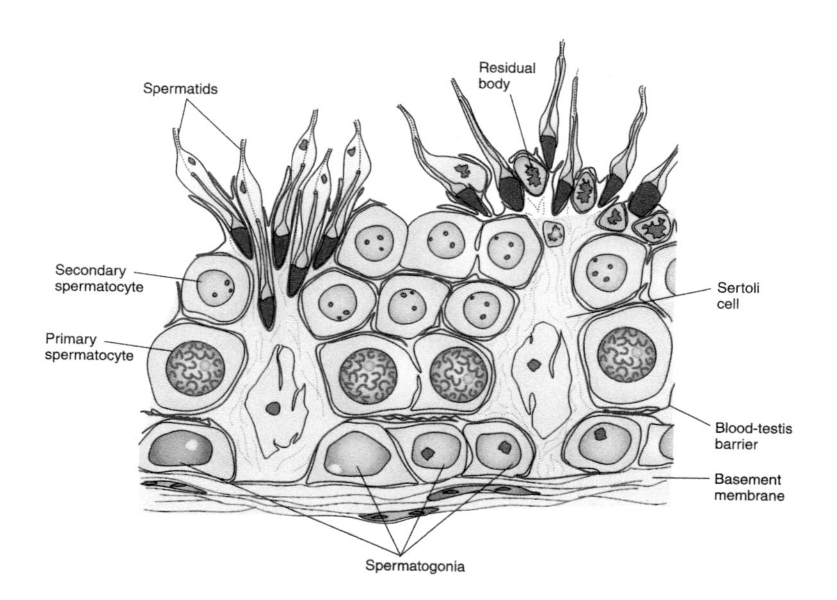

FIGURE 5.3 Cell types in the seminiferous tubules. A section through a portion of the wall of a seminiferous tubule. The Sertoli cells are somatic cells; all the other cell types are germ cells derived from primordial germ cells.

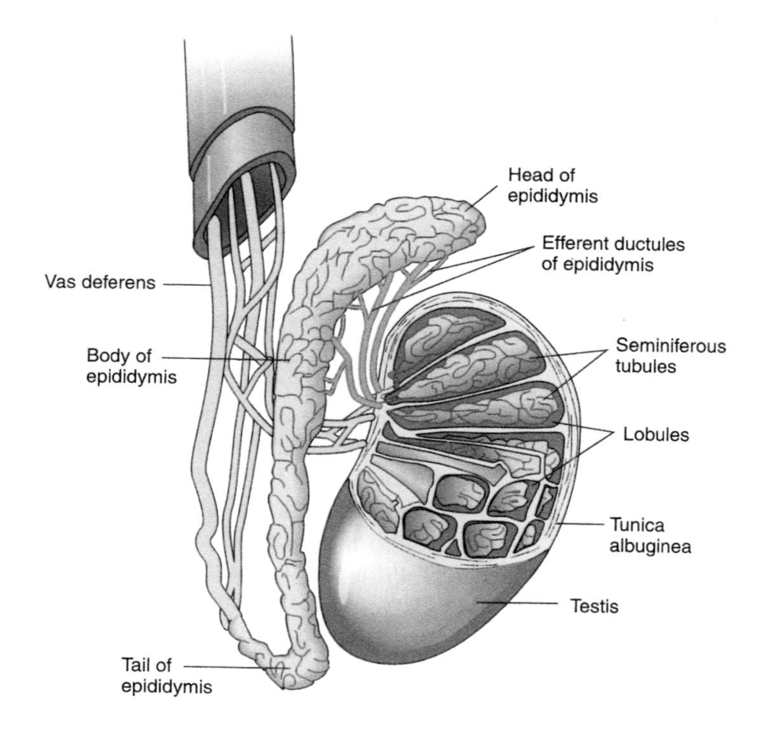

FIGURE 5.4 The epididymis. The efferent ductules of the epididymis convey sperm from the testis to the duct of the epididymis.

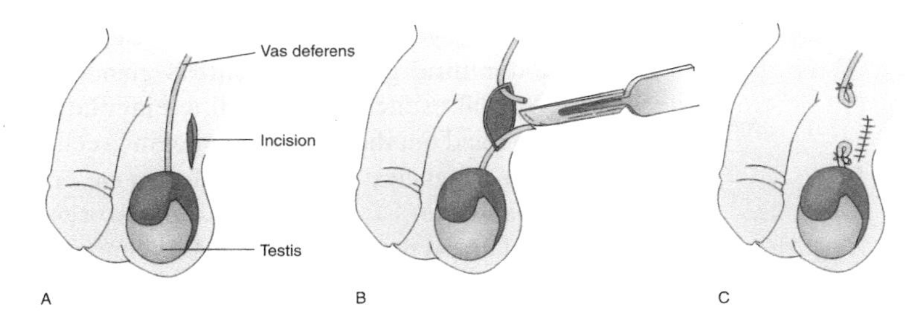

FIGURE 5.5 Vasectomy. By cutting the vas deferens (A, B) and tying off the cut ends (C), semen is prevented from reaching the urethra of the penis.

together with other structures (blood vessels, lymphatic vessels, and so on), make up the **spermatic cords**, which pass out of the scrotum, through the **inguinal canal**, and into the abdominal cavity.

Once inside the abdominal cavity, the vasa deferentia leave the other components of the spermatic cord, turn backward, and enter the prostate gland as the highly muscularized **ejaculatory ducts**. The prostate gland surrounds the **urethra**, where it emerges from the urinary bladder. Within the substance of the prostate gland, the ejaculatory ducts open into the urethra (here called the prostatic urethra), which emerges from the prostate gland and becomes the lumen (space inside of a tube) of the **penis** (penile urethra). The urethra opens on the tip of the **glans penis** as the **urogenital opening**, allowing the sperm to exit the body.

Accessory sex glands

Four types of accessory sex glands are found in the male reproductive system: two **seminal vesicles** (small fluid-containing sacs), one **prostate gland**, two **bulbourethral glands** (**Cowper's glands**), and multiple **glands of Littre**. Their function is to provide the liquid portion of the **semen** (**seminal plasma**), which nourishes and protects the spermatozoa suspended in it (see Figure 5.2).

Semen

Throughout their short lifespan, spermatozoa are suspended in fluid; the bulk of fluid in semen is derived from the seminal vesicles and prostate gland. The seminal vesicles produce thick, yellowish secretions, and the prostate gland produces thin, opaque secretions. Seminal plasma is a complicated chemical mixture, and we will discuss only a few characteristics as they pertain to sperm motility.

Seminal plasma's alkaline (basic) pH helps to neutralize the acidic environment of the vagina, since acidic pH inhibits sperm movement. Seminal plasma also provides energy. It contains the sugar **fructose**: sperm use fructose as their energy source, whereas most cells use glucose as their energy source. The most interesting constituent of seminal plasma might be the **prostaglandins**. These hormones stimulate smooth muscle contraction in the *female* reproductive tract, thus aiding the transport of sperm to the site of fertilization.

Mucus secretions

Other accessory sex glands play a role in the movement of sperm. Both the bulbourethral glands and Littre's glands produce mucus secretions during sexual excitement preceding **ejaculation**. These secretions into the urethra and out the urogenital opening seem to assist with lubrication, aiding either penetration during intercourse or the urethra's preparation for the **ejaculate**.

In total, the typical ejaculate has a volume of only 3.5 milliliters (mL) of fluid, but it contains about 100 million sperm per milliliter. Thus, a single ejaculate has about one-third of a billion sperm, even though only one is required to fertilize an egg!

Female reproductive system

The female reproductive system has the cardinal role in human reproduction and development. In addition to producing the female gametes and conveying them to the site of fertilization, it has the responsibility of assimilating the embryo and nurturing and protecting the embryo and the fetus. Furthermore, at the appropriate time, it has the responsibility of expelling the fetus into the world.

External genitalia

Collectively, the parts that make up the female external genitalia are called the **vulva** (Figure 5.6). The external genitalia include the **mons veneris, clitoris, labia majora, labia minora, fourchette**, and **vestibule**. At the center of the genitalia is the vestibule, an almond-shaped space bounded cephalically by the mons, laterally by the labia (majora and minora), and caudally by the fourchette. The vagina, urethra, Skene's glands, and Bartholin's glands all open into the vestibule. Note that the female urethra, unlike in the male, is not a urogenital duct, as it carries only urine from the urinary system. Rather, the vagina carries the products of the female reproductive system, whether that be the menses monthly or an infant during childbirth. The vagina is also the female organ of intercourse.

Gonads

The paired female gonads, the **ovaries**, are found in the **pelvic cavity**, which is the posteriormost portion of the abdominal cavity. Like most mammals (but unlike some other vertebrates), humans have a compact ovary. The ovaries are suspended in the pelvic cavity by a membrane called the **mesovarium**. Just inside the mesovarium is the **cortex** of the ovary, in which are found the germ cells. Inside the cortex is the **medulla** of the ovary, which contains a variety of housekeeping tissues such as blood vessels, connective tissue, and lymphatic vessels. The bulk of the ovary is composed of somatic cells, which support the development of eggs. Some somatic cells are endocrine cells that secrete the ovarian steroid hormones, **estrogen** and **progesterone**. Approximately once a month—by a process called **ovulation**—an egg is released from the surface of the ovary into the pelvic cavity. From here, the egg must enter the ducts of the female reproductive system (Figure 5.7).

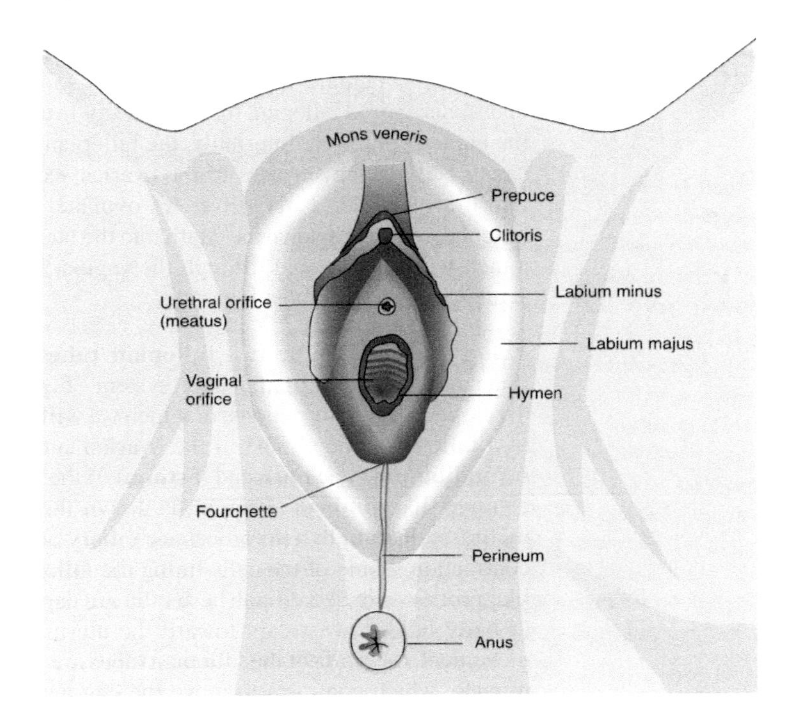

FIGURE 5.6 The female external organs of reproduction. (From Eastman, N.J. and L.M. Hellman. *Williams Obstetrics*, 13th edition. Appleton & Lange, 1966: Figure 1, p. 19 with permission from the McGraw-Hill Companies.)

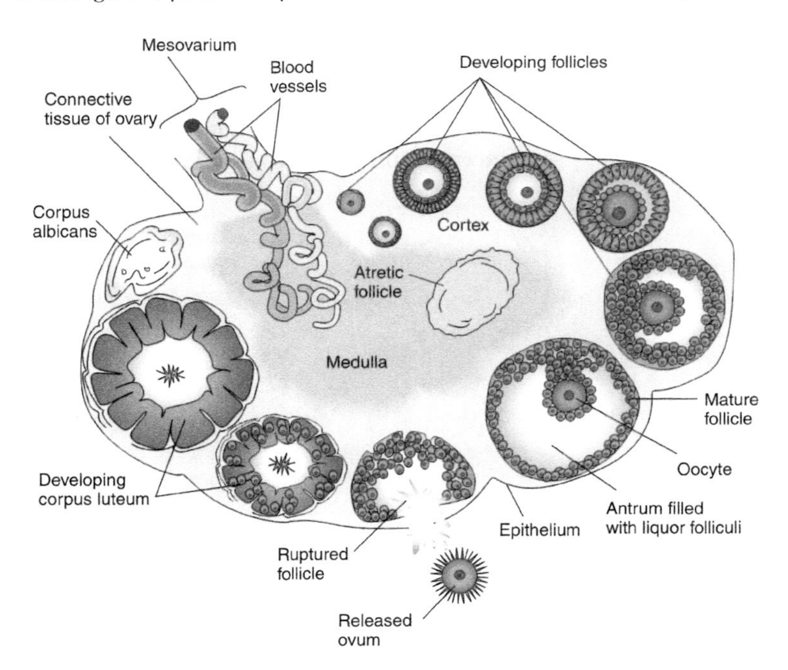

FIGURE 5.7 The ovary and ovulation. A diagrammatic section of the ovary, showing parts of the ovary, follicle development, and ovulation.

Reproductive ducts

The human female reproductive ducts consist of fallopian tubes, uterus, and vagina. Female reproductive ducts show great diversity among mammals; paired fallopian tubes and a single uterus and vagina constitute the human condition. Internally, the fallopian tubes open into the pelvic cavity close to the surfaces of the ovaries; externally, the vagina opens into the vestibule of the vulva. An ovulated egg will enter the closest fallopian tube, pass down its length into the uterine cavity and—assuming a menstrual cycle—exit through the vagina. If a pregnancy occurs, the same trip will take about 9 months.

Fallopian tubes The two **fallopian tubes** (**uterine tubes**) are the innermost portions of the duct system (Figure 5.8), and anatomists divide each into four regions, of which we will consider two. The portion of a fallopian tube closest to the ovarian surface is the funnel-shaped **infundibulum**. The open end (**ostium**) of the funnel is fringed by a ring of finger-like **fimbria**. The egg is drawn through the ostium and into the infundibulum by two processes: ciliary activity and smooth muscle contraction. Some of the cells lining the fallopian tubes have little hair-like processes called **cilia**. These cilia are capable of directing a current of body fluid downstream toward the uterus. To further aid the egg's movement, the walls of the fallopian tubes are made up mostly of **smooth muscle**, which contracts to move the egg away from the ovary. At the time of ovulation, hormones stimulate both the beating of cilia and the contraction of smooth muscle. As a result, the fimbriated infundibulum sweeps over the surface of the ovary and draws the egg in through the ostium, somewhat like the action of a vacuum cleaner.

The next region of the fallopian tube after the infundibulum is the **ampulla**. This is the site of **fertilization**, where sperm moving upstream meet the egg moving downstream. If fertilization does occur, the earliest stages of human development take place in the fallopian tube, while the **conceptus** (product of conception) moves downstream toward the uterus.

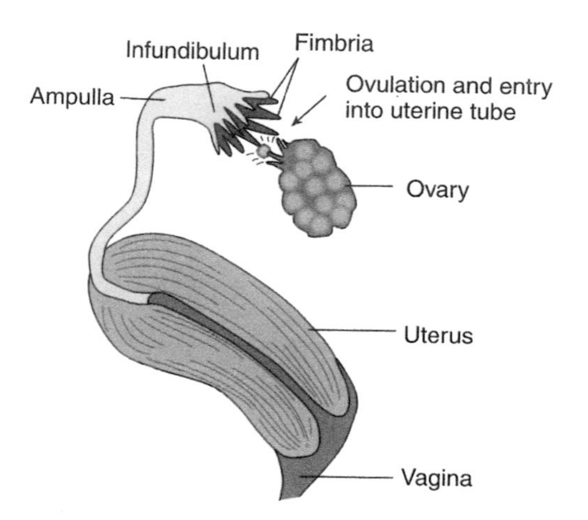

FIGURE 5.8 A fallopian tube and its relationship to the ovary and uterus.

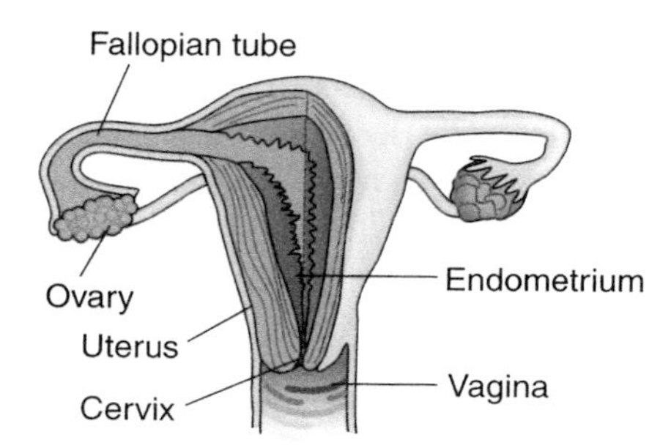

FIGURE 5.9 The uterus. A partially dissected uterus shows the surface and interior of the organ.

Uterus The destination of the conceptus is the **uterus**. The fallopian tubes connect to the upper portion of the body of the uterus, and the region between the fallopian tubes' attachment points is the **fundus**. Basically, there are two parts of the uterus: the **corpus** (body) and the **cervix** (neck). The cavity of the corpus is called the **uterine cavity**, and this is where most of the development of the conceptus takes place. The uterine cavity is lined by a mucous membrane called the **endometrium**, much of which is shed during the menses of the menstrual cycle (Figure 5.9).

The typical **menstrual cycle** has a duration of 28 days—its length is under the control of ovarian hormones (Figure 5.10). Days 1 through 5 are occupied by **menstrual flow**, commonly known as a woman's "period." Subsequently, and up until the 14th day of the cycle during the **proliferative stage**, the endometrium is rebuilt under the influence of estrogen from the developing ovarian **follicles**. On day 14 of the menstrual (or uterine) cycle, the egg is released from the surface of the

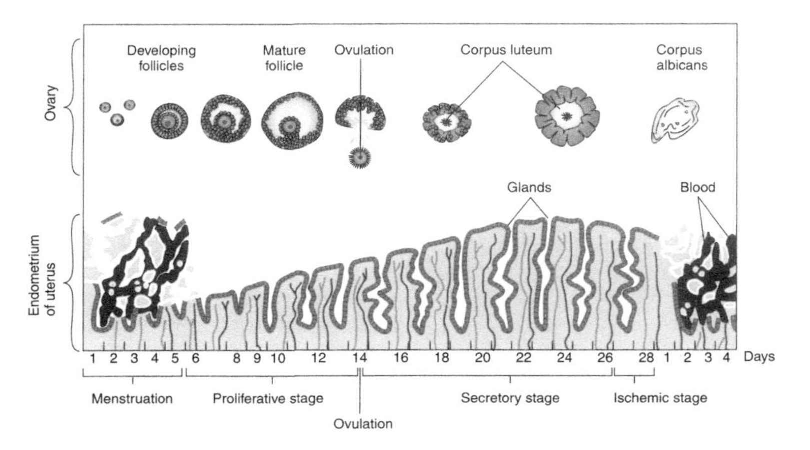

FIGURE 5.10 The ovarian and menstrual (uterine) cycles. The top of the figure shows the events of follicle development in the ovary. The bottom of the figure shows the corresponding events of the menstrual cycle.

ovary as the endometrium enters the **secretory stage**, under the influence of estrogen and progesterone from the ovarian **corpus luteum**.

The secretory stage, when the endometrium is synthesizing and secreting nutrients in preparation for a possible pregnancy, lasts until the 26th or 27th day of the menstrual cycle. If a pregnancy has not begun, the endometrium will enter the **ischemic stage** and begin to undergo degenerative changes in preparation for the next menses.

Cervix The lower part of the uterus is the cervix, which projects into the **vault of the vagina**. It has an opening into the uterine cavity (internal os) at its upper end and an opening into the vagina (external os) at its lower end (Figure 5.11). Between these openings is the lumen of the cervix, the **cervical canal**. The cervical canal is filled with a

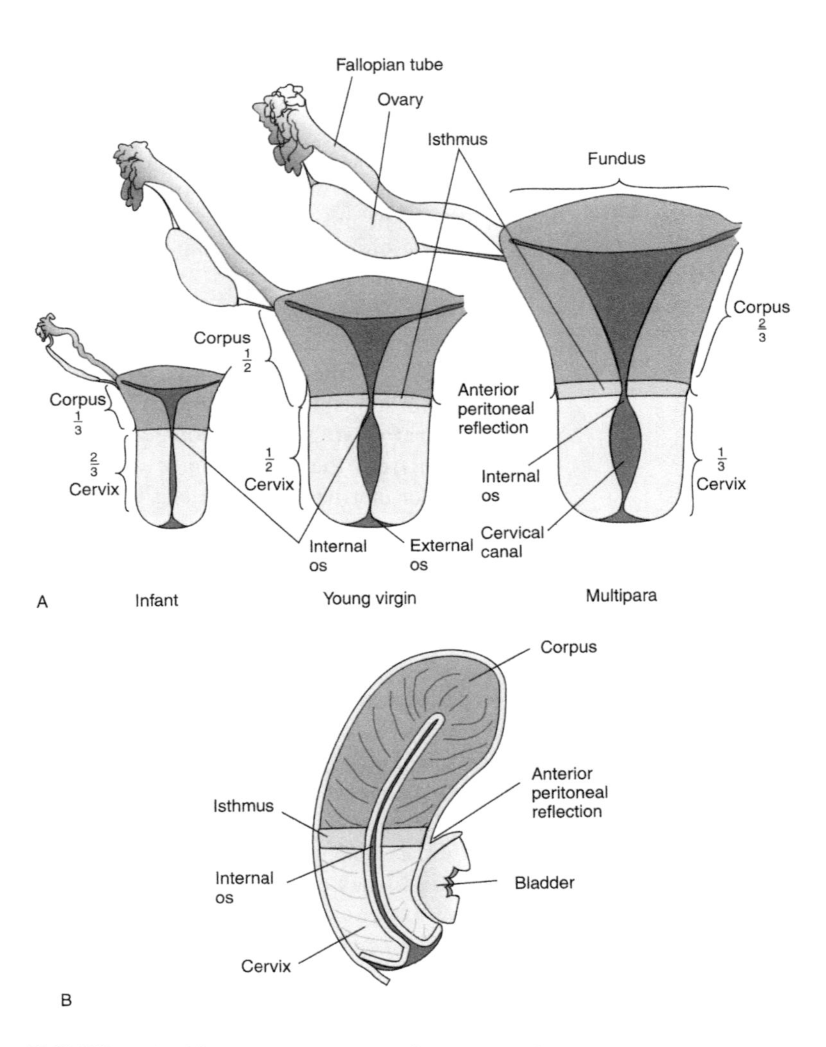

FIGURE 5.11 The uterus. (A) Frontal sections of the infant, young virgin, and multipara uteri. (B) Sagittal section of the uterus.

fluid, the **cervical mucus**. Hormones control the consistency of the cervical mucus as the menstrual cycle progresses. The mucus is more liquid (and therefore less of a barrier to the passage of sperm) at about the time that the egg is ovulated from the ovary. During pregnancy, the cervical mucus provides a barrier (the **cervical plug**), which prevents germs from moving up the cervical canal from the vagina and causing a uterine infection.

At the interior end of the vagina, called the "vault," the vagina connects to the cervix of the uterus. The cervix subdivides the vault into a number of spaces called fornices. The deepest of these is the **posterior fornix** and is worthy of mention for two reasons. First, it is an accessible pathway for surgical access to the pelvic cavity, and, second, it is the site where semen usually pools after intercourse and the site from which semen gains access to the cervical canal (Figure 5.12).

Accessory sex glands

The female accessory sex glands are of two kinds: the paired **Skene's glands** and the paired **Bartholin's glands**, both pairs of which release their secretions into the vestibule of the external genitalia. These secretions apparently play a role in lubricating the female genitalia in preparation for the penetration of intercourse.

A little more cell biology In order for development to be normal, it is necessary for cells to communicate, as indeed it is necessary throughout life. Cells may communicate in a variety of ways; for example, by way of nerve impulses (through the nervous system) or by way of hormones

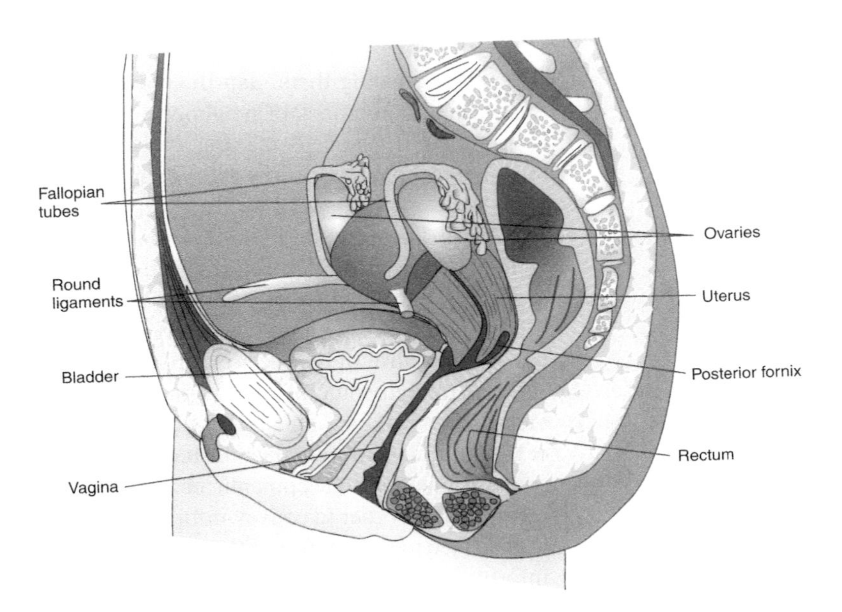

FIGURE 5.12 The female pelvis. A sagittal section of the female pelvis showing the relationships between the reproductive organs and between the reproductive and surrounding organs.

(through the endocrine system). Here we focus on hormones, and specifically steroid hormones (not all hormones are steroids; e.g., insulin is a protein hormone). We will start at the head of the body and proceed down to the pelvic cavity. A portion of the base of the brain is the hypothalamus, the cells of which produce hormones called gonadotropin-releasing hormones (GnRHs); these hormones affect cells of the anterior lobe of the pituitary gland, also found at the base of the brain. In a woman, cells of the anterior lobe of the pituitary gland respond to the GnRHs by producing two gonadotropins, follicle stimulating hormone (FSH) and luteinizing hormone (LH). These gonadotropins act on cells in the ovaries; in turn, these cells produce the steroid hormones estrogen and progesterone, which act on the cells that line the uterine cavity, converting these cells from endometrial cells to decidua cells, which make the uterine lining receptive to the implantation of an embryo.

Cell communication In addition to cells communicating over relatively long distances, for example, by nerve impulses and neurotransmitters of the nervous system or by hormones of the endocrine system, cells may communicate over shorter distances. Cell communication over shorter distances occurs by paracrine, juxtacrine, or autocrine signaling. If a cell has receptors for signal molecules it itself produces, this is called autocrine signaling, a relatively rare kind of signaling, but exemplified by the cytotrophoblast cells of the placenta, which make and secrete platelet-derived growth factor (PDGF), the receptors for which are found on the very same cells, this results in the explosive growth of the cytotrophoblast, which is instrumental in the implantation of the embryo into the lining of the uterus, whereby a pregnancy is initiated. If the signaling molecules are embedded in the plasma membranes of the cells producing them, and the receptors for them are embedded in the plasma membranes of neighboring cells, then juxtacrine signaling is occurring. If the cells making and secreting the signaling molecules attach to receptors of nearby cells, then paracrine signaling is occurring.

Intracellular signal transduction Developmentally important cellular activities include cell division (required to make a multicellular organism) and differential gene expression (required to make different kinds of cells); in a multicellular organism, these cellular activities are under the control of the cell's environment, including signals from other cells. Although some of these signals, for example, steroid hormones, are nonpolar and are able to pass through the plasma membrane of the cell and attach to cytoplasmic receptors, other signals are polar, for example, the hormone epinephrine, and attach to receptors in the plasma membrane. In order to convey information carried by the signal from the cell surface to the interior of the cell, for example, the nucleus, where the information is acted upon, the cell makes use of cascades of chemical reactions, which make up intracellular signal transduction pathways. For a specific example of intracellular signal transduction, see Chapter 7 on fertilization.

Study questions

1. What are the components of the male external genitalia? What is the role of each?
2. Which cells in the testes produce testosterone?
3. What are the three parts of the duct of the epididymis?
4. What is the major storehouse of sperm in the male?
5. List the four types of male accessory sex glands and their general functions.
6. What do sperm use as their energy source?
7. Approximately how many sperm are found in a single ejaculate?
8. What is another name for the female external genitalia? What structures make up the female external genitalia?
9. What is the mesovarium?
10. What part of the ovary contains the germ cells?
11. What is ovulation and how often does it occur? What two processes aid movement of the egg through the fallopian tube?
12. What are the two basic parts of the uterus? What is the role of each in development?
13. What are the four stages of the menstrual cycle and what is happening to the endometrium during each stage?
14. What are the two kinds of female accessory sex glands and what is the role of their secretions?
15. Distinguish, in terms of signaling distance, between nerve and endocrine signaling on the one hand and autocrine, juxtacrine, and paracrine signaling on the other.
16. Where are the cellular receptors for nonpolar and polar signals located, respectively?

Critical thinking

1. It has been proposed that menstruation prevents ascending infections of the female reproductive tract. Menstruation, of course, does not occur during pregnancy. What substance formed during pregnancy may take over this role of menstruation? Where is this substance found (i.e., in what and between what two openings)?
2. What are the roles of estrogen and progesterone in the uterine (menstrual) cycle?
3. Trace the pathway of sperm from its origin in the seminiferous tubule of the testis to its contact with the corona radiata (see Chapter 6) in the ampulla of the fallopian tube.

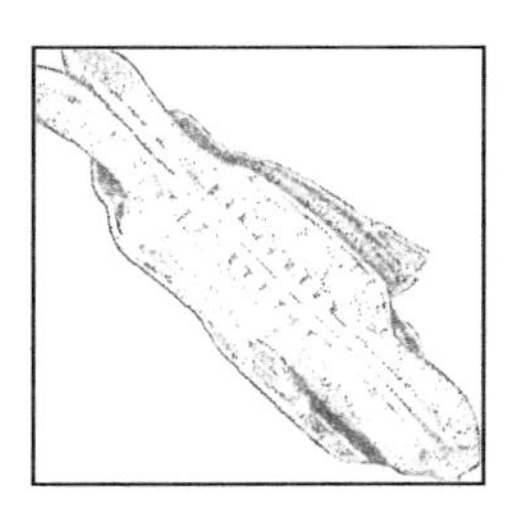

CHAPTER SIX

Gametogenesis

CHAPTER OBJECTIVES

After studying this chapter, you should be able to:

1. Explain the role of gametogenesis in human reproduction and development.
2. Describe the process of spermatogenesis, including the role of spermiogenesis.
3. Explain what a stem cell population is and state its role in spermatogenesis.
4. Describe the structure of the sperm and how this structure assists in its function—fertilizing the egg.
5. Describe the process of oogenesis.
6. Describe the structure of the egg and how this is suited to its role of being fertilized and eventually giving rise to the embryo.
7. Describe the similarities and differences between the male and female gametes, including lifespans, replenishment, relative number of gametes, and structure (i.e., size, motility, and reserve food).

The cells of the body may be classified into two categories: **somatic cells** and **germ cells**. Somatic cells are not directly involved in reproduction, and include muscle cells, skin cells, and bone cells. By contrast, germ cells are directly involved in *reproduction*. The subset of germ cells directly involved in *fertilization* are called **gametes** or sex cells and include spermatozoa and ova (or sperm and eggs). The formation of gametes is **gametogenesis** and comes in two varieties: **spermatogenesis** in men and **oogenesis** in women. Gametogenesis is the only process in the body involving **meiosis** (all other cell divisions are mitotic) and is confined to the gonads. However, during development, the germ cells arise outside the embryonic body and subsequently migrate into the developing **gonads**.

Spermatogenesis (also, see Chapter 21)

Sperm are produced in the **seminiferous tubules** of the testes beginning at the time of **puberty**. Until this time, there are two kinds of cells in these tubules: **spermatogonia** (germ cells) and **Sertoli cells** (somatic cells) (see Figure 5.3). The **primordial germ cells**, the first germ cells to appear in our development, are detected on the yolk sac of the 21-day-old embryo; by the 38th day, these cells have reached the developing testes and are then called spermatogonia (singular, spermatogonium) (Figure 6.1).

With puberty, the spermatogonia begin to produce a variety of other kinds of germ cells as well as additional spermatogonia. In other words, this **stem cell population** of spermatogonia produces some cells that stop dividing and differentiate, and it also produces cells that remain undifferentiated and capable of dividing.

Spermatogonia to spermatids

Let us now follow the fate of one spermatogonium that goes down the pathway of cell differentiation (also called cytodifferentiation) (Figure 6.2). It has 46 chromosomes and grows to become a larger cell (called a **primary spermatocyte**), which also has 46 chromosomes. The primary spermatocyte undergoes meiosis I (see Chapter 3), including anaphase I, in which 23 chromosomes (two chromatids each) move to each pole of the spindle. Meiosis I results in the formation of two smaller daughter cells (called **secondary spermatocytes**), each containing 23 chromosomes.

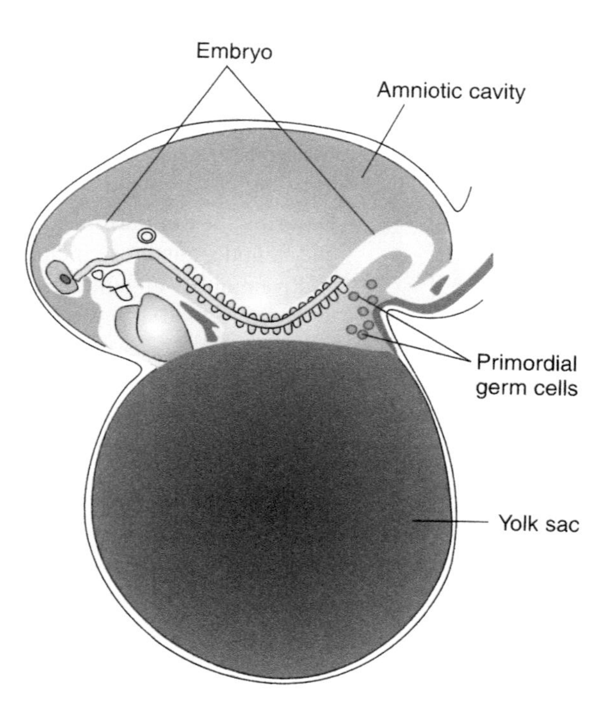

FIGURE 6.1 Primordial germ cells. Appearance of the primordial germ cells on the wall of the embryonic yolk sac.

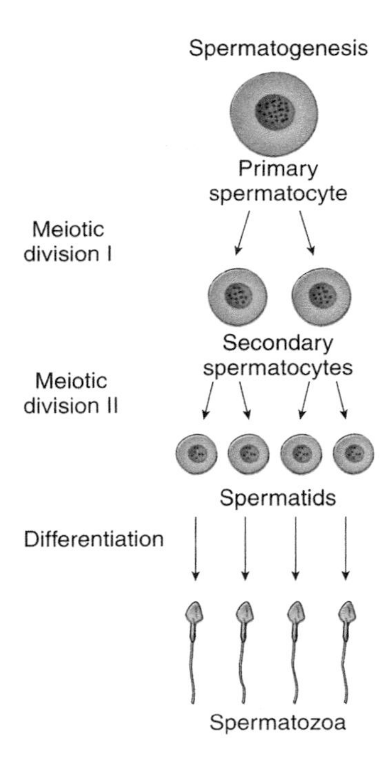

FIGURE 6.2 Spermatogenesis. A single spermatogonium undergoes spermatogenesis to produce four highly differentiated spermatozoa (sperm).

Each of the two secondary spermatocytes then undergoes meiosis II. Because there is no replication of DNA (chromatids) leading up to this second division, in each cell, 23 chromosomes (one chromatid each) are moving toward each spindle pole. This results in the formation of two smaller daughter cells (called **spermatids**) from each of the two secondary spermatocytes produced during meiosis I.

Spermiogenesis

The original spermatogonium is now four cells. The four haploid spermatids formed from the single diploid spermatogonium undergoing spermatogenesis look like typical cells at this point. However, they will now undergo a process of cell differentiation that results in the formation of cells that look anything but typical (see Figure 6.2). As the four spermatids undergo this process of **spermiogenesis**, they become intimately associated with the Sertoli cells, which appear to carry out some sort of "nursing" function for the differentiating cells. Note that spermiogenesis is a *part* of spermatogenesis—the part during which cell differentiation (spermatids to spermatozoa) occurs.

Stem cells

Unlike females, human males can produce gametes from puberty into advanced old age because the testes contain stem cell populations. These stem cells not only supply cells for spermatogenesis but also replenish themselves. Also, note that for each spermatogonium undergoing spermatogenesis, four functional gametes (sperm) are produced.

Sperm

The **spermatozoon** is a differentiated cell that is destined to fertilize an egg or die. From the time it is deposited in the female reproductive system, it has about 48 hours to find and fertilize the egg, or else it will die.

Structure

Perhaps more than any other cell, the spermatozoon reflects the correlation between cell structure and cell function. It is a small, motile cell without reserves of energy-rich molecules. It has three basic parts—the **head**, the **midpiece**, and the **tail**—a structure atypical of human cells (Figure 6.3). During spermiogenesis and while in close contact with Sertoli cells, the spermatids undergo a dramatic metamorphosis as part of their differentiation.

Development

During spermiogenesis, the Golgi apparatus of the spermatid gives rise to the **acrosome** of the head of the spermatozoon. The acrosome, a cellular organelle containing digestive enzymes, is a structure necessary for penetrating the egg. The spermatid nucleus condenses to become the highly **compact nucleus** in the head of the spermatozoon (Figure 6.4). The two **centrioles** of the spermatid's centrosome position themselves at the posterior pole of the sperm nucleus. One of the two centrioles acts as a **microtubule-organizing center**, providing the microtubular core of the sperm tail (**flagellum**), while the other centriole persists as the **centriole** of the spermatozoon, destined to play an important role after fertilization.

The **mitochondria** of the spermatid take up a position around the base of the core of the flagellum to form the middle piece of the sperm.

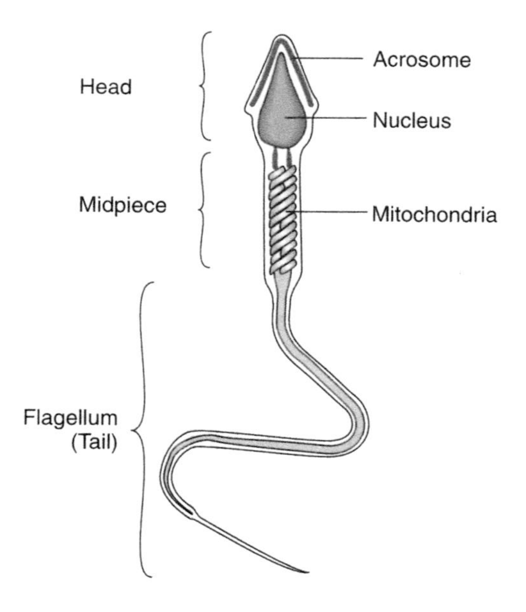

FIGURE 6.3 Anatomy of the spermatozoon.

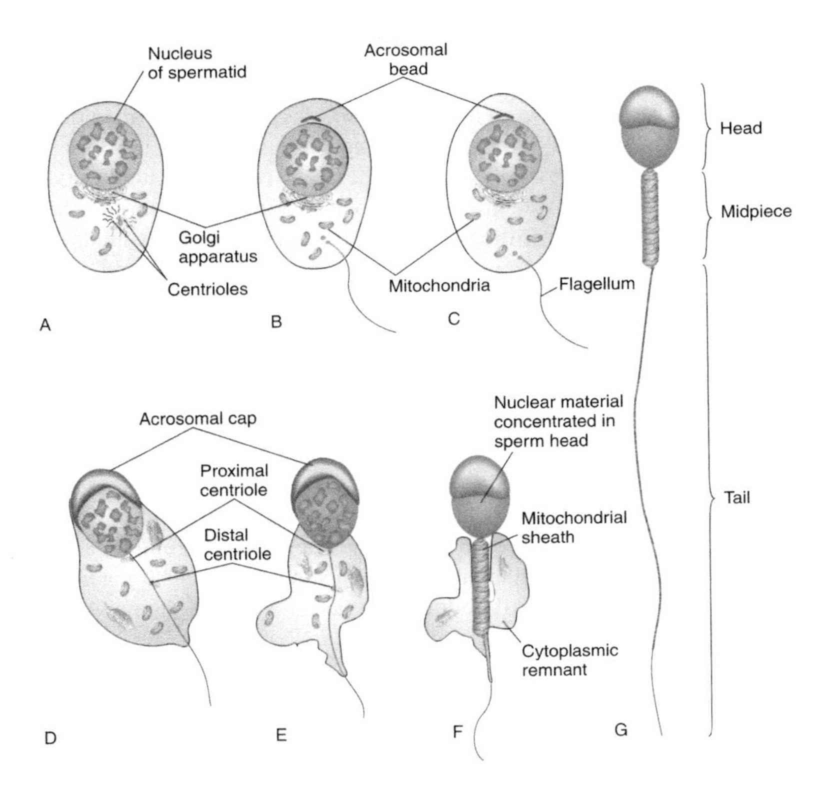

FIGURE 6.4 Spermiogenesis. During this type of cell differentiation, a morphologically ordinary cell, the spermatid, is transformed into a highly specialized, flagellated spermatozoon. Spermiogenesis is part of spermatogenesis.

So as not to impede sperm function, the bulk of the cytoplasm (in the form of a cytoplasmic remnant) is pinched off from the forming spermatozoon.

Functions

Once formed, the functions of the spermatozoon are (1) to reach the egg; (2) to fuse with the egg, thereby delivering the 23 paternal chromosomes with their paternal genes; and (3) to activate the egg to complete meiosis II and begin development of the new conceptus. The structure of the spermatozoon is exquisitely appropriate to serve these functions.

The sperm head contains both the acrosome, a bag of digestive (hydrolytic) **enzymes** that aids the sperm's passage through the egg coats, and the **sperm nucleus**, the sperm's genetic payload. The midpiece of the sperm is packed with energy-producing mitochondria that power the sperm's tail (its motility apparatus) so that it may reach the egg, which is slowly flowing down the fallopian tube. The motility of the sperm is affected by the pH of its environment, and the slightly alkaline pH of the female reproductive ducts above the vagina aids the sperm's motility. Of course, the movement of the sperm is also aided by muscular contraction of the female reproductive ducts.

Oogenesis (also, see Chapter 21)

As in the male, primordial germ cells arise on the yolk sac, then migrate into the developing ovaries on about the 38th day of development. Once in the ovaries, the germ cells are called **oogonia**. The ovaries contain the greatest number of germ cells before birth, when they number in the millions. At birth, the number is in the hundreds of thousands; at puberty (**menarche**) in the tens of thousands; and at **menopause**, the number is essentially zero. The number of germ cells in a woman's ovaries is in constant decline. A relatively small number of germ cells is lost through ovulation, and a much larger number is lost through normal cell death (Figure 6.5). In addition, the ovaries "do not have stem cell populations" to renew the supply of germ cells; however, see a dogma challenged at the end of this chapter.

Human females also differ from human males in that meiosis begins in the female before birth. While still a fetus, all of a female's oogonia become **primary oocytes** and remain as such (in prophase I; see Chapter 3) until, in some instances, menarche and, in other instances, until about menopause. Note that some cells may remain in meiosis I for as long as 50 years!

Oogonia to ootids

Let us follow the fate of a single diploid **oogonium** as it undergoes oogenesis (Figure 6.6). The oogonium begins to grow and then is called

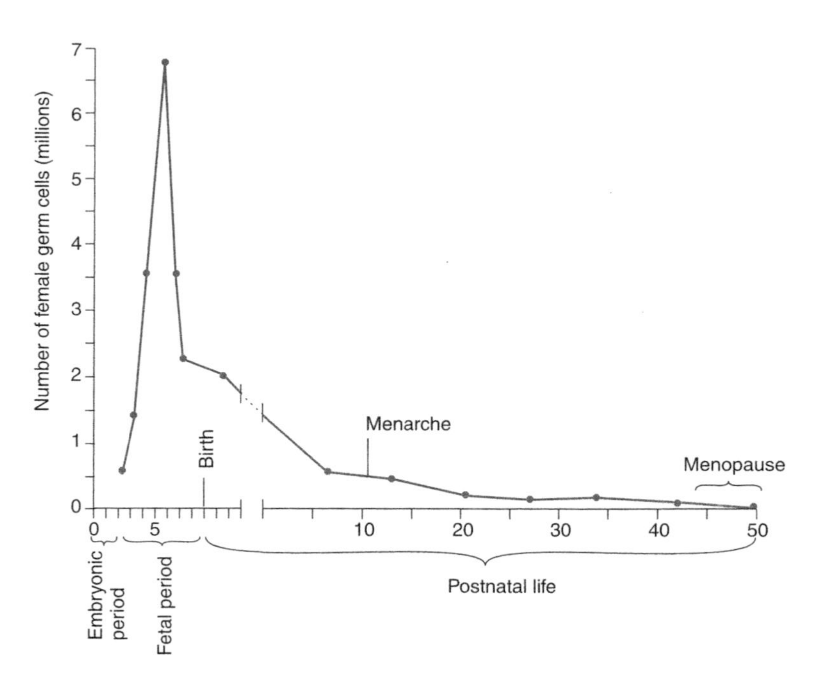

FIGURE 6.5 Decline in the number of germ cells during the life cycle of a woman. Note that the maximum number of germ cells is found in the fetal ovaries.

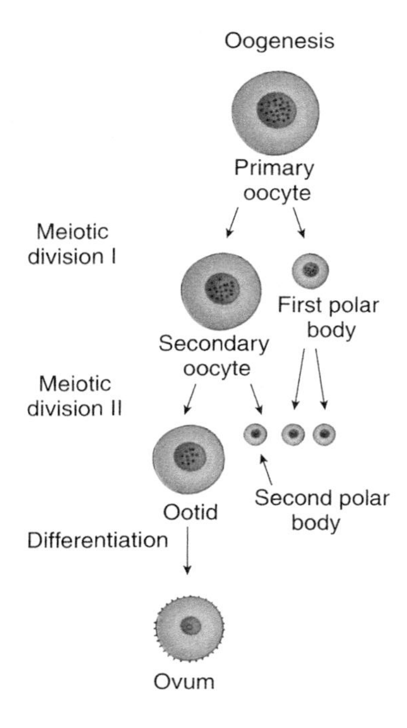

FIGURE 6.6 Oogenesis. A single oogonium undergoes oogenesis to produce a single large ovum (egg) and three tiny cells (polar bodies) that play no further role in normal development.

a **primary oocyte**. As such, it will be arrested in prophase I until at least menarche. Just before ovulation from the adult ovary, the primary oocyte completes meiosis I, producing two haploid cells: one large **secondary oocyte** and one small **first polar body**. The secondary oocyte proceeds through meiosis II until it reaches metaphase II, in which it remains until it is fertilized or dies. If fertilization occurs, the secondary oocyte completes meiosis II. As in meiosis I, cytokinesis is very unequal: it produces one large, haploid **ootid** (also known as the ovum or egg) and one small, haploid **second polar body**.

Polar bodies play no role in normal development and are probably just a means of reducing the chromosome number. For each oogonium that undergoes oogenesis, one functional egg is produced. During a human female's reproductive lifetime (barring pregnancies), the millions of germ cells once present are estimated to result in about 400 "eggs" being produced at the rate of about 1 per 28 days from menarche to menopause.

Follicles

As the oocyte develops in the ovary, it becomes associated with a number of somatic cells called **follicle cells** (or granulosa cells). Collectively, the germ cell and follicle cells make up an **ovarian follicle** (Figure 6.7). The oocyte surrounded by a single layer of follicle cells is called a **primary follicle**; two or more layers of follicle cells result in a **secondary follicle**. When fluid-filled spaces (**antral vacuoles**) begin to appear among the

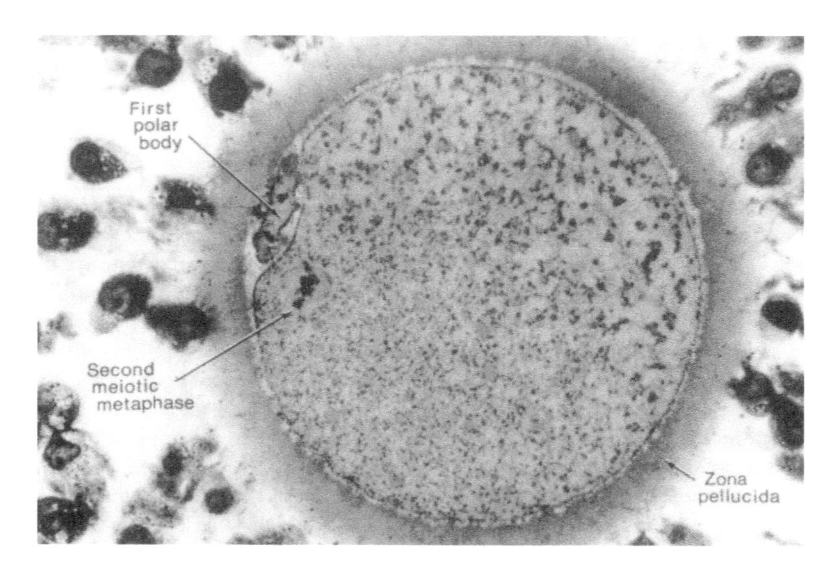

FIGURE 6.7 Ovulation and "the egg." What is actually released from the ovary at the time of ovulation is a secondary oocyte, a first polar body, a noncellular zona pellucida, and cellular corona radiata. (L. Zamboni and D. Fawcett/Visuals Unlimited.)

follicle cells, it is called a **tertiary follicle**. As the follicle matures, it moves to the surface of the ovary and causes the ovarian surface to bulge outward in a papilla. At this stage, we have a **mature follicle** (or **Graafian follicle**) containing a primary oocyte about to complete meiosis I and undergo ovulation (see Figure 5.10).

Egg

The egg is a differentiated cell that is destined to be fertilized by a spermatozoon or die. From the time it is released from the ovary, it has about 12 hours to be fertilized, or else it will die.

Characteristics

Unlike the spermatozoon, the female gamete (1) is relatively large: with the possible exception of fat cells, it is the largest human cell; (2) is nonmotile and is instead moved by forces in its environment; and (3) does contain some reserve food. The egg does not undergo the dramatic metamorphosis undergone by the developing sperm. Nevertheless, the egg does undergo a dramatic **biochemical cell differentiation** during oogenesis.

Prophase I

Recall that the primary oocyte becomes arrested in prophase I of meiosis. Prophase I is the most complicated part of meiosis. For the sake of study, it is subdivided into five substages, the details of which are beyond the scope of this book. Let it suffice to say that it is during the **diplotene**

stage of prophase I that the primary oocyte becomes arrested during its progress through meiosis I (as previously considered). However, it is during this stage that the oocyte is very active metabolically and synthesizes a variety of important biochemicals. Part of this synthetic activity is obvious in the growth of the size of the oocyte. At a more subtle level, this synthetic activity is stockpiling substances (such as maternal ribosomes and **maternal messenger RNA [mRNA] molecules**), which will be used to support development before the conceptus's own genes are able to produce ribosomes and mRNA.

Egg coats

As a cooperative venture between the primary oocyte and its surrounding follicle cells in the ovarian follicle, a noncellular coat (**zona pellucida**) is laid down around the outside of the egg. At about the time that the egg is to be released (ovulated) from the ovary, it completes meiosis I, resulting in formation of the secondary oocyte and first polar body. Consequently, the first polar body is retained beneath the zona pellucida in the fluid-filled **perivitelline space**.

When **ovulation** occurs, the ovary releases a secondary oocyte, a first polar body, a surrounding zona pellucida, and a number of follicle cells (collectively called the **corona radiata**), which adhere to the outside of the zona pellucida (see Figure 6.7). At the time of fertilization, the fertilizing spermatozoon encounters this collection of cellular and noncellular material in the ampulla of the fallopian tube.

A dogma challenged: For a good part of the twentieth century, the prevailing dogma was that the ovaries of mammals, including humans, did not contain stem cells after birth. By the time a baby girl was born, all of her fetal oogonia had progressed to being primary oocytes; in other words, all of her fetal ovarian stem cells had ceased being stem cells and had begun to differentiate. Early in the twenty-first century, researchers began to accumulate evidence that this was not the case, and postnatal human ovaries may indeed contain stem cells. For a recent review of postnatal oogenesis in humans, see https://www.ncbi.nlm.nih.gov/pmc/articles/PMC4376261/.

Stem cells: For a general background on stem cells, see Chapter 3 on cell division. We consider stem cells here, again, because gametogenesis involves two important kinds of stem cells. (1) The male spermatogonia, which divide by mitosis to give rise to additional spermatogonia or to primary spermatocytes, which divide by meiosis and undergo differentiation to give rise to sperm, and (2) the female oogonia, which divide by mitosis to give rise to additional oogonia or to primary oocytes, which divide by meiosis and undergo differentiation to give rise to ova. This is to say that spermatogonia and oogonia constitute stem cell populations. Figure 6.6 also shows two asymmetric divisions of non–stem cells, that is, division of a primary oocyte to produce a larger secondary oocyte and a smaller first polar body, and division of a secondary oocyte to produce a larger ootid and a smaller second polar body. Polar bodies are small cells.

Types of stem cells*

Totipotent Stem Cells	Capable of forming an entire organism
Pluripotent Stem Cells	Capable of forming most, but not all, tissues in an organism
Multipotent Stem Cells	Limited to giving rise to specific populations of cells
Induced Pluripotent Stem Cells (iPSCs)	Pluripotent stem cells derived from differentiated cell types

Examples of types of stem cells*

Totipotent Stem Cells	Embryonic stem cells (ESCs)
Pluripotent Stem Cells	Embryonic germ cells
Multipotent Stem Cells	Umbilical cord blood and tissue stem cells
Induced Pluripotent Stem Cells (iPSCs)	Pluripotent stem cells derived from differentiated cell types

Types of perinatal stem cells* (Prenatal and Postnatal Stem Cells)

Stem cell type	Developmental stage/source/potency
Human embryonic germ cells (hEGCs)	Prenatal/gonadal ridges/pluripotent; give rise to all 3 germ layers
Fetal stem cells	Fetuses/various fetal organs/pluripotent
Cord blood stem cells	Postnatal/newborn umbilical cord blood/multipotent
Cord tissue stem cells	Postnatal/umbilical cord tissue matrix cells/multipotent
Placental stem cells	Postnatal/cells derived from "placental tissue," that is, amniotic and chorionic membranes, chorionic villi, umbilical cord, and maternal decidua/multipotent

* Tables created from Hildreth, C. Definitive Guide to Perinatal Stem Cells, August 23, 2017 https://www.bioinformant.com/perinatal-stem-cells/.

Study questions

1. Cells of the body may be classified into what two categories? How are they functionally different?
2. What are gametes and what are the two kinds of gametes? What is their formation generally and specifically called?
3. Where and when do the primordial germ cells of both sexes first appear?
4. What are the primordial germ cells called when they reach the developing testes? The developing ovaries?
5. Which cells of the testes undergo meiosis I? Meiosis II? Which cells undergo spermiogenesis?

6. A single spermatogonium undergoing spermatogenesis gives rise to what (how many cells, cell type, and number of chromosomes)?
7. Do adult ovaries have stem cell populations? When do the ovaries contain their greatest number of germ cells?
8. About how many germ cells are found in the ovaries at birth, at menarche, and at menopause?
9. A single oogonium undergoing oogenesis gives rise to what (how many cells, cell type, and number of chromosomes)?
10. What are the three basic parts of a spermatozoon, and what is the function of each part?
11. What are the three roles of the spermatozoon in reproduction and development?
12. What factor in its environment affects the motility of the spermatozoon?
13. Compare and contrast the sperm and egg as to size, motility, and nutrient storage.
14. When an egg is released from the surface of the ovary at the time of ovulation, what, in detail, is actually released?
15. Do postnatal human ovaries contain stem cells?
16. Which type of gametogenesis involves asymmetric cell divisions and which type involves only symmetric cell divisions?

Critical thinking

1. Which human gamete is bigger, which is self-propelled, and which carries the greater responsibility for early development? What is meant by the greater responsibility for early development?
2. How are germ cells found in ovaries and testes at birth fundamentally different? (Hint: Think of meiosis and stem cell populations.)
3. Unlike the forming spermatozoon, which undergoes a dramatic morphological differentiation, the developing oocyte undergoes a dramatic biochemical cell differentiation. Explain.

CHAPTER SEVEN

Fertilization

CHAPTER OBJECTIVES

After studying this chapter, you should be able to:

1. Discuss the role of fertilization in human reproduction and development.
2. Describe the barriers to fertilization.
3. List the consequences of fertilization.
4. Describe the chromosomal basis of human sex determination.
5. Explain sex-linked inheritance and give examples of characteristics inherited in this way.

Two cells on the verge of death are the participants in **fertilization**, one of the most thought-provoking events in biology. If these two cells undergo fertilization, a new individual may result, who may in turn reproduce, leading to a long lineage of descendants as the process repeats itself. After the egg is released from the ovary during ovulation, it has a lifetime of about 12 hours; the spermatozoon has a lifetime of about 2 days in the female reproductive ducts. If the two cells do not meet during this time limit, that is the end of it.

Barriers to fertilization

A typical ejaculate of sperm contains about 350 million sperm, but only 50–100 reach the vicinity of the egg. This is because the spermatozoa face obstacles every step of the way—from the posterior fornix of the vagina to the ampulla of the fallopian tube. These obstacles include:

1. Vaginal acidity, which impedes sperm motility.
2. Traversing the mucus-filled cervical canal.
3. A relatively lengthy passage through the uterine cavity.

4. Selecting the correct fallopian tube because the wrong one leads to oblivion.
5. A vigorous swim to the immediate vicinity of the egg.

When the sperm finally encounters an egg, it is surrounded by two physical barriers: the outer, cellular **corona radiata** and the inner, noncellular **zona pellucida**. Both barriers must be breached before the plasma membranes of the two gametes can come into contact, an event necessary for cell fusion.

Despite all the obstacles, the sperm's unique structure makes fertilization possible. The acrosome of the sperm's head is a bag of digestive enzymes, which, together with the physical activity of the sperm, enable the sperm to get through the egg coats and into the fluid-filled **perivitelline space** between the zona pellucida and the egg plasma membrane (Figure 7.1). Once the sperm is in the perivitelline space, the plasma membrane of the posterior part of the sperm head contacts the plasma membrane of the egg. Subsequent fusion of these membranes provides a channel through which the contents of the sperm may enter the cytoplasm of the egg. The important contents to do so are the **sperm nucleus** and the **sperm centriole**.

Consequences of sperm penetration

The sperm's penetration of the egg triggers a series of changes.

1. The egg (secondary oocyte) completes meiosis II and becomes a bona fide egg (ovum), and the **second polar body** and **female pronucleus** are formed.
2. The sperm nucleus swells up, and its contained chromatin disperses, forming the **male pronucleus**.

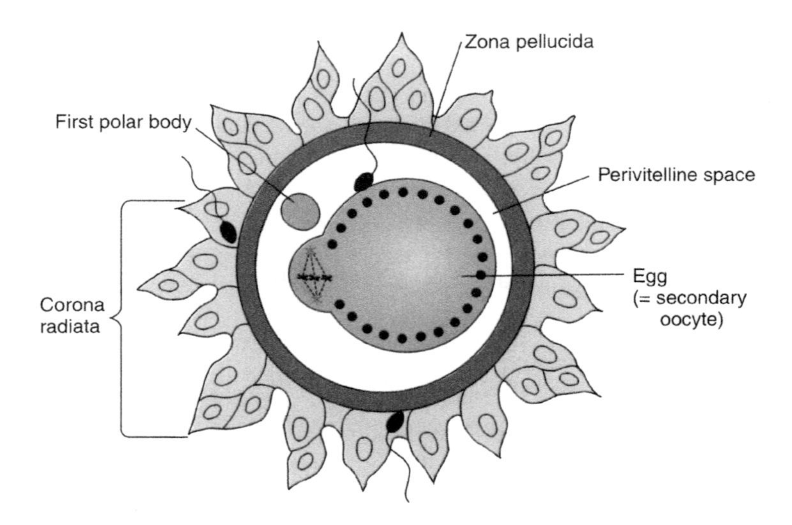

FIGURE 7.1 Breaching the egg coats. To enter the perivitelline space and come into contact with the surface of the egg, the successful spermatozoon must breach the cellular corona radiata and the noncellular zona pellucida.

3. As the **pronuclei** (male and female, each with 23 chromosomes) approach the center of the egg, the sperm's centriole begins to organize the spindle (see Chapter 3) of the first cleavage division.
4. The pronuclei each deposit their 23 chromosomes on the equator of the newly formed spindle, and the human diploid number of 46 chromosomes is reestablished.
5. Finally, the **egg is activated**: biochemical changes trigger DNA and protein synthesis, and the egg begins to develop into a new individual.

Significance of fertilization

At the time of fertilization, a unique genotype is created. This new and unique collection of genes results from the fusion of two cells. The spermatozoon brings the father's set of genes and the ovum brings the mother's set of genes to the zygote (fertilized egg). Thus, fertilization is an integral part of reproduction—the process by which parents produce children.

In addition to the transmission of genes from the parent to the offspring generation, fertilization also reestablishes the diploid number of 46 chromosomes. As far as the number of chromosomes is concerned, fertilization offsets the reduction of chromosome number that resulted from meiosis during the formation of the gametes.

Since children resemble their parents to a greater or lesser extent—after all, they are children and not puppies or kittens—it was once thought that children inherited their parents' characteristics. We now understand that what is inherited by children is the *potential* to develop their parents' characteristics. This potential resides in the genetic information encoded in the DNA molecules making up the genes inherited from the parents. When this information is transmitted to the zygote, fertilization has done its *genetic* job. By activating the egg to begin development, fertilization does its *developmental* job.

With the activation of the egg, numerous processes are set in motion and development begins. Very early in human development, the new embryo's genetic information, funneled from two parents to the offspring by fertilization, begins to be expressed (see Chapter 4 for details of gene expression), the preamble is over, and our story of development begins.

Sex determination

The 46 human chromosomes may be classified into two categories: two **sex chromosomes**, which determine genetic sex, and 44 **autosomes**, which do not (each autosomal chromosome contains thousands of genes). The sex chromosomes are designated *X* and *Y*, and the autosomes are designated by numbers 1 through 22 (Figure 7.2). At fertilization, we normally receive two sex chromosomes—one from each parent. A normal egg contains one X chromosome. However, sperm come in two varieties: those with one X chromosome and those with one Y chromosome.

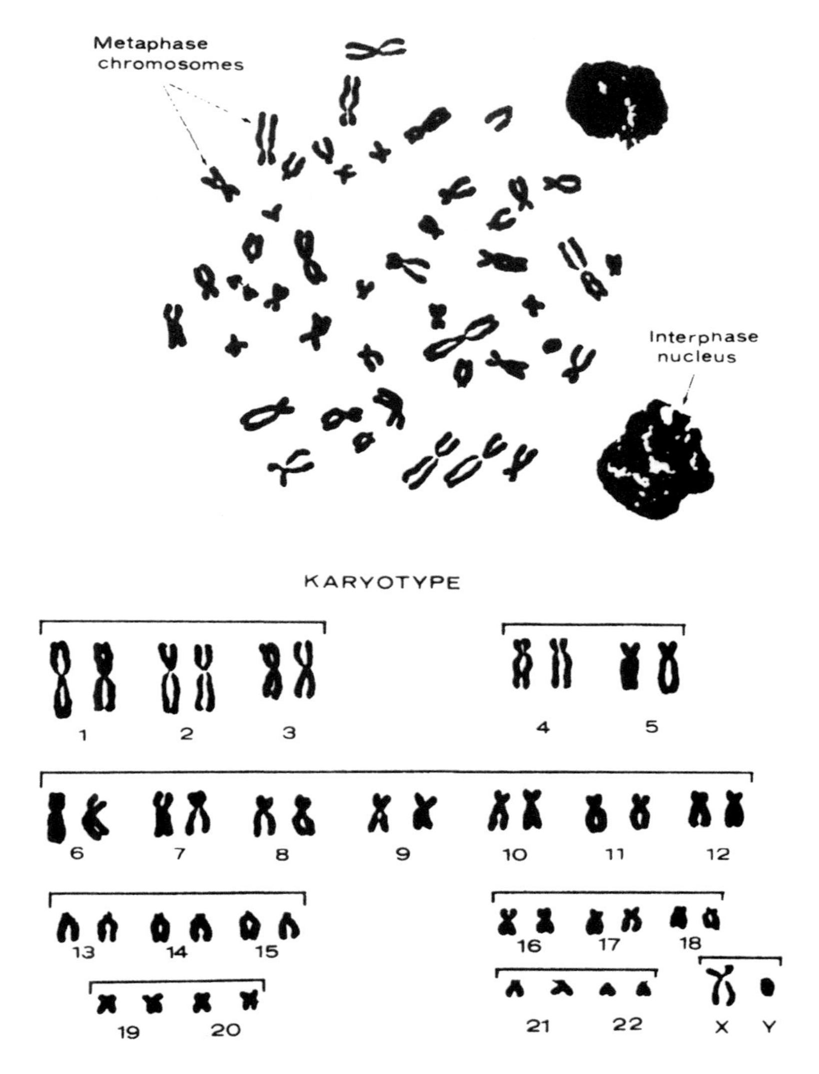

FIGURE 7.2 Human chromosomes. At top of the figure, a metaphase spread of human chromosomes obtained by using a mitosis-inhibiting drug. At bottom, the 46 chromosomes arranged as a karyotype. The sex chromosomes are the two chromosomes in the lower right-hand corner of the karyotype. (Don Fawcett/Visuals Unlimited.)

The human egg plays a passive role in sex determination because it normally contains one X chromosome. If the egg is fertilized by a Y chromosome-bearing sperm, the conceptus will have an ***XY*** sex chromosome makeup and will be a male genetically. If the egg is fertilized by an X chromosome-bearing sperm, the conceptus will have an ***XX*** sex chromosome makeup and will be a female genetically. In rare cases, an individual may be genetically a male and phenotypically (outwardly) a female, or vice versa.

Sex-linked inheritance

Note that **sex determination** should not be confused with **sex-linked inheritance** (also called **X-linked inheritance**). Sex-linked inheritance refers to those traits that are determined, and therefore inherited, by genes located on the X chromosome (**sex-linked genes**). The genes on the Y chromosome are expendable for the individual because no normal human female has them. This is not true of the vital genes found on the X chromosome because no living human lacks an X chromosome.

Autosomal chromosomes occur in pairs (Figure 7.3). Each member of a pair carries a gene for a given characteristic; that is, we have a pair of genes for characteristics carried on autosomal chromosomes. The members of the pair of genes may be identical (homozygous condition) or nonidentical (heterozygous condition). Sex chromosomes occur in pairs, but unlike autosomal chromosomes, they are not similar morphologically (the X chromosome is larger than the Y chromosome) or genetically (the X chromosome is gene rich; the Y chromosome is gene poor). Genes on the X chromosome do not have counterparts on the Y chromosome. Therefore, females, with two X chromosomes, may be homozygous or heterozygous for genes carried on X chromosomes. Males, having only one X chromosome, are said to be hemizygous for genes carried on X chromosomes.

Since a male has only one X chromosome, inherited from his mother, he has only one copy of the sex-linked genes (see Figure 7.3). If a male receives a "bad" gene (gene determining an unfavorable or unhealthy characteristic) on his X chromosome, it will be expressed and may give him color-blindness, hemophilia, or a very early death. On the other hand, if a female receives a bad copy of a gene on one of her X chromosomes, she still has a chance that her other X chromosome will have a good copy of the gene, so the bad copy of the gene will not be expressed. Statistically, a female is less likely to have color-blindness, hemophilia, and other sex-linked conditions.

Types of inheritance

The type of inheritance most people are familiar with is genetic inheritance through the genes on the chromosomes that we receive from our parents. A second type of inheritance, **maternal inheritance**, is controlled by extrachromosomal determinants (factors in the cytoplasm placed there when the egg was developing in the ovary).

Cell fusion: In humans, cell fusion is not a common event; three examples of cell fusion in humans occur during development: (1) during fertilization; (2) when cells called myoblasts fuse together to form multinucleated skeletal muscle fibers (a multinucleated mass of cytoplasm in animals is called a syncytium); and (3) during the formation of a part of the human placenta called the syncytiotrophoblast—this part

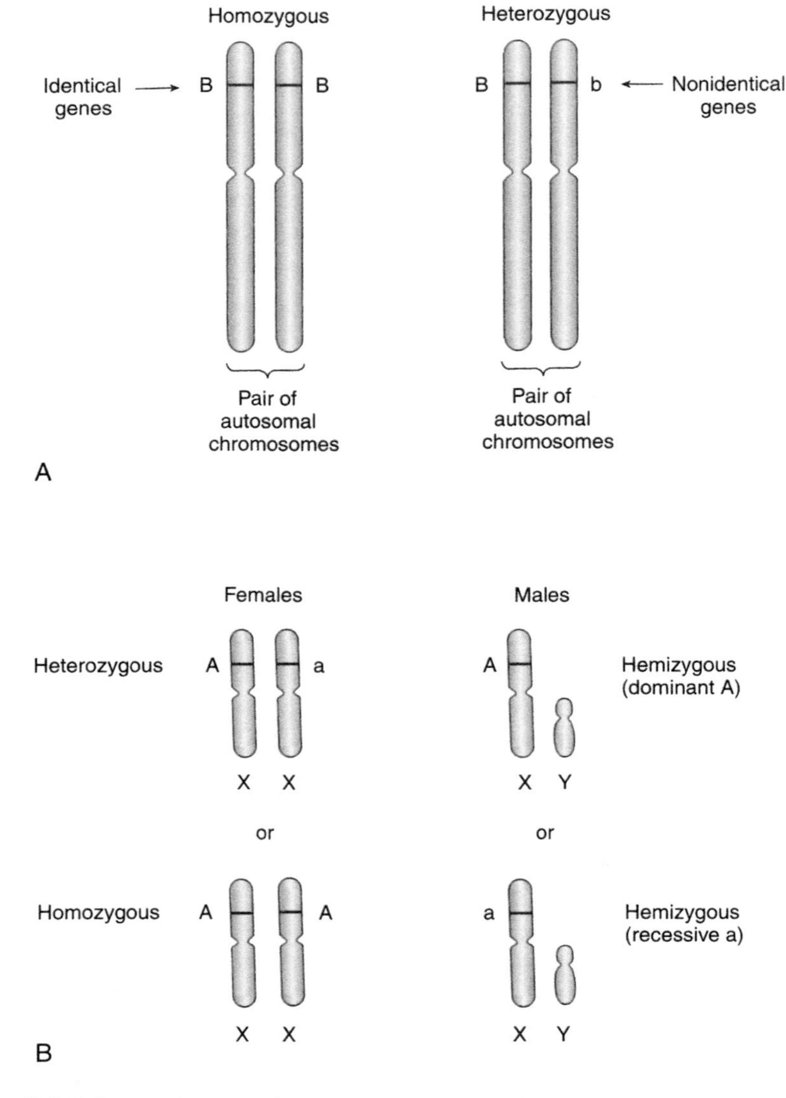

FIGURE 7.3 Autosomal genes and genes on the sex chromosomes. (A) Autosomal genes occur in pairs because they are found on autosomal chromosomes. (B) Genes on the sex chromosomes in males are generally not found in pairs because the X chromosome is gene rich and the Y chromosome is gene poor.

of the placenta is quite invasive and enables the conceptus to implant into the lining of the uterus, thereby initiating the pregnancy. Cell fusion may also be brought about, in cell culture, artificially through the use of chemicals (polyethylene glycol) or by the use of viruses (inactivated Sendai virus). It is quite remarkable that the human body is composed of trillions of cells, living in close quarters, and yet cell fusion is not a common event.

Study questions

1. What are the approximate lifetimes of the egg and sperm in the female reproductive ducts?
2. What two physical barriers to fertilization surround the egg?
3. The fusion of what two membranes is necessary for fertilization to occur?
4. What two important sperm contents enter the egg's cytoplasm?
5. The sperm's penetration of the egg triggers what five changes?
6. Human chromosomes are classified into what two categories? Which play a role in sex determination?
7. As far as sex determination is concerned, which parent's gamete plays the decisive role?
8. Where are the genes for sex-linked inheritance located?
9. In human development, if a Y chromosome is not present, what sex will the individual be?
10. Give three examples of normal cell fusion that occur during development.

Critical thinking

1. If the sperm and egg do not meet, what is the fate of both gametes? If the sperm and egg do meet, what may be the fate of the resulting zygote?
2. For individual human development, the Y chromosome is dispensable, but the X chromosome is not dispensable. Explain.
3. Statistically, why is a human female less likely to have color-blindness, hemophilia, and other sex-linked conditions?

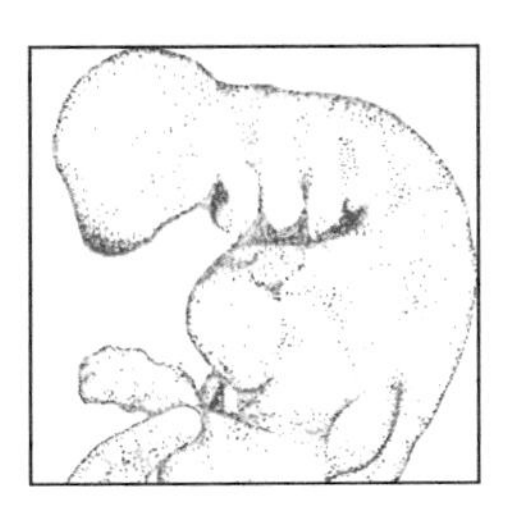

CHAPTER EIGHT

The embryonic period

CHAPTER OBJECTIVES

After studying this chapter, you should be able to:

1. Define the embryonic period and explain its importance in human development.
2. Describe cleavage and explain its contribution to development.
3. Describe the cell differentiation that takes place during the blastocyst stage.
4. Describe the process of implantation, including the following terms: implantation, nidation, syncytiotrophoblast, cytotrophoblast, lacunae, placenta, intervillous spaces.
5. Name the layers of the two-layered embryonic disc.
6. Discuss the significance of gastrulation in development and describe the origin of each of the following: primitive streak, primitive node, notochord, endoderm, mesoderm, ectoderm.
7. Discuss the significance of the notochord in animal classification and development, including the following terms: chordata, chordate, vertebral column, vertebrates, embryonic induction.
8. Define and describe neurulation, including its relation to anencephaly and spina bifida.
9. Explain the role of the neural crest in development.
10. Describe the organization of mesoderm into different regions and state to what each region gives rise.
11. List the kinds of body folds and their contributions to the morphogenesis of the embryo.
12. Describe the 4-week human embryo, including the following: forebrain protuberance, cardiac prominence, liver bulge, somite bulges, lens thickenings, branchial (gill) arches, limb buds, tail.
13. Describe, in a general way, the 8-week embryo.

During the 8 weeks after fertilization of the egg, a cell continues a journey begun, we believe, by a single cell billions of years ago. If successful, this journey will extend into the unimaginable future. Our concern in this chapter is with just 8 weeks—the embryonic period—during which the single cell, the zygote, creates millions of cells of 200 different kinds molded into the form of a recognizable human being.

We first look at the creation of multicellularity by the process of cleavage and the initiation of cell differentiation with the blastocyst. Next, we watch a mass of cells, the inner cell mass, organize into an embryonic disc, which is composed of two layers of cells. This is followed by the process of gastrulation. As the embryologist Lewis Wolpert said, "It is not birth, marriage, or death, but gastrulation which is truly 'the important event' in your life." It produces an embryonic disc composed of three germ layers, which collectively constitute the "Rosetta stone" for understanding early embryonic development. Neurulation follows on the heels of gastrulation and we see the beginning of the formation of our central nervous system, the brain, and spinal cord. During the fourth week, the cell layers we have been following for 2 weeks are folded into tubes—the tubes of which we humans are composed. During the second half of the embryonic period, weeks 5 through 8, the embryo transforms from a fishlike creature to a recognizable human. What have you done in the last 2 months?

Cleavage (cell proliferation)

During the process of embryo formation, the single-cell stage of the human life cycle is transformed into the multicellular condition that characterizes all other stages. This change is begun through a process called **cleavage**. The period of cleavage is characterized by the series of mitotic cell divisions by which the organism acquires the property of **multicellularity** (being many-celled).

The manner in which the egg undergoes cleavage (the **pattern of cleavage**) is determined by two things: yolk and genes. Human eggs, which have a small amount of yolk (**microlecithal**), uniformly distributed (**homolecithal**), undergo a type of cleavage pattern (rotational, holoblastic, equal cleavage) resulting in cells that are of equal size (Figure 8.1). You have correctly guessed that the term lecithal means yolk; the word root "lecithal" is derived from the Greek word *lekithos*, meaning yolk.

The human egg

The fertilized egg (**zygote**) is a unique cell. As much as any other stage in the human life cycle, this cell is human, although obviously not a human being. Of all stages of human development, only the fertilized egg is a single cell (**unicellular**). This zygote contains human genes and carries information from one generation to the next. This information is of three types: (1) the **genome** or genes, (2) materials deposited by the mother while the egg is in the ovary (leading to **maternal inheritance**), and (3) information in the form of **genomic imprinting** (see Chapter 10).

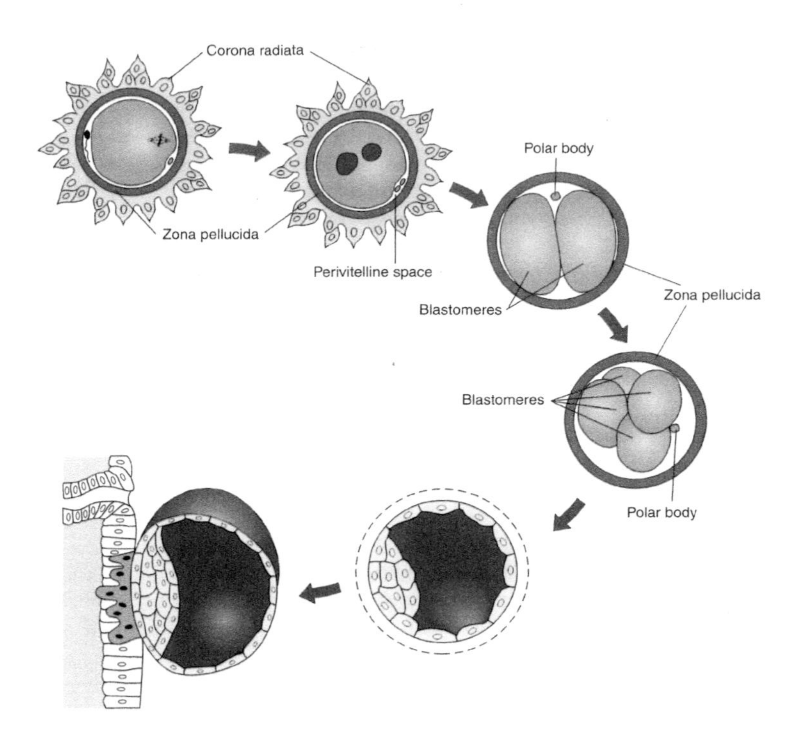

FIGURE 8.1 Early development of the egg. The early development of the egg, including cleavage and blastocyst formation, occurs in the fallopian tube. Cleavage occurs within the confines of the zona pellucida.

The informational content of this single cell is mind-boggling, but the amount of physical matter that makes up the zygote is microscopic. Nevertheless, even with a diameter of only about 100 μm (0.00004 inch), this is one of the largest human cells. To make an embryo, the zygote must undergo a transformation from its unicellular state into a multicellular animal. This transformation includes three fundamental developmental processes; **cell proliferation, cell differentiation**, and **morphogenesis**.

Comparative embryology

Earlier in this text, it was pointed out that comparative embryology is a powerful approach to the analysis of development. In this approach, rather than just observation of development in a single species, development is compared in several species. This provides greater insight into developmental processes. We will adopt this approach for the first time in considering cleavage (Table 8.1).

Yolk is stored food that the embryo requires to support its early development. Because it is physically inert, it has an inhibitory effect on cleavage. Therefore, the amount and distribution of yolk possessed by different kinds (species) of eggs have a great effect on the cleavage patterns undergone by these eggs. Microlecithal, mesolecithal, and macrolecithal eggs have small, intermediate, and large amounts of yolk, respectively. Isolecithal and telolecithal eggs have uniform and nonuniform distributions of yolk, respectively (Table 8.2).

Table 8.1 A classification of man

Taxon	Members of taxon		
Kingdom	Animalia	Animals	All animals
Phylum	Chordata	Chordates	All animals with a notochord
Subphylum	Vertebrata	Vertebrates	All chordates with a backbone (vertebral column)
Class	Mammalia	Mammals	All vertebrates except fish, amphibians, reptiles, and birds
Order	Primates	Primates	A group of mammals including monkeys, apes, and humans
Genus	*Homo*	Men	All humans who have ever existed
Species	*Homo sapiens*	Man	All living humans

Note: This table puts humans in the context of the animal kingdom.

Table 8.2 Types of eggs

Yolk amount (terminology)	Yolk distribution (terminology)	Vertebrate examples[a]
Large (macrolecithal)	Nonuniform (telolecithal)	Fish, reptiles, birds
Moderate (mesolecithal)	Nonuniform (telolecithal)	Amphibians
Small (microlecithal)	Uniform (isolecithal)	Mammals

[a] Eggs are classified according to their amount and distribution of yolk.

The yolk that most people are familiar with is the yolk of a hen's egg. In fact, this yolk is the egg and—enormous as it is—therefore a single cell. The other parts of the egg, the egg white, membranes, and shell, are accessory egg coats designed (by natural selection) to nourish or protect the developing chick. As you might imagine, the hen's egg (yolk) is of the macrolecithal type. The small whitish disc found on the surface of the yolk is a concentration of the active part of the egg (cytoplasm with contained nucleus), as opposed to the inert yolk. The yolk of the hen's egg is not uniformly distributed because most of the active part of the egg is isolated from the inert yolk. Thus, the hen's egg is telolecithal.

Unlike macrolecithal, telolecithal eggs produced by chickens, humans, and most other mammals (except the duck-billed platypus and the spiny anteater, which have birdlike eggs) produce microlecithal, homolecithal eggs. These eggs have little yolk (microlecithal), which is uniformly distributed (homolecithal) throughout the egg.

Patterns of cleavage

A number of different cleavage patterns are seen among animals. These are partially determined by yolk and partially by genes. These patterns may be categorized into two general types, **holoblastic** and **meroblastic**, according to whether the entire egg undergoes cleavage or not, respectively.

Table 8.3 Types of cleavage patterns

Yolk amount (terminology)	Yolk distribution (terminology)	Cleavage pattern	Vertebrate examples
Large (macrolecithal)	Nonuniform (telolecithal)	Meroblastic, discoidal	Fish, reptiles, birds
Moderate (mesolecithal)	Nonuniform (telolecithal)	Holoblastic, unequal	Amphibians
Small (microlecithal)	Uniform (isolecithal)	Holoblastic, equal	Mammals

Note: The type of cleavage pattern is determined by yolk and genes. Note the effects of the amount and distribution of yolk on cleavage patterns. Note also that genes have an effect on cleavage patterns.

Holoblastic and equal cleavage Human eggs undergo holoblastic cleavage, resulting in cells that are of equal size. More precisely, microlecithal, homolecithal human eggs undergo holoblastic, equal cleavage (Table 8.3).

Across the spectrum of the animal kingdom, a variety of animal eggs undergo holoblastic cleavage, but there are several subcategories of this type. A variety of microlecithal, isolecithal animal eggs undergoing different kinds of holoblastic equal cleavage indicates that something other than just yolk plays a role in determining the specific cleavage patterns. Indeed, the bilateral, spiral, radial, and rotational types of holoblastic cleavages are determined by the genes of the involved species. Mammals, including humans, undergo holoblastic, equal, rotational cleavage.

Meroblastic cleavage Because the hen's egg has an enormous amount of yolk, and since yolk inhibits cleavage, during chick development, the bulk of the egg (yolk) does not undergo cleavage. The bulk of the yolk is instead enveloped by the developing chick. Thus, the macrolecithal, telolecithal egg of the hen undergoes meroblastic cleavage.

Blastomeres

The cells produced by cleavage are called **blastomeres**. As the human zygote undergoes cleavage, an increasing number of ever-smaller blastomeres are produced. The "home" of the human embryo undergoing cleavage (which occurs as it is carried along) is the interior of the **fallopian tube**. In addition, the process occurs within an egg covering, the **zona pellucida** (see Figure 8.1), which has surrounded the egg (and later the early embryo) from before the time it was ovulated from the ovary.

Almost universally in the human body, cell division alternates with cell growth. Cleavage is an exception to this generalization. During cleavage, blastomeres alternate duplication of their genetic material (S phase of the cell cycle) with cell division (M phase of the cell cycle). Moreover, because growth is eliminated between divisions, as cleavage progresses, blastomeres get smaller and smaller. Obviously, there is a limit to how long this can continue. However, it does continue throughout the cleavage period. Blastomeres are uniquely able to do this because they are using material deposited by the mother in the egg. As these materials are exhausted, it becomes necessary for growth to be introduced (G1 and G2 of the cell cycle) between successive divisions. This is an example of the egg's uniqueness. Note that sometimes the early embryo is referred to as the dividing egg.

Some of the egg's uniqueness is reflected in the blastomeres, produced from the zygote by cleavage. In other animals (e.g., sea urchins or mice), if the early blastomeres (those of the two-cell or four-cell stages) are separated from each other, each one is able to give rise to a complete animal! This exemplifies a type of development referred to as **regulative** development and is also probably the mechanism of at least one means of the origin of human identical **twins**. As cleavage continues, later blastomeres lose this ability and, in this sense, become less and less unique. Related to regulative development is a type of human prenatal diagnosis wherein a single blastomere is removed from an eight-cell human embryo for genetic analysis, with the remaining seven-cell embryo regulating for the removal of the one blastomere (see Chapter 21).

The earliest human embryos are described by the number of blastomeres they contain. Thus, we have two-cell embryos, four-cell embryos, eight-cell embryos, and so on. Actually, this geometric progression, though accurate for many lower animals, is not accurate for mammals, since there are interspersed three-cell, five-cell, and so on stages.

Compaction

Before the human embryo leaves the fallopian tube to enter the uterine cavity, it undergoes a stage referred to as the **morula**. The morula is a stage in animal development characterized by a solid ball of blastomeres (Figure 8.2). During mammalian development, blastomeres undergo a process called **compaction**. Until this stage, the individual blastomeres are easily discernible—like members of a cluster of grapes. During compaction, however, the association of the blastomeres becomes so intimate that it is no longer possible to make out the individual blastomeres using an ordinary light microscope. While the cells are intimately associated, they form among themselves two types of connections, or **cell junctions**. These cell junctions allow the next stage of development, the blastocyst, to emerge.

Blastocyst (initiation of cell differentiation)

As the embryo leaves the fallopian tube to enter the cavity of the uterus, it transforms into the embryonic form called the **blastocyst**. As the embryo becomes the blastocyst, a fluid-filled cavity appears among the blastomeres, and two kinds of cells are discernible: trophoblast cells and embryoblast cells.

The outer cells of the embryo collectively make up the **trophoblast** and are joined to each other by **tight cell junctions**. The tight junctions between these cells and the cells themselves partition the

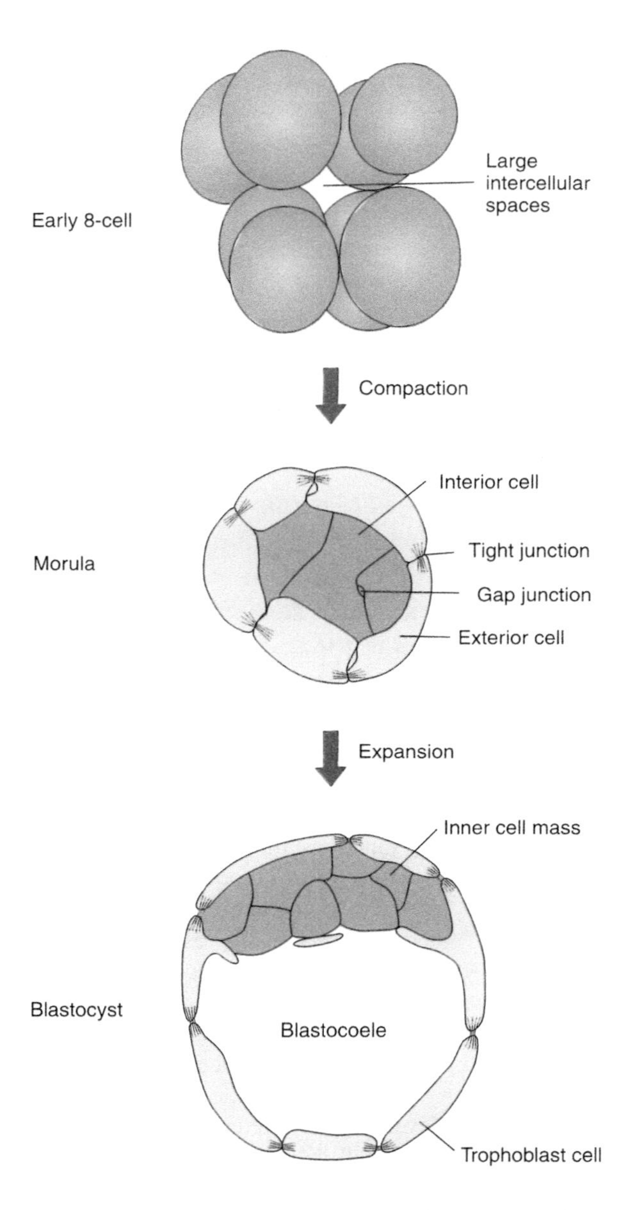

FIGURE 8.2 The morula, compaction, and blastocyst. Cells making up the trophoblast are joined together by tight junctions. These seal off the blastocyst cavity (blastocoele) from the surrounding environment. Gap junctions are formed between inner cells. These junctions are generally involved in the metabolic coupling of the cells that form them.

outside environment from the interior of the embryo and allow for the accumulation of fluid into the **blastocyst cavity** (**blastocoele**).

The remaining, inner blastomeres of the embryo make up the **inner cell mass** (**ICM** or **embryoblast**). These cells share cell junctions called **gap junctions**. In recent years, it has been found that tight junctions

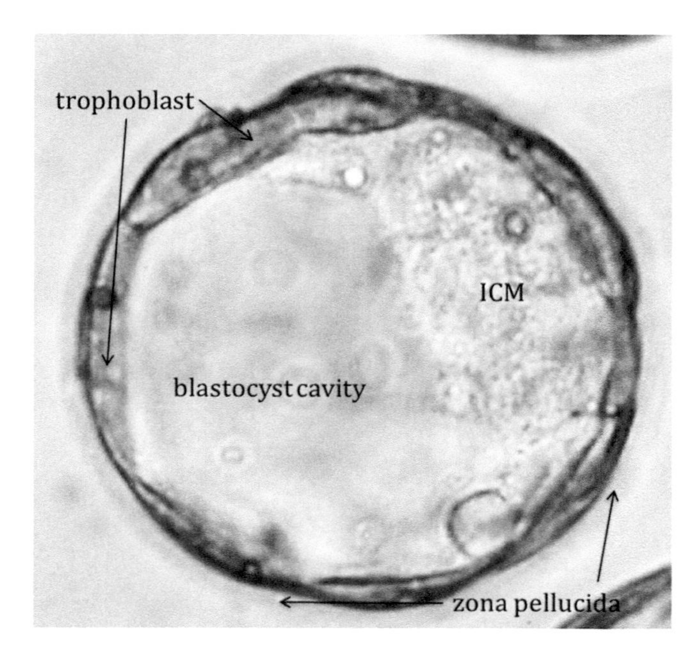

FIGURE 8.3 The blastocyst. This photograph shows a mouse blastocyst still within the confines of the zona pellucida. The inner cell mass is at the right. (Photo by the author.)

contribute to the formation of partitions between compartments, whereas gap junctions allow for communication between cells that share them.

The two cell populations of the blastocyst, trophoblast and ICM, give rise to **extraembryonic membranes** and the developing organism (e.g., embryo, fetus), respectively (Figure 8.3). The origin of these two populations is the first visible instance of cell differentiation during human development. Because an adult human is made up of more than 200 different kinds of cells, it is obvious that many more instances of cell differentiation will occur during human development.

We are now in the latter half of the first week (after fertilization) of development. The embryo is developing in its private fluid-filled space (perivitelline space), the boundary of which is the zona pellucida. Before the embryo can attach to the lining of the uterus (**endometrium**), it must **hatch** from the confines of the zona. This hatching is *not* very similar to a chick hatching out of its egg in which the egg shell is more or less broken into two pieces. During hatching of the blastocyst, a small opening is made in the zona by cells of the trophoblast, and the blastocyst squeezes out of its "shell" (Figure 8.4). This takes place on about the fifth or sixth day after fertilization.

Experimentation with early mouse embryos has shown that zonae are not sticky, but embryos from which zonae have been removed are quite sticky. After the embryo has hatched, its sticky trophoblast comes into contact with the endometrium, and the embryo attaches to it. A useful term to explain at this point is **conceptus**, which refers to the products of

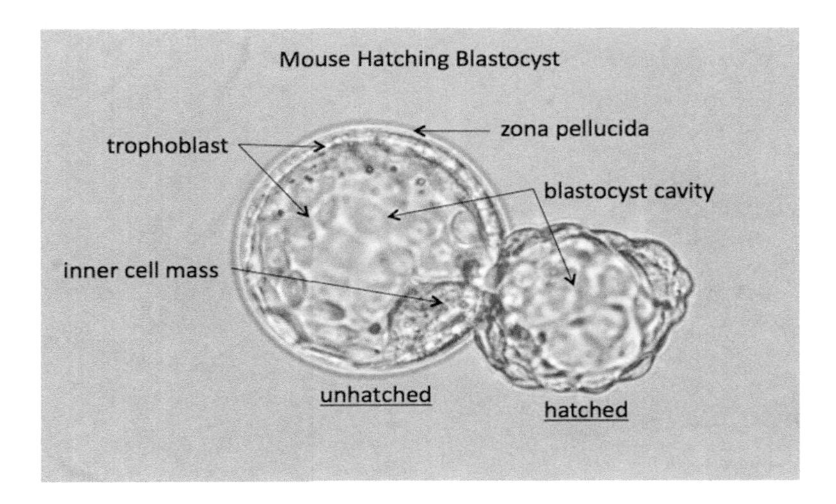

FIGURE 8.4 Hatching. Photograph of a mouse blastocyst escaping from the confines of the zona pellucida. Where part of the blastocyst has escaped from the zona pellucida (lower right), its surface is rough. The part of the blastocyst that has not yet hatched (upper left) is still within the confines of the smooth zona pellucida. (Photo by the author.)

conception, that is, both the developing organism (embryo or fetus) and its associated membranes. Usually, when the conceptus attaches to the endometrium, it does so by its embryonic pole, which is the region of the trophoblast at which the ICM is found.

After the trophoblast contacts the lining of the uterus, it becomes active, which is necessary for invasion of the endometrium by the conceptus (a process called **implantation** or **nidation**). Very soon, two recognizable regions of the trophoblast are apparent: (1) an outer portion, the **syncytiotrophoblast**, which has a syncytial rather than a cellular nature (i.e., individual nuclei are not packaged into separate cells, but share a common mass of cytoplasm), and (2) an inner portion, the **cytotrophoblast**, which, as its name indicates, does have a cellular nature (Figure 8.5).

It is the syncytiotrophoblast that is especially invasive. As it erodes its way into the endometrium, fluid-filled spaces (**lacunae**) appear between its strands. These lacunae are initially filled with fluid derived from tissue destruction caused by the invasive syncytiotrophoblast. As endometrial blood vessels and glands are eroded, maternal blood and glandular secretions fill the lacunae (Figure 8.6). With progressive development, an extraembryonic structure, the **placenta**, forms from a portion of the trophoblast and a portion of the endometrium. As this happens, the lacunae give rise to the **intervillous spaces**, which will contain the maternal blood to nourish the developing conceptus.

By the end of the first week of development, the conceptus is advanced in its implantation into the endometrium, but most women are unaware that they are pregnant because their next expected period is still a week away. During the second week, the inner cell mass transforms into a two-layered embryonic disc.

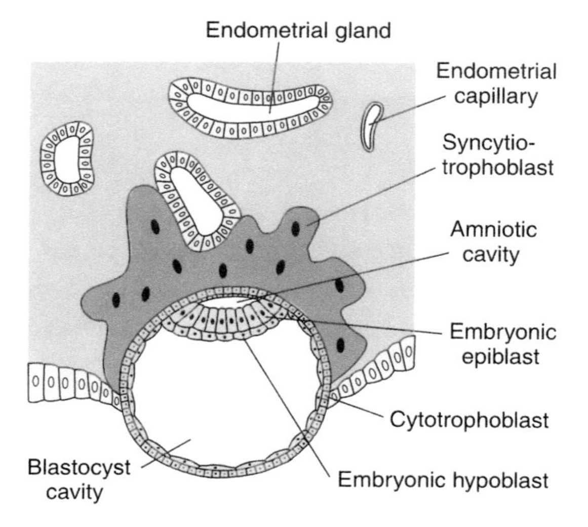

FIGURE 8.5 The trophoblast. Early in implantation, the trophoblast gives rise to an outer, highly invasive syncytiotrophoblast and an inner cytotrophoblast.

Two-layered embryonic disc

Recall that three vital processes in early development are cell proliferation, cell differentiation, and morphogenesis. We have already considered cell proliferation (during cleavage) and cell differentiation (trophoblast and ICM). Early in the second week, dramatic **morphogenesis** joins these two continuing processes. An example of morphogenesis is the

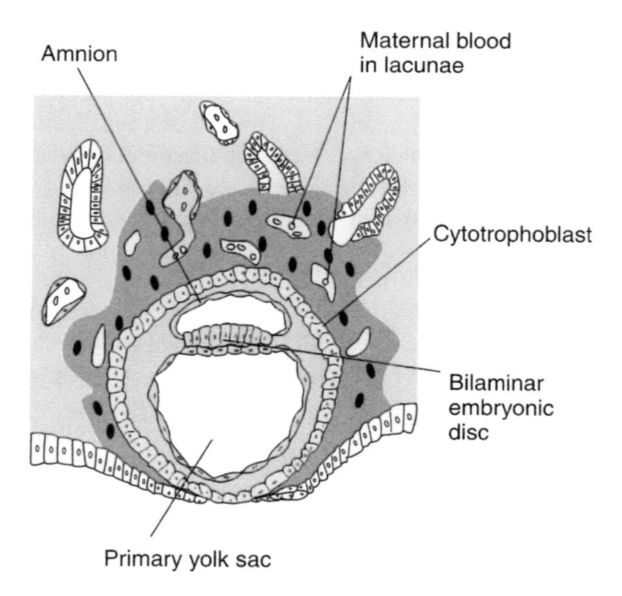

FIGURE 8.6 Lacunae filled with maternal blood.

formation of the two-layered embryo from the ICM. This two-layered embryo comes to occupy a position between two fluid-filled cavities: the **amniotic cavity** above it, and the **yolk sac** beneath it.

The amniotic cavity has a roof and walls of cytotrophoblast and a floor of the upper layer (**epiblast**) of the **embryonic disc** (see Figures 8.5 and 8.6). The contained **amniotic fluid** eventually fills an "aquarium" in which the embryo and fetus carry on their successive aquatic existences. This fluid is also an important source of liquid and cells for prenatal diagnosis later in pregnancy (see Chapter 19).

The roof of the yolk sac consists of the lower layer (**hypoblast**) of the embryonic disc; the rest of it is derived from trophoblast (see Figures 8.5 and 8.6). Although human yolk sacs do not contain yolk and have only a temporary existence, they are the source of two important kinds of cells: (1) **primordial germ cells**, which give rise to the gametes of adult life, and (2) the first **blood cells** formed by the conceptus.

The **two-layered embryonic disc**, amniotic cavity, and yolk sac all are contained within the fluid-filled **exocoelom** (or **extraembryonic coelom**), which has the **chorion** as its boundary. The chorion is the outermost layer of the conceptus, derived from the trophoblast, and, in turn, is in contact with uterine tissues (Figure 8.7).

The **amnion** (boundary of the amniotic cavity), yolk sac, and chorion, three of the four extraembryonic membranes formed during human development, all are composed of two layers. In addition to a layer derived from trophoblast, each also has a layer derived from what is called mesoderm.

One of the unifying concepts in animal development is that of **germ layers**. Three of these appear during development: **ectoderm,**

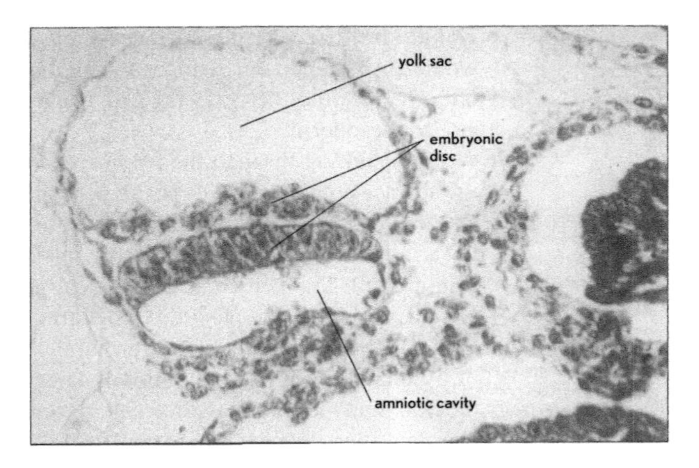

FIGURE 8.7 The chorionic vesicle. The embryonic disc, suspended between the amniotic cavity and the yolk sac, is within the fluid-filled extraembryonic coelom. The outer wall of this exocoelom is made up of the chorion derived from trophoblast and extraembryonic mesoderm. The chorion is in contact with the uterine tissues. (From England, MA: *Color Atlas of Life Before Birth*. Year Book Medical Publishers, Chicago, 1983, Figure D, p. 36.)

mesoderm, and **endoderm** (outer, middle, and inner skin, respectively). During the second week, the epiblast of the bilaminar embryonic disc gives rise to the endoderm, and during the third week, it gives rise to the mesoderm and ectoderm of the embryo. However, the mesoderm of the extraembryonic membranes is **extraembryonic mesoderm** and comes from the trophoblast.

At about the end of the second week, a woman notices that she is not having her expected menstrual period.

Gastrulation and the three-layered embryonic disc

Gastrulation, which occurs very early in the developmental process, is critical for human development. If something goes wrong during this process, development will be significantly abnormal. Gastrulation is most dramatic during the third week of development and consists of profound morphogenetic movements.

If we were somehow miniaturized and we entered the amniotic cavity of a beginning third-week human conceptus, we could look down on top of the two-layered embryonic disc, the epiblast, which would have a somewhat circular outline. Early in the third week, a thickened ridge of cells would appear on the midline of the epiblast. This ridge is called the **primitive streak** and is a characteristic of early developing higher vertebrates (reptiles, birds, and mammals). At the (future) head end of the primitive streak is a mound of cells, the **primitive node**.

As we watched what transpires during this third week, we would observe cells in the epiblast moving toward both the primitive streak and primitive node. As the cells arrive at these two locations, they would sink beneath the surface epiblast. Some of the sinking cells would sink deeper and displace the hypoblast to give rise to **embryonic endoderm**; other cells would sink to an intermediate position to give rise to mesoderm.

Cells sinking through the primitive streak move out laterad beneath the surface epiblast to give rise to mesoderm and endoderm. Those cells sinking through the primitive node would move out *beneath* the epiblast toward the future head end of the organism to give rise to the **notochord** (Figure 8.8). The notochord is characteristic of a group of animals that are classified into the phylum **Chordata**; this is to say that what makes an animal a **chordate** is the formation during its early development of a notochord. Most such animals later develop a **vertebral column** (backbone); such animals are called **vertebrates**. All vertebrates are chordates, but the converse is not true.

We humans are vertebrates, as are the most familiar animals—fishes, amphibians, reptiles, birds, and mammals.

The notochord is considered a region of the **embryonic mesoderm**. When all the cells destined to give rise to embryonic endoderm and mesoderm have moved out of the epiblast, what remains is **embryonic ectoderm**.

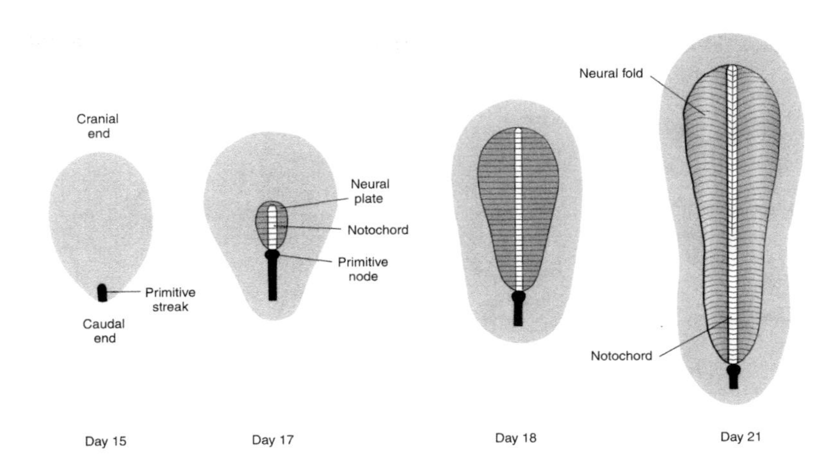

FIGURE 8.8 Origin of the notochord. The notochord arises from the primitive node at the cranial end of the primitive streak.

Neurulation

To the events of cell proliferation, cell differentiation, morphogenesis, and gastrulation, which occur during the third week after fertilization, we must now add the process of **neurulation**. This refers to the formation of the **neural tube**, a very important structure that gives rise to the **central nervous system**, which is composed of the **brain** and **spinal cord**.

Neural tube

The first indication of neurulation is the formation of a thickened region, the **neural plate**, in the ectoderm *above* the notochord. After its appearance, the midregion of the neural plate sinks inward, converting the plate into a **neural groove** flanked by two **neural folds**. Gradually, these folds come together in the midline and fuse to begin the formation of the neural tube (Figure 8.9A).

Although the process of neural tube formation starts during the third week, it is not completed until sometime during the fourth week. In the meantime, there are openings at the head end and tail end of the tube, referred to as the **anterior neuropore** and **posterior neuropore**, respectively (see Figure 8.9B and C). This means that the interior of the neural tube (the **neurocoele**) is in direct continuity with the fluid-filled amniotic cavity. If these anterior and posterior openings do not close, they will contribute to birth defects known as **anencephaly** and **spina bifida**, respectively (see Chapter 19). As you might expect, it is the head end of the neural tube that gives rise to the brain; the balance of the neural tube gives rise to the spinal cord (Figure 8.10).

Neural crest

When the neural folds fuse to form the neural tube, a region of ectoderm (**neuroectoderm**) disappears from the surface and the balance of the ectoderm (**epidermal ectoderm**) closes over it. Actually, a small population of ectodermal cells disappears from the surface but is not

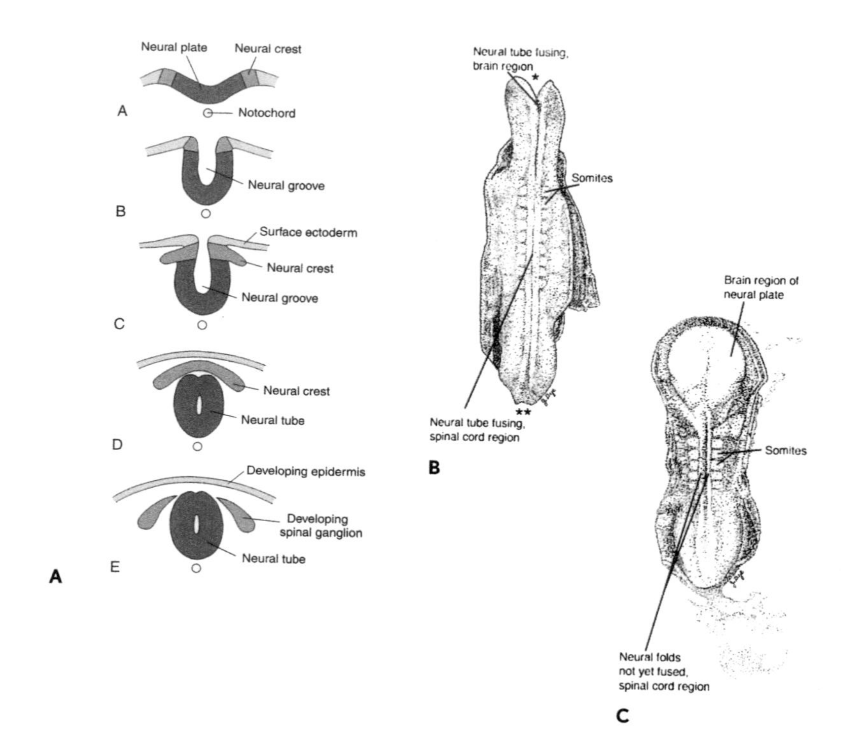

FIGURE 8.9 Neurulation. (A) Diagrammatic transverse sections through 3-week human embryos undergoing neurulation. (B and C) Surface views of human embryos at the beginning of the fourth week, showing the neural tube progressively closing, with open anterior (*) and posterior (**) neuropores.

incorporated into the neural tube; this population of cells is called the **neural crest** (see Figure 8.9A). Neural crest cells migrate from their initial position and give rise to a variety of structures in the developing embryo; for example, **ganglia** (collections of nerve cells outside the CNS), **pigment cells** (some of which provide us with a tan when we are exposed to the sun), and some cells (**medulla**) of the **adrenal glands** (which produce the hormones that prepare us for "fight or flight").

Embryonic induction

Many experiments (**experimental embryology**) with vertebrates (especially amphibians and birds) have shown that it is the presence of the notochord that causes the formation of the neural tube from the ectoderm. If the notochord is removed, a neural tube does not form; in place of the neural tube, ordinary skin ectoderm forms. In other words, the presence of the notochord changes the fate of cells. This is an example of **embryonic induction**, wherein one part of the embryo determines the fate of a different part of the embryo. Many embryonic inductions occur during our development, and they must occur during specific,

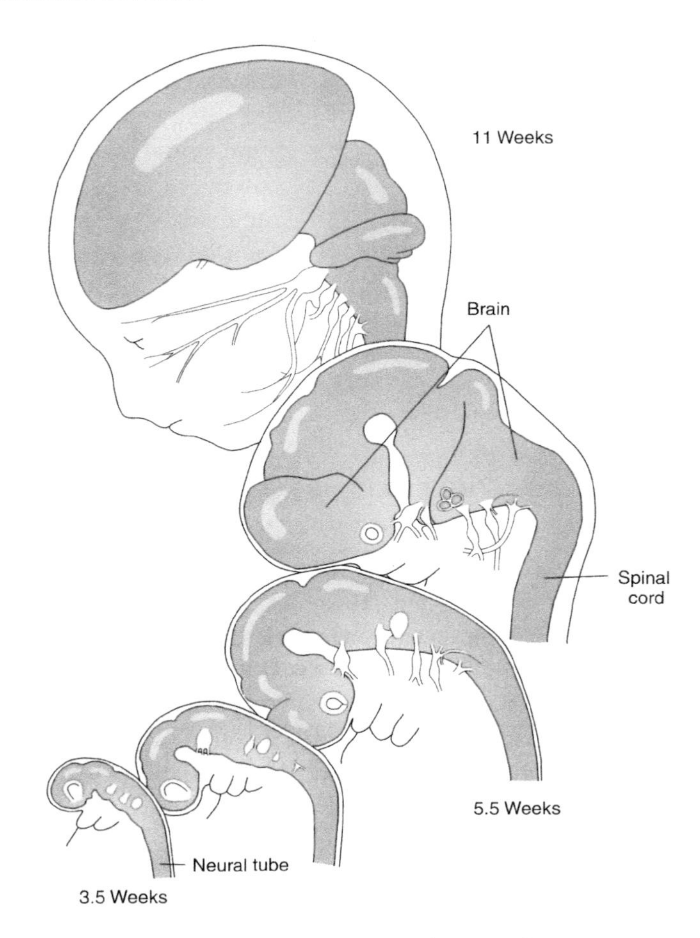

FIGURE 8.10 The neural tube gives rise to the central nervous system. The head end of the neural tube gives rise to the brain; the rest of the neural tube gives rise to the spinal cord.

narrow "windows of opportunity." Abnormal embryonic inductions mean abnormal development.

The changing mesoderm

Previously, we referred to the notochord as a part of the mesoderm; in fact, there are several parts of the mesoderm. To understand their spatial arrangement, it is necessary to review the concept of **bilateral symmetry** (see Chapter 1).

Somitic, intermediate, and lateral mesoderm

Imagine that the developing notochord occupies the same relative position in the embryo that your spinal column occupies in your body. If you move away from it toward either the left or right side of your body, you are moving laterad (in a lateral, or side, direction). In the embryo, the notochord is flanked on both sides by what is called **somitic mesoderm**

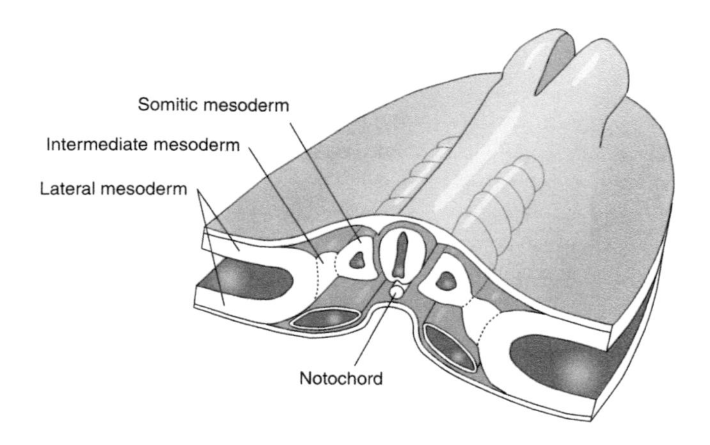

FIGURE 8.11 Regions of embryonic mesoderm.

(also called paraxial, segmental, or dorsal mesoderm). This is a thickened region of mesoderm and is called somitic because it will soon give rise to paired blocks of tissue (cell aggregates) called **somites**. Lateral to somitic mesoderm on both sides is **intermediate mesoderm**, and beyond this on both sides is **lateral mesoderm** (Figure 8.11).

To anticipate some embryonic shape changes (morphogenesis) that we are going to consider later, suffice it to say at this point that (1) much of the notochord will be replaced, but it will give rise to the **nuclei pulposi** (centers) of the adult intervertebral discs; (2) the somites will give rise to **vertebrae** of the spinal column, **dermis** of the skin, and some **muscles**; (3) a portion of the intermediate mesoderm will give rise to the **kidneys**; and (4) the lateral mesoderm will give rise to muscles of the gut and the body wall.

Cardiogenic mesoderm

Recall that the mesoderm arises from cells of the epiblast that move through the primitive streak. A portion of the lateral mesoderm, once it has moved laterad from the primitive streak, swings cephalad (toward the head region). This mesoderm comes to occupy a region beyond the head and is called **cardiogenic mesoderm** because it gives rise to the heart.

The heart initially arises from a pair of lateral mesodermal tubes that come together in the midline of the embryo and fuse to give a single mesodermal tube. This simple tube undergoes dramatic morphogenetic movements to eventually give rise to the four-chambered, valved **heart**. Because human eggs are not provided with sufficient yolk to support their entire development, it is necessary that the developing human conceptus quickly tap into a source of nourishment that can be delivered to all cells of the developing body.

Note that the first system to "come online" (begin to function) is the cardiovascular system. It begins to function at the end of the third week when the embryo is barely visible to the naked eye. As the embryonic period proceeds, a portion of the yolk sac of the embryo will be incorporated into the embryo as the early **gut**; nevertheless, nourishment provided by the mother reaches the embryo through its circulatory system, *not* through its digestive system, which does not begin to function until after birth.

From layers to tubes: The fourth week

During the fourth week after fertilization, the embryo undergoes dramatic changes in body shape. Afterward, it is much easier to see the **cephalic** (head) end and the **caudal** (tail) end, the **dorsal** (back) surface and the **ventral** (belly) surface, and the sides of the embryo.

Four body folds cause this dramatic conversion just about simultaneously: the **head fold**, the **tail fold**, and the two **lateral folds**. In addition to forming the head, the head fold brings the developing heart and **mouth** into their proper (adult) positions. The tail fold brings the developing **anus** into its proper position and, yes, human embryos do develop **tails**. As these two folds progress, portions of the yolk sac are incorporated into the body as the **foregut** and hindgut (Figure 8.12, left).

Simultaneously, the two lateral body folds complete the ventral (belly) body wall and mark the boundaries of the embryonic and extraembryonic regions of the conceptus (Figure 8.12, right). Even though the ventral body wall is "completed" during the fourth week, it will have protruding from it first a **yolk stalk** and a **connecting stalk**, then a single **umbilical cord** (see Figure 8.13B).

The changes in the external appearance of the human embryo during the fourth week can be thought of as the appearance of surface bulges and depressions. During this period, the embryo is quite transparent, and it is possible to relate the surface features to underlying developmental changes. Conspicuous surface features of this embryo are: **forebrain protuberance, cardiac prominence, liver bulge, somite bulges, lens thickenings, branchial (gill) arches**, limb buds, and tail.

At the end of the fourth week, the embryo does not look human, but, perhaps, somewhat fishlike. At this point, the study of a model of the embryo would be very instructive. If a model is not available, a photograph is a reasonable substitute (Figure 8.13).

Becoming a recognizable human: Weeks 5–8

During the second half of the embryonic period, the embryo undergoes a transformation resulting in what appears to be a diminutive human being. By the end of the eighth week, enough species-specific genes have been expressed to impose on the developing embryo the stamp of recognizable humanity. Still not capable of writing a book, composing a symphony, or running a wind sprint, this embryo will nonetheless evoke from the observer some degree of affinity. No longer is the embryo the "fish" of the middle of the embryonic period.

During this period of time, the branchial arches participate in the formation of the face and ears, and the limb buds give rise to arms and legs,

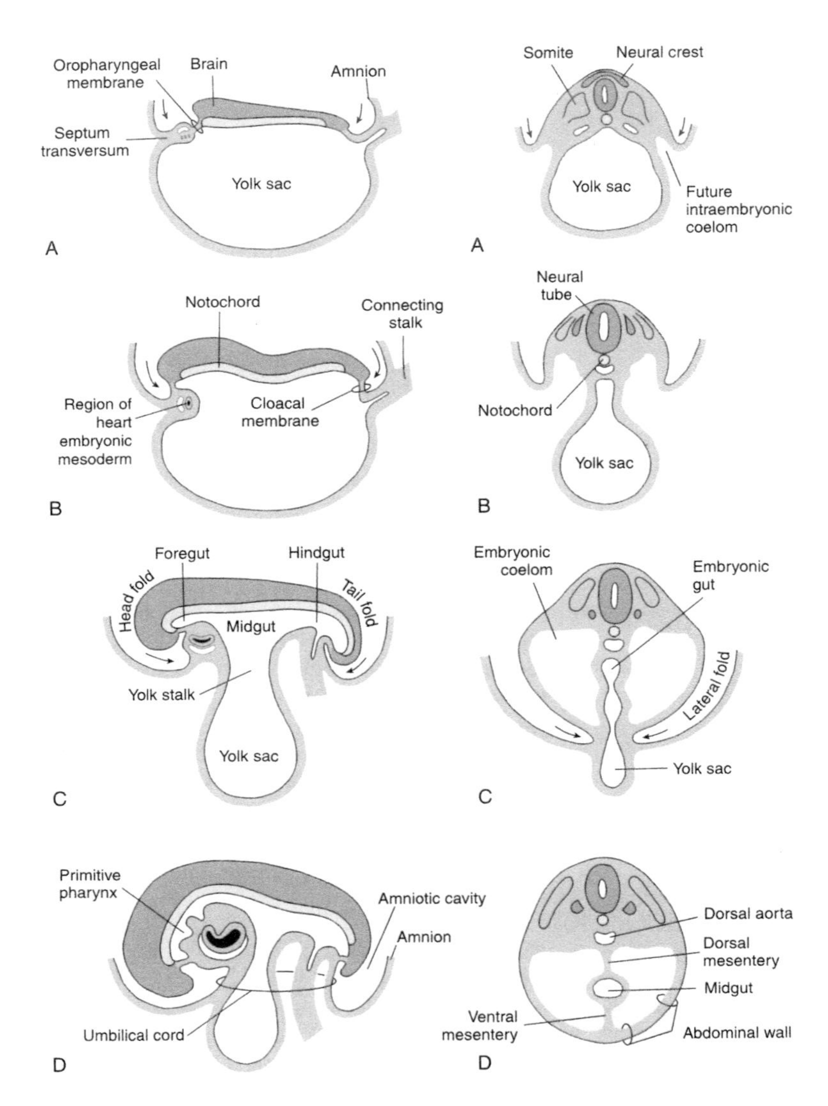

FIGURE 8.12 The body folds. (Left) Diagrammatic sagittal sections of embryos during the fourth week, showing the contributions of the head and tail folds to the formation of the tubular embryonic body. (Right) Diagrammatic transverse sections of embryos during the fourth week, showing the contributions of the two lateral folds to the formation of the tubular embryonic body.

hands and feet, fingers and toes. The various regions of the developing brain are no longer visible in surface view, but the disproportionately large head hints at the dramatic development occurring within. The somites will have disappeared from surface view as they, also, go about their genetically dictated tasks. The heart and liver have also disappeared from surface view, but the beating heart has begun a regimen of activity that must not cease for 70, 80, 90, or 100 years. The liver is on its way to

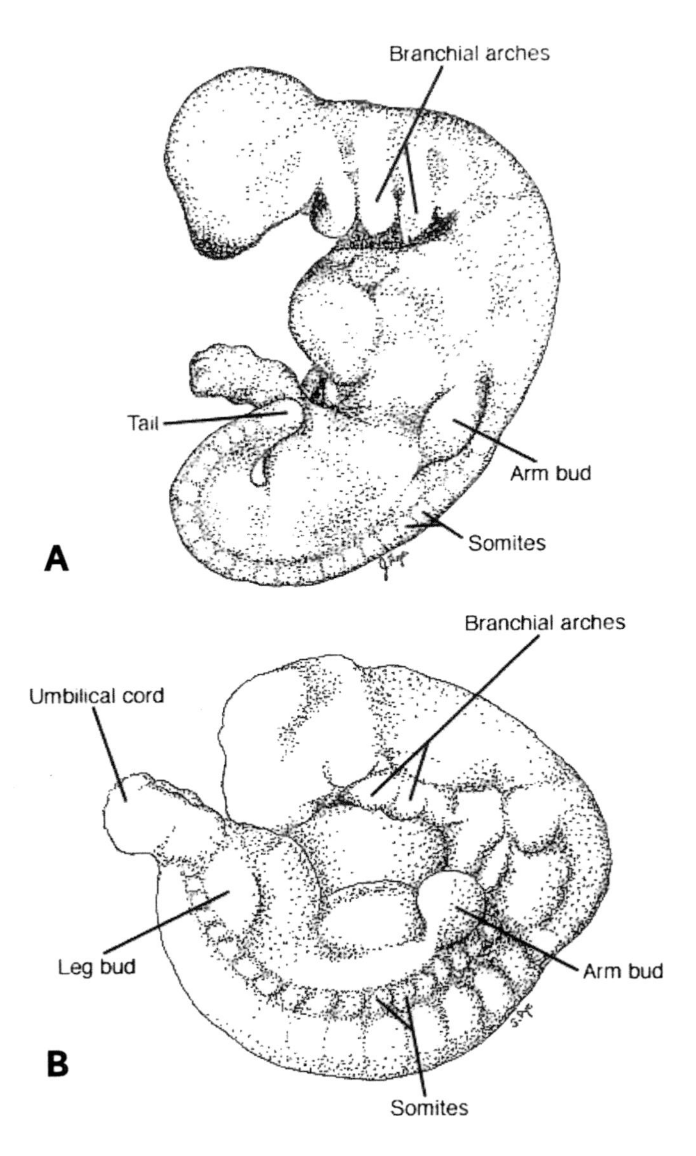

FIGURE 8.13 The human embryo at the end of 4 weeks. (A) Note presence of arm buds; the head is off the chest. (B) Note arm and leg buds; the head is on the chest.

becoming a most sophisticated biochemist, which is destined to produce vital biochemicals and detoxify poisons for a lifetime.

At the end of the eighth week, the embryo resembles a human, and it is now referred to as a **fetus** (Figure 8.14). Although this fetus is only slightly longer than the width of your thumbnail, every major system of the body has been laid down! It is during the fetal period that these systems grow and come to functional activity (begin to work). Between now and birth, the fetus grows and develops into a baby that must successfully make the transition from an aquatic existence to a terrestrial existence.

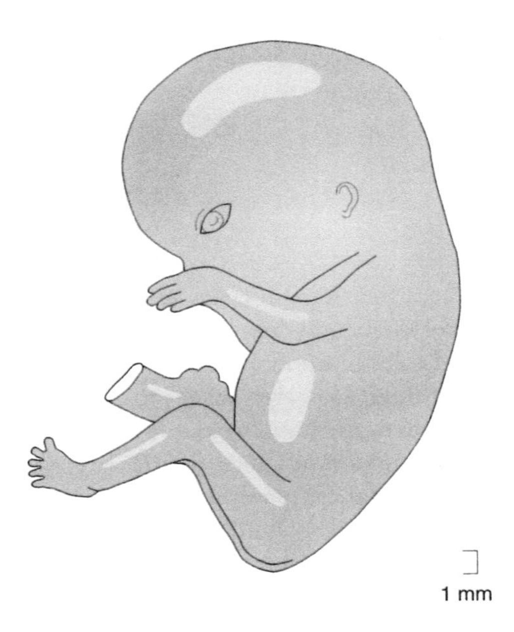

FIGURE 8.14 The human fetus at the end of 8 weeks.

Cell movements

Gastrulation and neurulation occur during the early embryonic period. Both of these processes involve cell movements, including invagination, evagination, epiboly, involution, convergence extension, migration, and ingression. Invagination is the movement of a sheet of cells into a preformed cavity. Evagination is the movement of a sheet of cells away from a preformed cavity, for example, the evaginations of the developing retinas from the lateral walls of the early forebrain. Epiboly is the spreading of cells upon a surface. Involution is the turning in of cells over a rim, for example, the involution of epiblast cells into the primitive streak. Convergence is the movement of cells toward each other. Extension is the elongation of a structure as a result of cell convergence. Migration is the movement of single cells, for example, the movement of neural crest cells throughout the developing embryo to give rise to a wide diversity of structures, such that the neural crest has been referred to as the **fourth germ layer**. Ingression is the movement of single cells out of a cell layer into a preformed cavity, for example, ingression of epiblast cells along the primitive streak to give rise to endoderm and mesoderm, or formation of the notochord by cells that ingress through the primitive (Hensen's) node.

Artificial embryos

In 2017, it was reported that scientists at the University of Cambridge had created an artificial mouse "embryo" for the first time. The embryo was created by using (1) embryonic stem cells, (2) trophoblast stem cells, and (3) a 3D scaffold (extracellular matrix) on which they could grow. Recall that the inner cell mass of the blastocyst, be it mouse or human, is the source of embryonic stem cells. The rest of the blastocyst consists of trophoblast stem cells (TSCs), which normally contribute to the formation of the placenta. The Cambridge researchers were able, by using these three components, to create a structure: (1) capable of assembling itself and (2) resembling the normally developing early embryo. Magdalena Zernicka-Goetz, who led the research, reported that: (1) the embryonic and extra-embryonic cells start to talk to each other, and (2) they become organized into a structure that looks and behaves like an embryo and has anatomically correct regions that develop in the right place and at the right time.

The Cambridge group, comparing their artificial "embryo" to a normally developing embryo, noted that its development followed the same pattern of development; that is, (1) the stem cells organized themselves with ESCs at one end and TSCs at the other end, (2) a cavity opened up within each cluster, and (3) these cavities joined together to become the large pro-amniotic cavity, in which the embryo develops.

http://www.cam.ac.uk/research/news/scientists-create-artificial-mouse-embryo-from-stem-cells-for-first-time

Study questions

1. What three types of information does the zygote convey to the new individual?
2. What process initiates the multicellular state of the new individual?
3. Microlecithal, homolecithal human eggs undergo holoblastic, equal cleavage. In your own words, what does this statement mean?
4. Yolk has what effect on cleavage?
5. Where does the human embryo undergo cleavage and within what membrane does cleavage occur?
6. When does the embryonic period of human development occur?
7. The embryo that arrives in the uterine cavity is in what stage of development? What is the fluid-filled cavity of the embryo called?
8. What are the two general parts of an early blastocyst and what type of cell junction is associated with each part?
9. From an embryologist's point of view, human development takes how many days? How many days for an obstetrician? What are the different points of reference used?
10. Before the embryo can attach to the endometrium, what must it do and from where?
11. What is implantation or nidation?
12. Which part of the trophoblast is especially invasive? What are lacunae and to what will they give rise?
13. The ephemeral human yolk sac is the source of what two important kinds of cells?

14. What are the names of the three germ layers formed in early development?
15. The profound morphogenetic movements of the third week of development collectively make up what process?
16. Through what do many cells move on their way to their destinations during the third week of development? What is the name of the cephalic end of this structure?
17. Humans are both chordates and vertebrates. What do these two terms mean?
18. What is neurulation?
19. What two birth defects are associated with abnormal neurulation?
20. Moving laterad, what are the three regions of mesoderm that bilaterally flank the early notochord?
21. Intermediate mesoderm gives rise to what? Lateral mesoderm gives rise to what?
22. Why is it necessary for human embryos to quickly tap into a source of nourishment?
23. What is the first system to come "online"?
24. What are the conspicuous external features of the 4-week human embryo?
25. When does the heart begin to beat?
26. Name six types of cell movements that occur during gastrulation and neurulation.
27. Artificial mouse embryos were recently created from what three things?
28. When the stem cells were joined together to form the artificial embryos, what did they start to do.

Critical thinking

1. The neural crest has been called the fourth germ layer. Why? List several derivatives (e.g., spinal ganglia) of the neural crest.
2. It has been said that gastrulation is the important event in your life. Why may this be true?
3. In what sense are embryonic inductions "windows of opportunity?" What is the relevance of this for birth defects?
4. Under normal conditions, if the brain is deprived of oxygen for 5–10 minutes, we die. What must the heart not do, even as it begins to form?
5. When does the human embryo become a fetus? What happens at this time?

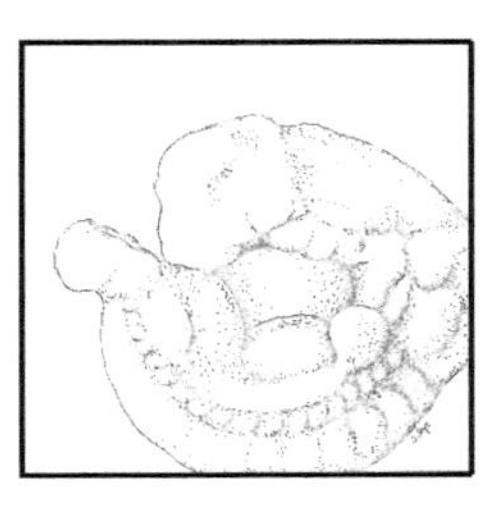

CHAPTER NINE

The fetal period

CHAPTER OBJECTIVES

After studying this chapter, you should be able to:

1. Define the fetal period.
2. Explain the importance of the fetal period in human development.
3. Give examples of structures that were established during the embryonic period, but come to functional maturity during the fetal period.
4. List the two major processes that occur during the fetal period.
5. Relate the size of the 8-week embryo and the fetus at the end of the first and second trimesters to the size of different parts of your own anatomy.
6. Describe the change in the relative size of the head during the fetal period.
7. State approximately when during the fetal period the following may be accomplished:
   - Gender of the fetus may be determined by external genitalia.
   - Fetal eyes close and open.
   - The umbilical hernia is retracted.
   - Fetus sucks on its thumb and swallows amniotic fluid.
   - The fetal body goes from lean to plump.
   - Fetal skin goes from wrinkled to smooth.
8. Describe the following: vernix caseosa, lanugo.

By the beginning of the fetal period, every major system of the body has been laid down! During the fetal period, these systems grow and begin to work, and the fetus grows and develops into a baby who must successfully make the transition from an aquatic to a terrestrial existence. The frog tadpole makes this transition over a period of weeks, whereas each of us has made this transition over a period of minutes. From an evolutionary point of view, this transition took much longer. Although the respiratory system begins to develop during the middle of the *embryonic period*, it is not ready to allow the change from water living to air living until the beginning of the third trimester at best. One of the most serious problems faced by the premature baby is respiratory distress syndrome (see Chapter 16).

Why the change in name from embryo to fetus? When the developing organism begins to resemble an adult of the species, it is referred to as a **fetus**, and the **fetal period** of development has begun. Beginning at the end of the eighth week of development (56 days after fertilization) and ending at birth, the fetus develops into a baby.

Developmental changes

Many significant developments occur during the fetal period. For example, sentience (having feeling) develops. During a significant part of the fetal period (from 9–26 weeks), the eyes are closed, but toward the end of the fetal period, the fetus can see light and hear sound. The heartbeat is affected by the level of light or the tempo of music to which the mother is exposed. This type of response might be a secondary response by the fetus to the mother's primary response. However, others point to the premature baby to support the theory that the responses are the baby's own.

A baby born prematurely can be thought of as a fetus outside the uterus, but such babies are certainly responsive to sensory input. Of course, the "fetus" lying in an air-filled hospital incubator must have sharper sensory input than the fetus inside the fluid-filled uterus. Before leaving this subject of premature babies (**extrauterine fetuses**), consider Portmann's theory proposed in the 1940s that human babies born at term are also extrauterine fetuses: on the basis of **comparative embryology**, human gestation should last 21 months rather than 9 months.

Structures established during the embryonic period come to functional maturity by birth. A good example is lung development. The laryngotracheal groove arises from the floor of the pharynx (throat) during the fourth week, and the resulting tracheal tube divides during the sixth week to form the lung buds. However, not until the 26th–28th week, during the fetal period, are the lungs mature enough to support life outside the uterus.

Another example of a structure that comes to functional maturity at birth is provided by the kidneys. The beginnings of the kidneys can be seen in the fifth-week embryo, but it is during the fetal period that the nephrons (functional units of the kidneys) are formed. The nephrons begin to excrete a dilute urine before birth.

Sex determination

Although the germ cells entered the developing gonads of the embryo during the 38th day after fertilization, we cannot determine the sex of the fetus by ultrasound until the early fetus sufficiently develops its external genitalia at about 12 weeks. The ability to use ultrasound depends on the presence of a sufficient amount of amniotic fluid. Although the fluid-filled amniotic cavity first appears during the eighth day after fertilization, it is some time later before ultrasound can be used as a prenatal diagnostic technique.

Growth

The major processes during the fetal period are growth and differentiation of previously established structures. An indication of the significant amount of growth that occurs during the fetal period is shown by the fetus's weight gain. From the beginning of the fetal period (at 9 weeks, when the fetus weighs 8 g) to the middle of the fetal period (at 23 weeks, when the fetus weighs 820 g), the fetus will have increased its weight 100-fold. By the end of the fetal period (38 weeks), the fetus will have quadrupled its midfetal weight to 3500 g!

Significant growth is also reflected in the increasing length of the fetus. At the end of the embryonic period, the embryo's length (27 mm) is about the width of an adult thumb. At the end of the first trimester, the fetus's length (92 mm) is about the width of the palm of a hand. By the end of the second trimester, the length of the fetus (250 mm) is about equal to the length of an adult hand span (from the tip of the thumb to the tip of the index finger of a stretched-out hand).

Differential growth of the fetus is shown by the dramatic morphogenetic changes that occur. At the beginning of the fetal period, the fetal head is half the length of the fetal body. By the end of the fetal period, the relative size of the head is reduced to one-fourth of the fetal length. On the other hand, during the same period of time, the legs have increased their relative size from one-eighth to one-half the fetal length.

Subdivisions of the fetal period

The 30-week fetal period is here divided into three stages of equal length, each consisting of two phases of equal length. However, keep in mind that all of development is a continuous process. For example, although ossification (bone formation) begins during the embryonic period—with the clavicles developing during the sixth week—it continues through the fetal period and into postnatal development.

Early fetal stage or "the fetus goes to sleep" (9–18 weeks)

At the beginning of the early fetal stage, the eyes are closing or closed and remain so well into the next stage. In this sense, the fetus is going to sleep. As the eyes close, eyelashes and eyebrows have not yet developed. During this stage, the cerebral hemispheres—derived from the forebrain—overgrow the midbrain. The pituitary gland assumes its characteristic morphology. By 10 weeks, the intestines have left the umbilical cord and are in the abdomen. Although not initially distinguishable, by 12 weeks the sex of the fetus is distinguishable externally.

Vernix caseosa (a whitish, cheesy coat on the skin of the fetus) and **lanugo** (fine, downlike hair confined to the skin of the fetus) are present on the fetal skin by the end of the early fetal period and persist until about term.

Phase 1 (9–13 weeks) (Figures 9.1 through 9.3): In the first phase of the early fetal period, the head is relatively large, the neck develops, and the chin is raised from the chest. The wide-set closed eyes have moved from the sides to the front of the head; the ears are low-set on the head. The mandible (lower jaw) grows rapidly, and the palate fuses, creating the roof of the mouth. The first permanent tooth buds form. The midgut herniation withdraws from the umbilical cord into the enlarged abdomen. The arms are well developed, and the fingernails are developing. The external genitalia have developed sufficiently by the end of this phase so that gender may be ascertained by observation. Blood vessels are easily seen through the thin skin.

The fetus may begin to suck on its thumb and toward the end of this phase begins to regularly swallow amniotic fluid (during the fetal period, the fetus also urinates into the amniotic fluid).

Phase 2 (14–18 weeks) (Figures 9.4 and 9.5): The external ears have moved from the upper neck onto the sides of the head. At the beginning of this phase, tooth enamel formation begins. The legs are well developed,

FIGURE 9.1 Week 9, beginning of the fetal period. The fetal head is almost half the length of the fetus.

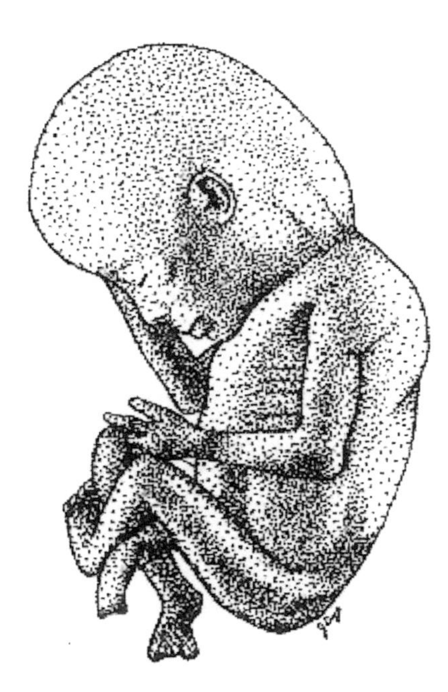

FIGURE 9.2 Week 12. The ears have moved from the neck onto the head, and the eyes have moved from the sides to the front of the head.

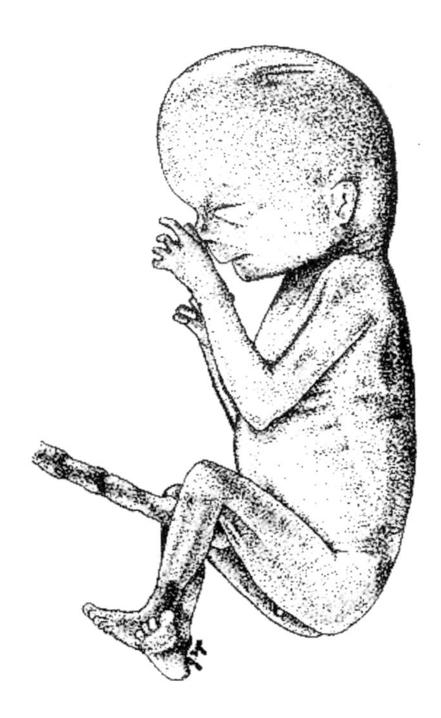

FIGURE 9.3 Week 13, end of the first trimester. In the living fetus, blood vessels are visible through the thin skin.

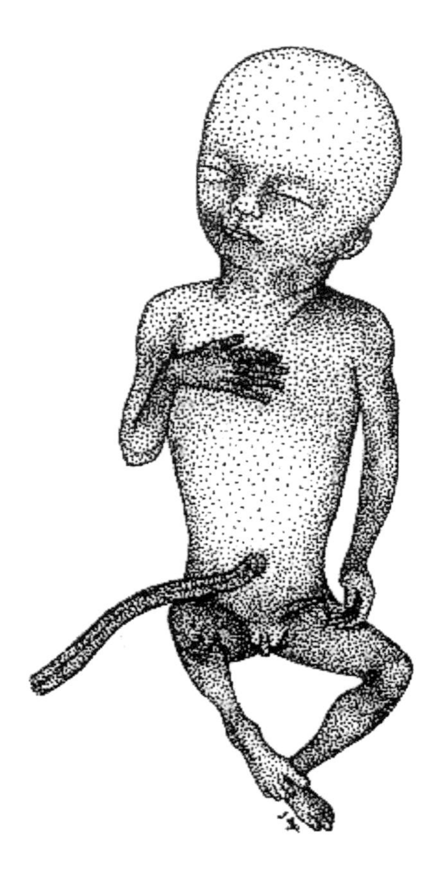

FIGURE 9.4 Week 16. The appearance of the external genitalia shows that this is a male fetus.

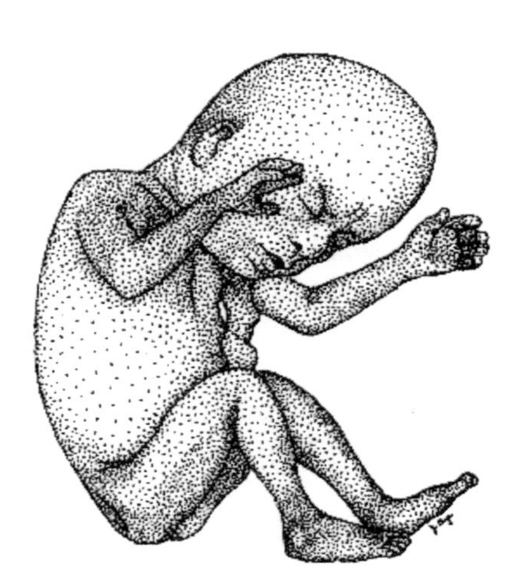

FIGURE 9.5 Week 17. Fetal movements are felt by the mother.

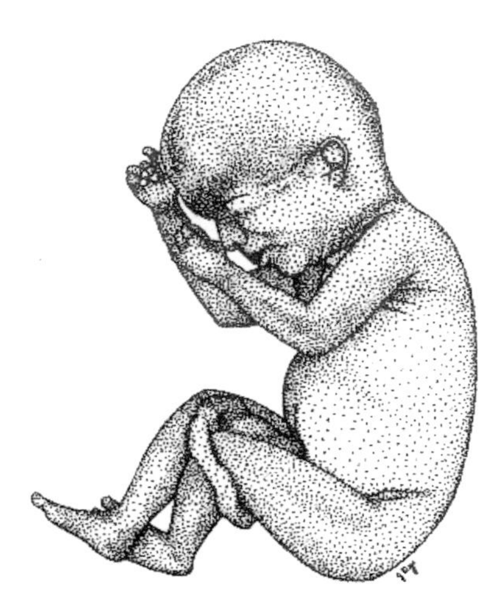

FIGURE 9.6 Week 20. Eyelids, eyebrows, and fingernails are well developed.

and toenails are developing. Brown fat (which will produce heat for the newborn infant) and vernix caseosa form. Fetal movements are felt by the mother (quickening).

Middle fetal stage or "an awkward stage" (19–28 weeks)

During this stage, the fetus is lean and the skin is wrinkled—an awkward stage (Figures 9.6 and 9.7). The fetus achieves viability (the ability to survive outside the uterus) but may still be **presentient** (not yet conscious). The eyes open and eyebrows and eyelashes are present. First, fingernails and then toenails appear, and a good head of hair develops. By the end of this period, the testes are descending.

Phase 1 (19–23 weeks): Eyelids, eyebrows, and fingernails are well developed during the first phase of the middle fetal stage (see Figure 9.6).

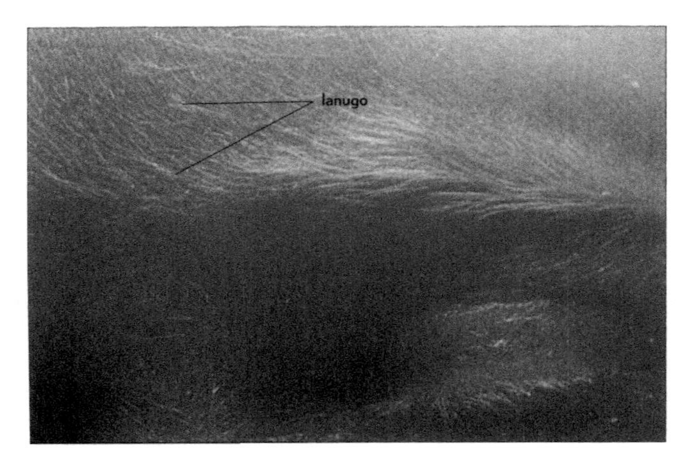

FIGURE 9.7 Week 15. Lanugo. By week 21, lanugo covers most of the body. (England, MA: *Color Atlas of Life Before Birth*. Year Book Medical Publishers, Chicago, 1983: Figure B, p. 203.)

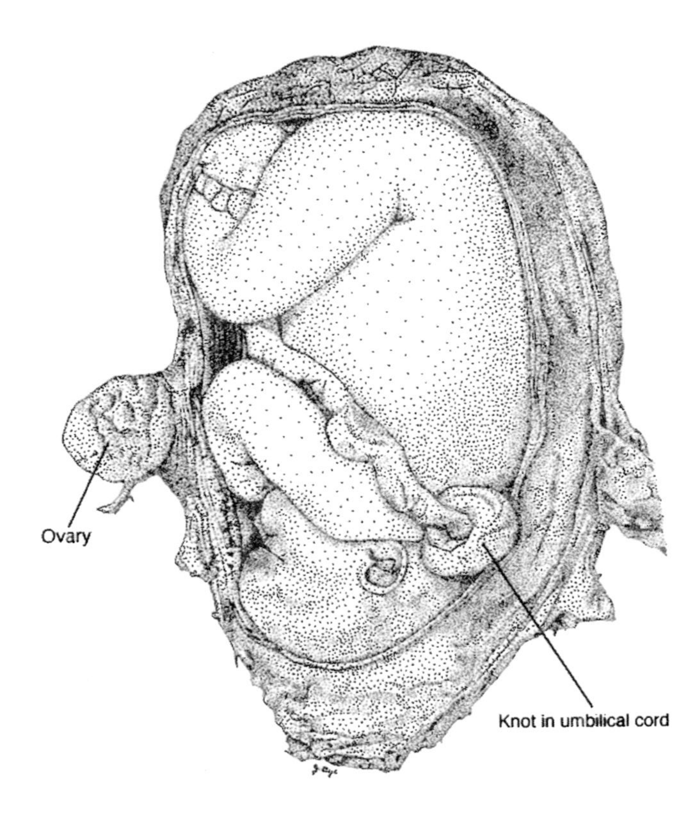

FIGURE 9.8 Week 28, third trimester. Although this fetus appears to be sleeping, the eyes by now are open again. Head is down, and there is a knot in the umbilical cord. With the deposition of subcutaneous fat, the body has become plump and round. Fetus is viable.

The wrinkled skin exhibits lanugo, which covers most of the body (see Figure 9.7).

Phase 2 (24–28 weeks) (Figure 9.8): The well-proportioned body is lean. By the end of this phase, the eyes are open and the fetus is likely to survive outside the uterus. The previously smooth surfaces of the cerebral hemispheres develop **gyri** (convolutions) and **sulci** (furrows). Secondary hair is longer than the lanugo. As this second phase ends, subcutaneous fat deposition begins.

Late fetal stage or "the swan emerges" (29 weeks to birth)

The swan emerges in the sense that the skin is smooth and pink and the body is plump compared with the earlier fetal stages. The chest is prominent with protruding breasts, and the testes are in or close to the scrotum.

Phase 1 (29–33 weeks): The body of the fetus becomes plump with smooth and pink skin. Nails reach the fingertips.

Phase 2 (34–38 weeks): Generally, lanugo is shed from the skin of the plump body during this phase. The testes are in the scrotum or descending into it, but the ovaries do not descend completely into the pelvic cavity until after birth. Nails reach the tips of the toes.

The characteristics previously enumerated for the fetal period are primarily external characteristics and particularly reflective of morphogenesis. However, keep in mind that enormous, though less conspicuous, contributions are being made by cell proliferation (increasing the number of cells) and cell differentiation (increasing the number of kinds of cells). For even though the organ systems are laid down during the preceding embryonic period and, in the instance of the cardiovascular system, begin to function during the embryonic period, it is during the fetal period that most of the developing systems begin to come "online" (begin to function) as far as their functional development is concerned.

Examples of the onset of biochemical functions include the following:

- Thyroxine appears in thyroid follicles.
- Urine is formed (8 weeks).
- The adrenal cortex produces corticosteroids.
- Adrenocorticotropic hormone (ACTH) is released (8 to 9 weeks).
- Gonadotropins are produced by the fetal pituitary gland (9 weeks).
- The neurohypophysis (a portion of the pituitary gland) produces small amounts of growth and lactogenic hormones (10 weeks).
- The parathyroid glands produce parathyroid hormones (12 weeks).
- Bile and bile pigments are produced (13–16 weeks).
- Rennin is present in the stomach (18 weeks).
- Thyrotropic hormone is released from the pituitary gland (19 weeks).
- Insulin is formed by the pancreas (20 weeks).

Some of the details of fetal development are considered in Part II.

Organoids During much of the twentieth century, researchers extensively used cell culture in biomedical research. Additionally, tissue, organ, and embryo culture were also used. The formation of an organ during development is called organogenesis. A certain amount of organogenesis could occur in culture; for example, the development of mouse tooth germs (embryonic mouse teeth) would undergo a certain amount of organogenesis *in vitro* (i.e., in culture); however, organ culture was limited by size restrictions related to the rate of diffusion of nutrients into the volume of the organ or organ fragments. Embryo cultures, especially of mouse embryos, were limited to early development because mammalian embryos must at some point implant into a uterus, so although these embryo cultures provided good insight into early mammalian development, development beyond the blastocyst stage would be hidden from view. Cell and tissue cultures for most of the twentieth century were generally two-dimensional (2D) cultures (see Figure 9.9), as the cells or tissues grew as a flat layer on their glass, or, increasingly, plastic, substratum. A relatively recent innovation involves the use of 3D cultures (see Figure 9.10). This is a more natural environment for the cultures, as cells grow *in vivo* (i.e., within the organism) in three dimensions.

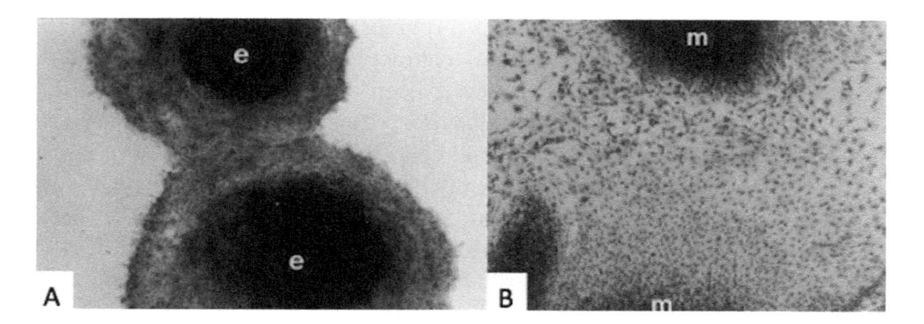

FIGURE 9.9 (A) Two enamel organ epithelia (e) derived from two 15-day molar mouse tooth germs growing as tissue cultures in 2D. (B) Two dental papilla mesenchymes (m) derived from two 15-day molar mouse tooth germs growing as tissue cultures in 2D. Preparations and photos by author.

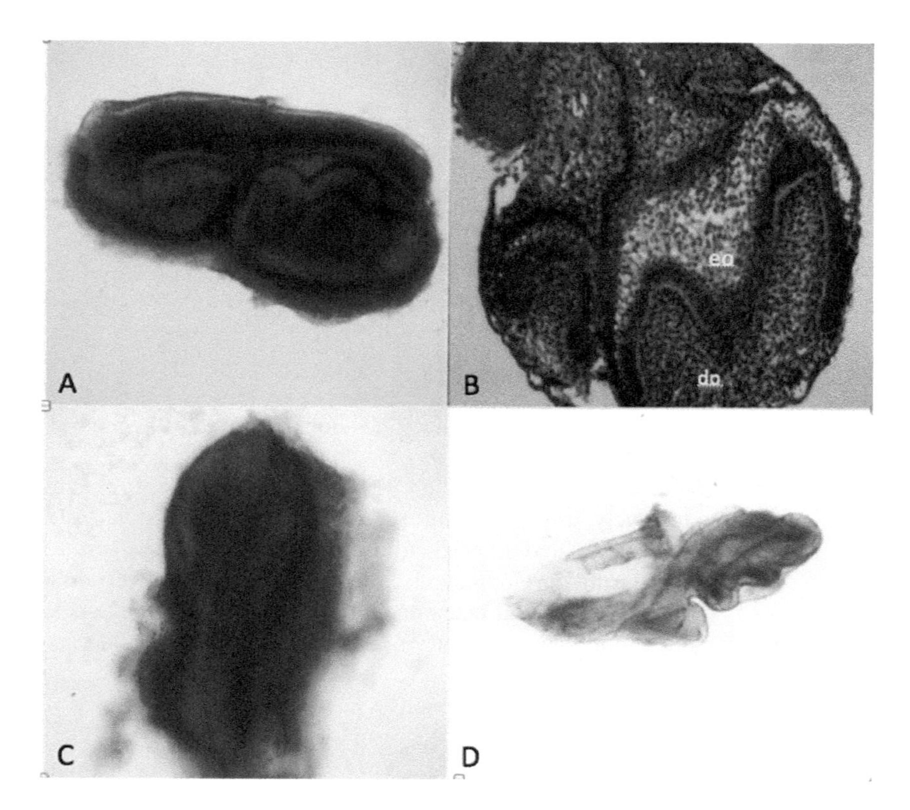

FIGURE 9.10 (A) First and second mouse molar explant from a 15-day mouse fetus, in lateral view; first molar is on the right. (B) First and second mouse molar fixed and stained sagittal section from a 15-day mouse fetus; first molar is on the right. (C) First and second mouse molar explant from a 15-day mouse fetus, in dorsal view; first molar is on the top. (D) First and second mouse molar enamel organ explant from a 15-day mouse fetus; first molar is on the right. The enamel organ portion of the explant was obtained by treating the intact molar explant with enzyme solution. A and C, when placed in culture will develop as 3D organ cultures. (D) When placed in culture, will grow as a 2D tissue culture (see Figure 9.9A). eo is the enamel organ; dp is the dental papilla. Preparations and photos by author.

Early in the twenty-first century, the creation and culture of organoids have been insightful innovations in *in vitro* culture. Organoids are small versions of organs derived from stem cells or tissue cells that have been converted into stem cells called induced pluripotent stem cells. The creation of organoids begins with such cells grown in 2D culture, where the cells are exposed to a "chemical cocktail" that directs them to differentiate (specialize) in a specific direction, that is, toward becoming nerve cells, heart cells, and so on. These cells, when transferred into 3D culture, begin to communicate with each other and self-organize into a miniature version of the organ the cells would be expected to be part of, that is, nerve cells into miniature brains, heart cells into miniature hearts, and so on. The organoids have a micro-anatomy closely resembling that of the corresponding organ.

Researchers are using organoids to study normal organ development, organ disease, and therapy for such diseases.

Study questions

1. What are the major processes during the fetal period?
2. Differential growth of the fetus is shown by what during the fetal period?
3. When during the fetal period do the eyes close? Open?
4. By when is the sex of the fetus distinguishable externally?
5. What is vernix caseosa and where is it found? What is lanugo and where is it found?
6. During the early fetal period, the eyes make what positional change?
7. What is meant by midgut herniation, and when is it usually resolved?
8. Beginning in the early fetal stage, the fetus may do what with its thumb?
9. During the fetal period, the fetus may do what two things with the amniotic fluid?
10. During the early fetal period, the ears make what positional change?
11. What is quickening and when is it first detected?
12. What special type of fat tissue begins development in the early fetal period? What of significance does it produce for the newborn?
13. At what stage does the fetus achieve viability?
14. During the early fetal stage, the cerebral hemispheres overgrow the midbrain. By the end of the middle fetal stage, what develop on the surfaces of the previously smooth cerebral hemispheres?
15. When do most of the developing systems come "online" as far as their functional development is concerned?

16. What is an organoid?
17. What cells are used to form organoids?
18. What must cells do to form an organoid?

Critical thinking

1. It has been suggested that human babies are actually extrauterine fetuses. It is true that human babies are quite immature. Why do you suppose it is necessary for us to be born so "premature"?
2. Developmentally, when do we become sentient?

CHAPTER TEN

The placenta and the umbilical cord

CHAPTER OBJECTIVES

After studying this chapter, you should be able to:

1. Explain the significance of implantation in human development.
2. Describe the development of the placenta, including the role of the decidua, chorion, and genomic imprinting.
3. List the functions of the placenta and its significance in human development as the organ of cooperation between mother and unborn child.
4. List the substances that can pass across the placenta—good and bad.
5. Explain the problems that can occur with the placenta, such as abruptio placentae and placenta previa.
6. Explain how the umbilical cord's anatomy allows interaction between the embryo or fetus and placenta through the process of umbilical circulation.

To survive, every living organism must be able to acquire nutrients and rid itself of waste products. To accomplish this, we adults interact directly with our environment. The embryo and fetus do not directly interact with the environment, but do so only through the intermediary of the mother. What the mother puts into her body—from oxygen and good food to alcohol, nicotine, and other drugs—affects the developing baby.

This relationship between embryo and mother begins even as the cleaving (dividing) embryo is tumbling down the fallopian tube. It becomes much more intimate by the end of the first week after fertilization, when the embryo is implanting into the uterine lining. With implantation, the placenta begins to form; the strands of the trophoblast and the blood-filled lacunae (spaces) between them give rise to the villi and the blood-filled intervillous spaces between them, respectively. As the inner cell mass gives rise to the embryonic disc, a connecting stalk forms between it and the chorion (membrane), developing from the trophoblast. This body stalk develops into the **umbilical cord** between the embryo and placenta.

Implantation

Recall that at the end of the first week of development, the blastocyst has two distinct cell populations: trophoblast and inner cell mass. After the blastocyst hatches from the zona pellucida, the trophoblast attaches to the lining of the endometrium and begins **implantation**. Implantation is the process by which the embryo burrows into the endometrial lining of the uterine cavity.

The trophoblast does not give rise to any part of the organism; rather, the trophoblast erodes into the endometrium. The trophoblast (review Chapter 8, "The embryonic period") carries the rest of the conceptus with it, organizing itself into strands separated by the blood-filled lacunae. These strands will become organized into more discrete, feathery structures called **villi** (Figure 10.1). As the trophoblast begins this process, it picks up a layer of extraembryonic mesoderm, transforming it into the chorion. Initially, the distribution of villi over the surface of the chorion is uniform. The lacunae, filled with maternal blood, give rise to the blood-filled **intervillous** spaces. Moreover, the villi, also soaking in maternal blood, come to contain fetal blood vessels. Thus, it is easy

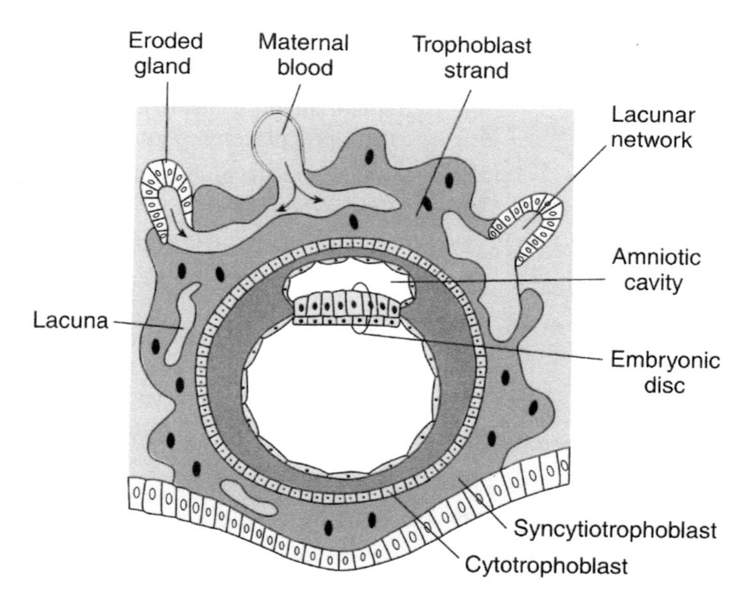

FIGURE 10.1 Lacunae (spaces) and strands of trophoblast. The syncytiotrophoblast erodes the uterine tissues, including blood vessels, resulting in blood-filled lacunae.

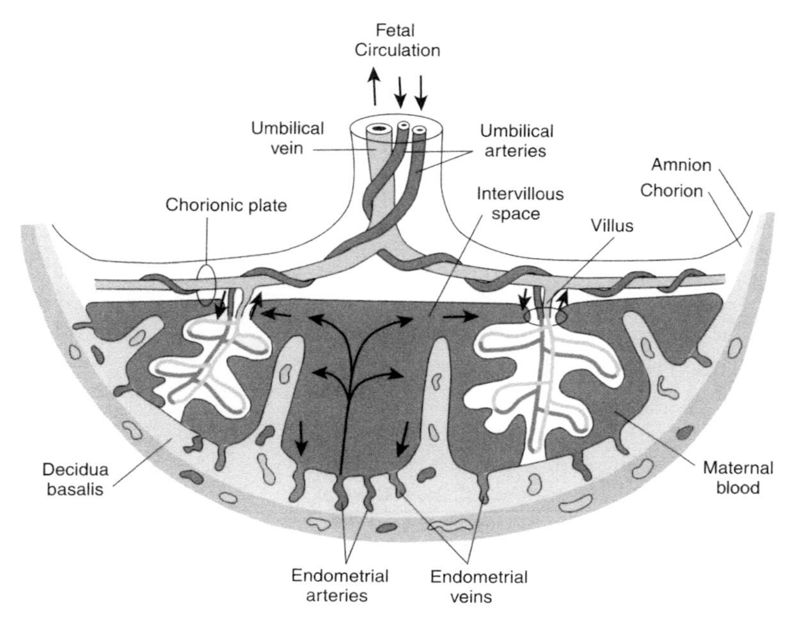

FIGURE 10.2 The placenta. The diagrammatic section through the fully developed placenta shows villi containing fetal blood vessels bathed by maternal blood in the intervillous spaces.

to understand that the villi are important sites of transfer of materials between mother and fetus (Figure 10.2).

Placenta

The human placenta is unique for several reasons. No other human organ provides such a wide variety of services to the organism it serves. The placenta serves as respiratory, digestive, excretory, and endocrine systems for the developing embryo and fetus. Perhaps the placenta is most unique because it is the only human organ that results from the cooperation of two separate individuals—mother and fetus.

Decidua and chorion

Recall that the chorion is a source of interchange between a mother and her unborn child. Therefore, you will not be surprised to learn that the chorion plays a role in the development of the placenta. Also playing a role is the **decidua**, which is the major part of the endometrium during pregnancy. Just as deciduous trees shed their leaves during autumn, the decidua is shed by uterine contractions at childbirth (a process called **parturition**).

A portion of the chorion and a portion of the decidua cooperate to form the placenta. The chorion is composed of two parts, and the decidua is composed of three. As already stated, initially, the chorion is uniformly

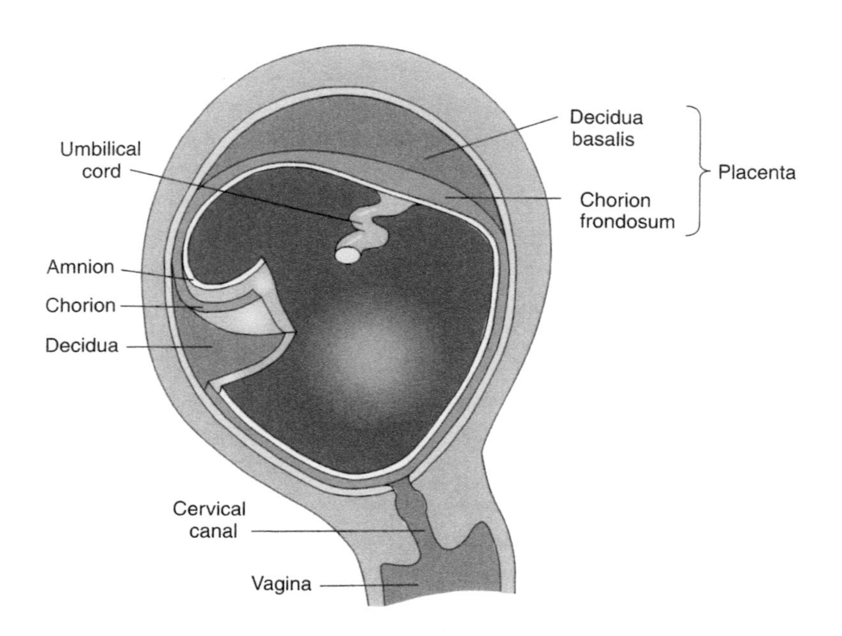

FIGURE 10.3 Fetal and maternal components of the placenta. The villated portion of the chorion and the decidua basalis contribute to the placenta—the only human organ formed by two individuals.

villated; however, as the conceptus grows, much of it loses its villi and is then referred to as the **chorion laeve (smooth chorion)**. The balance of the chorion is the **chorion frondosum (bushy chorion)**. As the conceptus implants into the decidua, three decidual regions are discernible: the **decidua parietalis** is initially farthest from the conceptus, the **decidua capsularis** is above the conceptus (between conceptus and uterine cavity), and the **decidua basalis** is the portion underneath the conceptus.

The placenta arises from the intimate relationship developed between the chorion frondosum (therefore, **placenta fetalis**) and the decidua basalis (therefore, **placenta materna**) (Figure 10.3).

Genomic imprinting

It might surprise you to learn that the placenta, the organ of cooperation between mother and fetus, may be the genetic responsibility of the father! Research in the area of genomic imprinting seems to indicate this. Recall that the chromosomes we receive from our mother and father are not equivalent (see Chapter 3). In the formation of sperm in the male and eggs in the female, each parent places a stamp of his or her sexuality on the chromosomes of the developing gametes—**genomic imprinting**. In other words, the chromosomes possess a chemical modification "saying" that they came from a female or a male.

Just as imprinting of chromosomes seems to be unique to mammals, the formation of a placenta is also unique to mammals. Experiments have demonstrated that normal mouse development requires a set of maternally modified chromosomes and a set of paternally modified chromosomes. More specifically, if a zygote is experimentally created with two sets of paternal chromosomes, a placenta will develop, but no fetus. On the other hand, if two sets of maternal chromosomes are provided, a placenta will not develop. Although not understood at a cause-and-effect level, paternally modified chromosomes seem to be necessary for the development of an organ resulting from the cooperation of the mother and the conceptus, both of which possess both maternally and paternally imprinted chromosomes.

Placental function

Vertebrate animals develop in watery environments. In general, fish and amphibians develop in bodies of water. Reptiles, birds, and two mammals (the duck-billed platypus and the spiny anteater) develop in eggs that are supplied with amniotic fluid and sufficient nutrients to get the animal to the hatching stage. The porous egg shell provides both the benefit of gas exchange (oxygen in and carbon dioxide out) and the problem of **desiccation** (water out).

On the other hand, most mammals, including humans, develop in watery amniotic fluid enclosed within the body of the mother. Materials must flow between the conceptus and the mother to provide for normal development; these materials cross the placenta. The placenta provides for embryonic and fetal nourishment (food in), respiration (oxygen in and carbon dioxide out), and excretion (waste out). However, not everything that crosses the placenta—especially in the modern world—is beneficial to the embryo or fetus.

We once thought that the placenta was a protective barrier between the conceptus and its environment, but we now know that many things are able to pass through the placenta in both directions. These substances and radiations are physical, chemical, and biological in nature.

Radiation is a physical factor that can pass through the placenta and affect the embryo or fetus. It is common knowledge that pregnant women should avoid x-rays, and birth defects were among the tragic consequences of the exposure of populations to the radiation of nuclear explosions. Sound waves (including **ultrasound**) are able to pass through to the embryo or fetus, as evidenced by the now routine use of ultrasound as a prenatal diagnostic procedure and by the effect of various types of music on fetal activity. From these examples, we can see that the effects of physical factors on the fetus can be either harmful or innocuous.

Passage of chemicals across the placenta can be absolutely essential or absolutely tragic. Chemicals such as **glucose** (a simple sugar) and **amino acids** (building blocks of protein) are essential for the nourishment of the developing fetus, and the availability of oxygen and water is absolutely required for human life. On the negative side, chemicals such as **thalidomide**, cancer **chemotherapeutic drugs** (if the mother also happens to be a cancer patient), and alcohol can cause birth defects of various degrees or even fetal death. Remember that it is possible to chemically induce an abortion. The passage of **antibodies** (protective proteins produced by the immune system) across the placenta may be a good thing (by providing the newborn with temporary **passive immunity** against some diseases) or a bad thing (by causing **erythroblastosis**

fetalis, a condition in which maternal antibodies destroy fetal red blood cells in response to Rh incompatibility between mother and fetus).

Rubella (German measles) virus, herpes simplex virus, and cytomegalovirus are examples of **biological agents** that are able to cross the placenta into the fetus, causing severe birth defects or fetal death. We are all painfully aware of the fact that infants may be born already infected with the HIV-1 (AIDS) virus. In addition to viruses, **protozoa** (one-cell organisms) may also cross the placenta: *Toxoplasma gondii* (found in the feces of infected cats) is the causative agent of **toxoplasmosis**, which can result in blindness, mental retardation, and other birth defects.

Placental problems

Normally, the placenta is the fetus's life support system, staying with the fetus until birth. However, in a condition called **abruptio placentae**, the placenta separates prematurely from the lining of the uterus (Figure 10.4). This condition threatens both mother and fetus, subjecting the mother to potential massive hemorrhage (bleeding) and the fetus to an interruption of its life support.

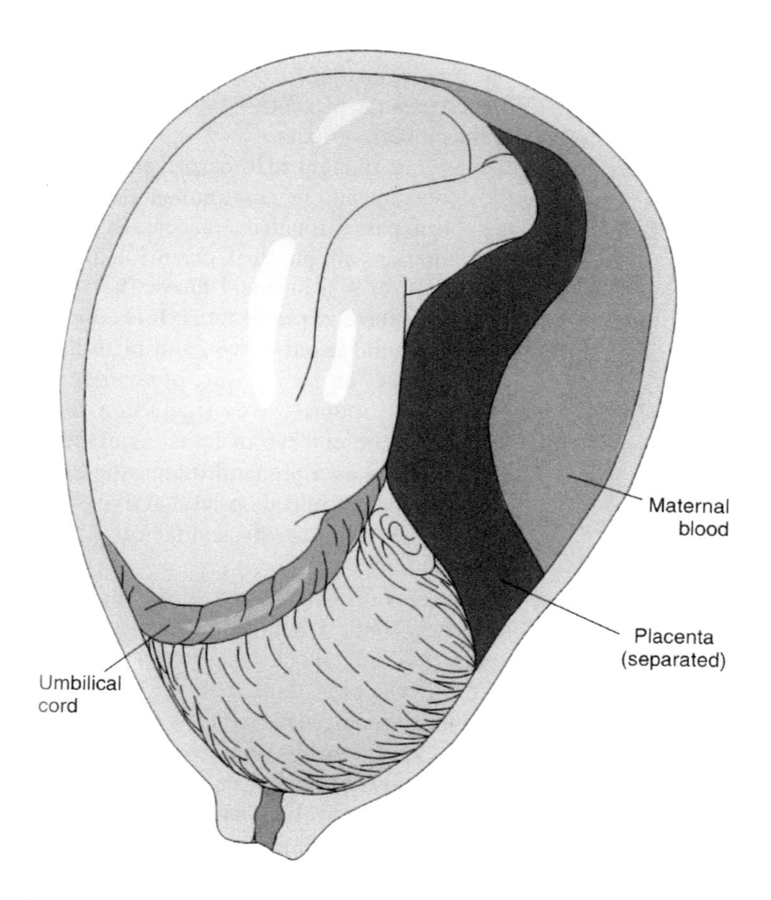

FIGURE 10.4 Abruptio placentae. Premature separation of the placenta from the lining of the uterus.

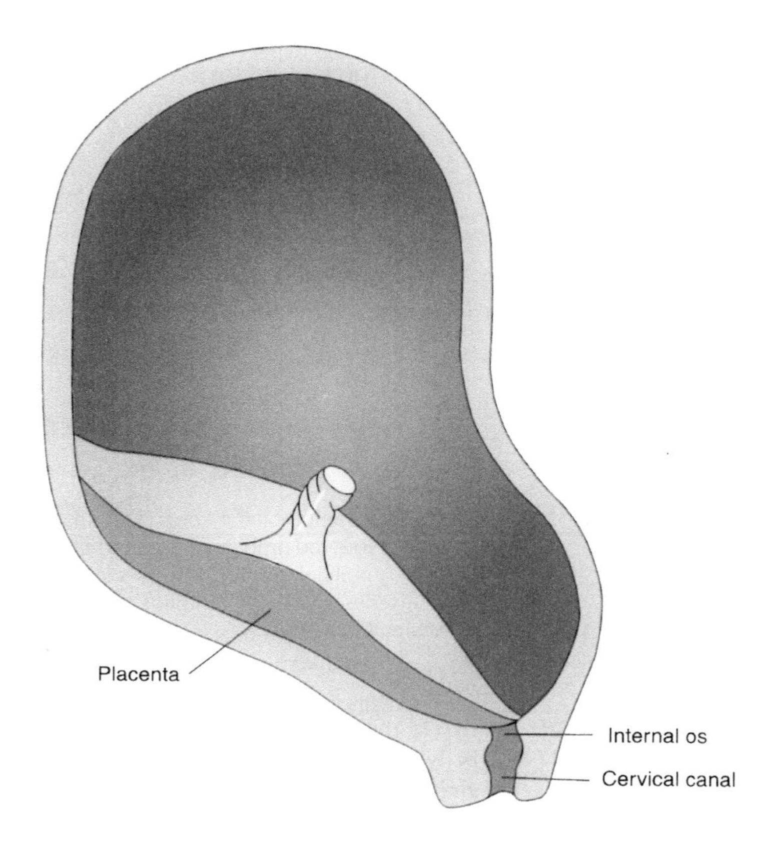

FIGURE 10.5 Placenta previa. An example of abnormal location of the placenta. Here, the edge of the placenta is blocking the internal os of the cervical canal of the uterus.

Another placental abnormality is **placenta previa**, in which the placenta is formed in an abnormal location (Figure 10.5). Normally, the placenta forms on the superior (upper), dorsal (back) wall of the uterus. In placenta previa, the placenta forms across the opening into the cervical canal, thus blocking the normal exit from the uterus during childbirth and necessitating a delivery by cesarean section.

The umbilical cord

Just as the brain and distant parts of the body must "connect" for normal activity to occur, the embryo and fetus must connect with the placenta for survival, let alone development, to occur. The connection between the embryo or fetus and the placenta is through the umbilical cord.

Structure

At full term, the fetus's umbilical cord has an average length of 50 cm and a diameter of 1.25–1.9 cm. The umbilical cord finds its origin in the connecting stalk of the early conceptus, a structure formed between the

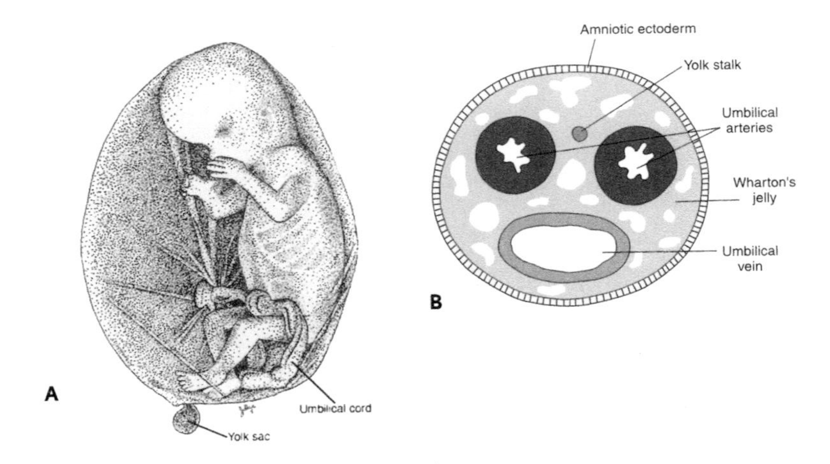

FIGURE 10.6 The umbilical cord. (A) Human fetus within its amniotic sac at 12 weeks. Note the umbilical cord and the yolk sac (the yolk stalk passes through the umbilical cord). (B) Cross-sectional view at term. Note the two umbilical arteries, the single umbilical vein, and the yolk stalk. The allantois is not visible at this level.

trophoblast and the forming amnion. It gradually changes its position relative to the developing embryo and comes to appear to arise from the belly of the embryo (see Figure 8.12, showing the early body stalk, the changing position of the forming umbilical cord, and the definitive umbilical cord attached to the belly of the embryo).

The surface of the umbilical cord is covered by a layer of amnion, which gives it a shiny appearance. Included within this outer layer is the long, attenuated yolk stalk. At the end of the yolk stalk at the level of the placenta is found the diminutive yolk sac (Figure 10.6A). A cross-section through the umbilical cord reveals the two smaller arterial cross-sections and the somewhat larger cross-section of the umbilical vein. In addition, the very tiny yolk stalk cross-section is present as well as that of the **allantois** (one of the four extraembryonic membranes). Although in humans the allantois is not the well-developed structure it is in the development of other mammals, its blood vessels become the umbilical blood vessels. Beneath the surface layer of amnion is found a special kind of connective tissue, **Wharton's** jelly, within which are embedded all the structures just mentioned in considering the cross-section of the cord (see Figure 10.6B).

Umbilical circulation

The umbilical cord, as mentioned, is the conduit between the embryo/fetus and the placenta through which flows the life-supporting blood of the embryo/fetus, confined to two umbilical arteries and a single umbilical vein. The surface of the umbilical cord at term is ridged by the arteries spiraling around the umbilical vein.

Blood from the fetus is carried out to the placenta through the two umbilical arteries. Unlike other arteries, these carry blood depleted of oxygen and laden with the wastes of metabolism. In the placenta, carbon dioxide and nitrogenous wastes are transferred into the mother's blood, and oxygen and nutrients are transferred into the fetal blood. The replenished fetal blood is carried through the umbilical vein back to the fetus. Unlike other veins, the umbilical vein carries blood with a high level of oxygen and nutrients.

Because the umbilical cord is a life-support system, its contained blood vessels must remain patent (open) or the baby's oxygen supply will be compromised—with disastrous consequences. The cord must not become compressed during the final stages of labor and childbirth.

Childbirth

It has become fashionable for the father of the baby to cut the cord between two applied clamps. One piece of the cord remains attached to the placenta, which is born about several minutes after the baby—thus, **afterbirth**. The other clamped piece of umbilical cord remains attached to the baby's belly, until it eventually shrivels up and falls off. The cord attached to the baby must be clamped or tied off because the baby's blood vessels are continuous with the umbilical blood vessels.

Umbilical cord blood

Umbilical cord blood is a life-saving gift! During the life of the fetus, an expansion of the blood-forming tissue occurs, associated with a large number of highly proliferative stem cells migrating in the peripheral (circulating blood, which, in the adult, does not contain blood-forming stem cells) blood. In the past few years, the umbilical cord, although still routinely discarded, has come to be valued as a readily available source of these blood stem cells—cells that are able to produce new blood cells as well as maintain themselves as a stem cell population.

Why save umbilical cord blood? Some parents are having their baby's umbilical cord blood (UCB) cryogenically stored (frozen) for possible future use, as in replacement of the hematopoietic (blood cell-forming) system after its elimination by cancer treatment. In addition, parents can choose to donate their infant's UCB at no cost in the hope of saving the life of someone with cancer. About 9,000 transplant candidates—one-third of them children—die each year before a suitably matched bone marrow donor is found. The availability of UCB will greatly improve the chances of locating a good match for patients who need a life-saving transplant.

Diseases treated The first UCB transplant was performed in France in 1989. Recently, it has been reported that nearly 80 diseases are treated with UCB (https://www.viacord.com/references/). These include malignancies (a number of leukemias, lymphoma, and neuroblastoma), blood disorders (including aplastic anemia and sickle cell anemia), immunodeficiencies (including severe combined immunodeficiency disorders [SCIDs]), and a number of inborn errors of metabolism (including Gaucher's disease and Hurler's syndrome). Treatment of many of these disorders otherwise requires a bone marrow transplant.

Collecting umbilical cord blood Collection of large volumes of UCB is a simple, safe, efficient, and *noninvasive procedure*. It takes 1–3 minutes and doesn't interfere with the normal delivery process. The cord blood may be made available worldwide for transplantation into patients with cancer and blood diseases.

UCB collections, with advanced parental permission, must be thoroughly prepared in advance. Expectant mothers must undergo certain blood tests, including but not limited to HIV testing. At the time of delivery, there is *prompt* clamping of the umbilical cord. This is critical because the sooner the clamping is performed, the larger and richer the UCB sample. Although UCB transplants had been limited to children, studies indicate that UCB samples have enough stem cells for transplantation to adult patients. Also, it has been shown that cryopreserved UCB stem cells may be increased in the laboratory by stimulating the stem cells with chemicals called **cytokines**.

Artificial placenta It has been reported that, "in the United States, extreme prematurity is the leading cause of infant morbidity and mortality, with over one-third of all infant deaths and one-half of cerebral palsy attributed to prematurity. Advances in neonatal intensive care have improved survival and pushed the limits of viability to 22–23 weeks of gestation. However, survival has been achieved with high associated rates of chronic lung disease and other complications of organ immaturity, particularly in infants born before 28 weeks."

https://www.nature.com/articles/ncomms15112

Consequently, it is not surprising that there is interest in developing an artificial placenta to bridge the gap between 23 weeks of gestation and term pregnancies. It has recently been reported ("An extra-uterine system to physiologically support the extreme premature lamb," *Nature Communications*, April 2017), that such an artificial placenta has been developed capable of supporting lamb development for 28 days; that is, the lamb fetuses were supported for 4 weeks in an extra-uterine artificial placenta. The objective of this research is to create a human artificial placenta capable of carrying otherwise preterm infants past 23–24 weeks and closer to full term, thus ameliorating those conditions associated with prematurity.

Researchers at the Children's Hospital of Philadelphia devised an artificial placenta for lamb fetuses, which consists of a fluid-filled bag with ports for oxygen and nutrient delivery, with a pumpless oxygenator for the fetus to circulate blood using its own heartbeat.

A significant feature of this artificial placenta is that the fetus remains in a fluid environment, mimicking a natural placenta. This is significant because it avoids the mechanical ventilator, wherein the fetal lungs are exposed to air, which can lead to lung disorders as well as other problems later in life. These researchers, starting with lambs whose lungs were at a similar stage of development as those of a 23-week-old human fetus, performed tests on lung and brain tissue and found they were able to produce lambs that showed no significant differences between the preterm lambs that developed in their artificial placenta and those that were able to develop in the womb. The gestation period for lambs is 142–152 (average 147) days, and the lamb is a common animal model for comparative lung function studies. The external components of their artificial placenta were fed through the lambs' umbilical cord, with the ports designed to preserve sterile conditions.

While their results are promising, these researchers point out significant challenges that need to be dealt with before application of this technology to human fetuses. For example, human fetuses are smaller; the fluid that they used, a mix of electrolytes, needs to be customized to humans; and the umbilical connections must also be human customized. One of the researchers predicted that a similar artificial placenta could see its first human applications within 3 to 5 years.

http://blogs.discovermagazine.com/d-brief/2017/04/25/artificial-placenta-lamb/#.WuB6QMgpBgo

Study questions

Placenta

1. What organ serves as respiratory, digestive, excretory, and endocrine systems for the developing embryo and fetus? In what way is this organ most unique?
2. As implantation progresses, the chorion comes to be uniformly covered with what feathery structures? These dangle in blood-filled spaces (derived from lacunae) called what?
3. Although the chorion is initially uniformly villated, it gradually becomes distinguishable as two regions. The part from which the villi disappear and the part where the villi persist and flourish are called what, respectively?
4. What is the endometrium of pregnancy called, and what three parts are discernible while the conceptus implants into it?
5. What does the placenta provide for the embryo and fetus?
6. What is the condition called abruptio placentae?
7. What is the condition called placenta previa?
8. What is the objective of developing a human artificial placenta?
9. Such an artificial placenta has been developed capable of supporting development of what organism and for how long?

Umbilical cord

10. The umbilical cord is the connection between what two things?
11. What is the origin of the umbilical cord? What covers the surface of the umbilical cord and what five structures are found in a cross-section of the umbilical cord?
12. Describe the umbilical circulation: the blood vessels involved, the directions of blood flow, and the general contents of its blood.
13. Umbilical cord blood (UCB) has been harvested for several years. It is a source of what valuable cells and what is the value of these cells?

14. UCB may substitute for what kind of transplant?
15. What kinds of diseases have been treated with UCB?

Critical thinking

1. If we need vitamins as adults, is it reasonable to expect that embryos and fetuses need them? If so, how are they obtained by embryos and fetuses?
2. From a functional point of view, how do the umbilical arteries differ from other arteries in the body? How does this make sense from the fetus's point of view?
3. From a functional point of view, how does the umbilical vein differ from other veins in the body? How does this make sense from the fetus's point of view?
4. Explain genomic imprinting.

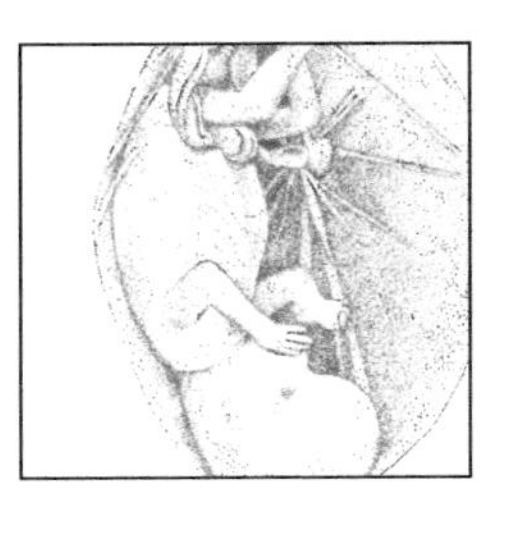

CHAPTER ELEVEN

The pregnant woman, childbirth, and multiple pregnancies

CHAPTER OBJECTIVES

After studying this chapter, you should be able to:

1. Describe changes undergone by the pregnant woman during pregnancy including gastrointestinal changes, changes in appearance, and disease like conditions.
2. Describe the three stages of labor and childbirth.
3. Describe the seven cardinal movements of labor.
4. Explain the roles of sutures and fontanelles in childbirth.
5. Distinguish among the following kinds of twins: monozygotic, dizygotic, identical, fraternal, normal, conjoined.

In describing a phenomenon—whether it is photosynthesis or human development—it is necessary to consider the context in which that phenomenon occurs. For the developing human conceptus, the context is the pregnant woman. Although we are here concerned with human life *before* birth, just as we began our story with gametogenesis, it seems appropriate to carry our fetus over the threshold that is childbirth. Most often, human life before birth is a story about the development of a single individual. However, there are instances in which the womb is occupied by more than one conceptus. People have traditionally been especially interested in multiple pregnancies.

Here, we first consider the pregnant woman, the *context*. Next, we look at what is involved in the birth of the baby. Finally, we look at some aspects of multiple pregnancies.

The pregnant woman

The life of a pregnant woman is to a great extent affected by two factors—hormones and the increasing size of her uterus. These causative factors result in a variety of discomforts, both major and minor. As discouraging as this list of pregnancy changes must seem to the woman contemplating motherhood, most mothers have more than one child and these changes are a necessary part of the miracle of human development. Because pregnancy is not a disease, it would be misleading to refer to the following as "symptoms," but here are *some* of the less desirable characteristics of pregnancy.

Gastrointestinal effects

A number of characteristic changes are associated with the gastrointestinal system. Perhaps the best known of these changes is "morning sickness," and, although generally confined to the first trimester, nausea may persist throughout pregnancy. We know the causes of some of these changes. For example, **constipation** results when hormones decrease the motility of the bowel (**progesterone** relaxes the intestinal musculature) and the enlarging uterus exerts pressure on the bowels. **Heartburn** (a burning sensation in the chest due to acid reflux into the esophagus) is a secondary effect of relaxation of the cardiac sphincter muscle. The pregnant woman may also be predisposed to **gallstone formation** because of delayed emptying of the gallbladder.

Appearance

The most obvious change in appearance is weight gain that ensues as the fetus, placenta, fluid-filled amniotic cavity, and uterus grow. Other changes occur to the skin. High levels of estrogen and progesterone are responsible for skin pigmentations; it is thought that they have a melanocyte-stimulating effect (melanocytes are pigment, melanin-producing, cells). Examples include hyperpigmentation of the nipples and areolae (circular areas surrounding the nipples) of the breasts; **linea nigra** (a dark pigmented line extending from the pubes upward); and chloasma ("mask of pregnancy"), patchy hyperpigmentation located chiefly on the forehead, temples, and cheeks).

The hyperestrogenemic state of the pregnant woman might also be the culprit of **vascular spiders** (minute reddened elevations of the skin from which there is branching of small blood vessels) and **palmar erythema** (bright red palms). Pregnant women also suffer from swelling, including **varicose veins** (veins that have become abnormally dilated and tortuous) of the legs and vulva and **edema** (swelling from excessive fluid accumulation) of the lower extremities. The enlarged uterus produces occlusion of the inferior vena cava in the supine and semirecumbent positions.

Other changes

Cardiovascular changes include increased heart rate, **heart murmurs** (physiologic and generally reversible), and **hemorrhoids**. Contributions from the urinary system include frequent **micturition** (urination) and **urinary tract infection** (due to relaxation of the bladder musculature and the resulting urinary stasis). To this litany of maladies may be added **backache**, a **waddling gait** (due to slackening of the pelvic joints and the resulting pelvic instability), **hyperventilation** (stimulated by progesterone), and **diabetes** (physiologic and generally reversible).

Labor and childbirth

Here, we come to the end of our story. Approximately 266 days after a tiny cell, initially deposited in the vagina, meets another tiny cell in the upper reaches of a fallopian tube, a 7-pound (more or less) baby emerges from the vagina. The complex process of human birth (parturition) is not completely understood. However, it seems to involve hormonal interaction among the fetus, placenta, and mother.

In each woman's life, childbirth is a milestone. Whether a pregnant woman chooses to give birth or not, a life-altering decision is made. There are many reasons for opting not to have children. However, if a woman takes the option to have a child, carry a pregnancy, and go through childbirth, other options regarding the birth process itself present themselves. Sometimes, the options are dictated by medical considerations. Should a cesarean section be done, or should labor and childbirth be induced? At other times, personal choice is possible. Should childbirth be natural; should analgesics (pain relievers) or anesthetics be used; do I want to deliver in a hospital, a birthing center, or at home? Do I want an obstetrician or a midwife in attendance? Do I want any family members or friends present? Do I want the birth videotaped?

Many women seek the advice of their obstetrician, who should lay out all the choices and critique them.

Childbirth is part of a larger process called **labor**. This series of events is divided into three stages: (1) the **dilatation stage**, (2) the **expulsion stage**, and (3) the **placental stage**. These stages vary in duration, but in general are shorter in the **multigravida** (a woman who has had previous pregnancies) than they are in the **primigravida** (a woman who is pregnant for the first time).

Stage 1: dilatation

The dilatation stage begins with the first "true" **contractions** of labor. These contractions begin to cause effacement and dilatation of the cervix. **Effacement** is the thinning of the cervix caused by movement of the amniotic sac into the cervical canal during contractions. During dilatation of the cervix, its **external os** (outer opening) is stretched from an opening of a few millimeters to about 10 centimeters. The duration of the dilatation stage is about 10–14 hours.

Stage 2: expulsion

The expulsion stage begins with full dilatation of the cervix and ends with the birth of the baby. How the baby is born is influenced by the attitude, lie, presentation, and position of the fetus. All four descriptions relate to the fetus's alignment in the uterus. The fetus is able to easily change its alignment in the uterus up until about 6 weeks before birth. That is, the fetal head may alternate between being at the top or the bottom of the

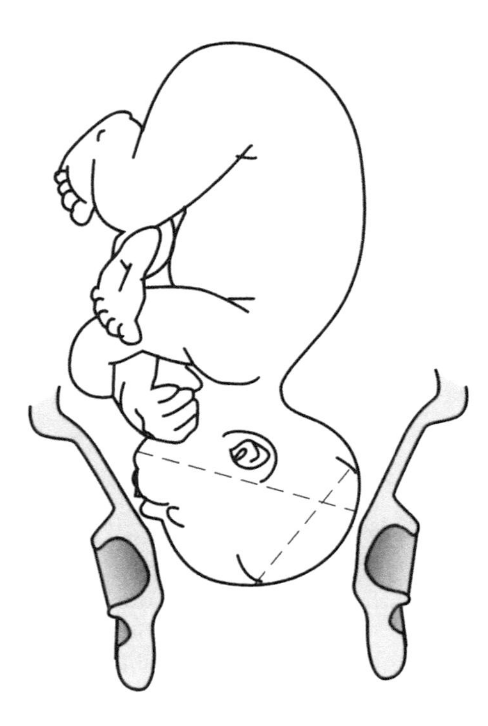

FIGURE 11.1 Alignment of the fetus in the uterus. Diagram shows examples of attitude (fetal position), lie (longitudinal), presentation (cephalic), and position (face right).

uterus. As the fetus grows, it must at some point settle into one position. An obstetrician must be familiar with these four characteristics and their variations (Figure 11.1).

The **attitude** is the posture that the fetus assumes in the uterus near the end of pregnancy. In common usage, this position is called "the fetal position," but it should not be confused with "position" described below. **Lie** of the fetus refers to the relationship of the long axis of the fetus to that of the mother. If the axes are parallel, a longitudinal lie exists; if the axes are perpendicular to each other, the lie is transverse.

The **presentation** of the fetus refers to the fetal part leading the way down the birth canal. In the **cephalic presentation**, the presenting part is the head, whereas the buttocks are the presenting parts in **breech presentation**. Most often, the fetus settles into the cephalic presentation about 6 weeks before birth. **Position** relates a chosen part of the fetus to the right or left side of the mother (e.g., face left or face right).

During the expulsion stage, the fetus moves down the birth canal (uterus and vagina), which varies in shape along its length. Because the shape of the canal varies, it is necessary for the fetus to change its posture as it descends through this passageway. These postural changes are known as the **seven cardinal movements of labor: engagement, descent, flexion, internal rotation, extension, external rotation**, and **expulsion** (Figure 11.2).

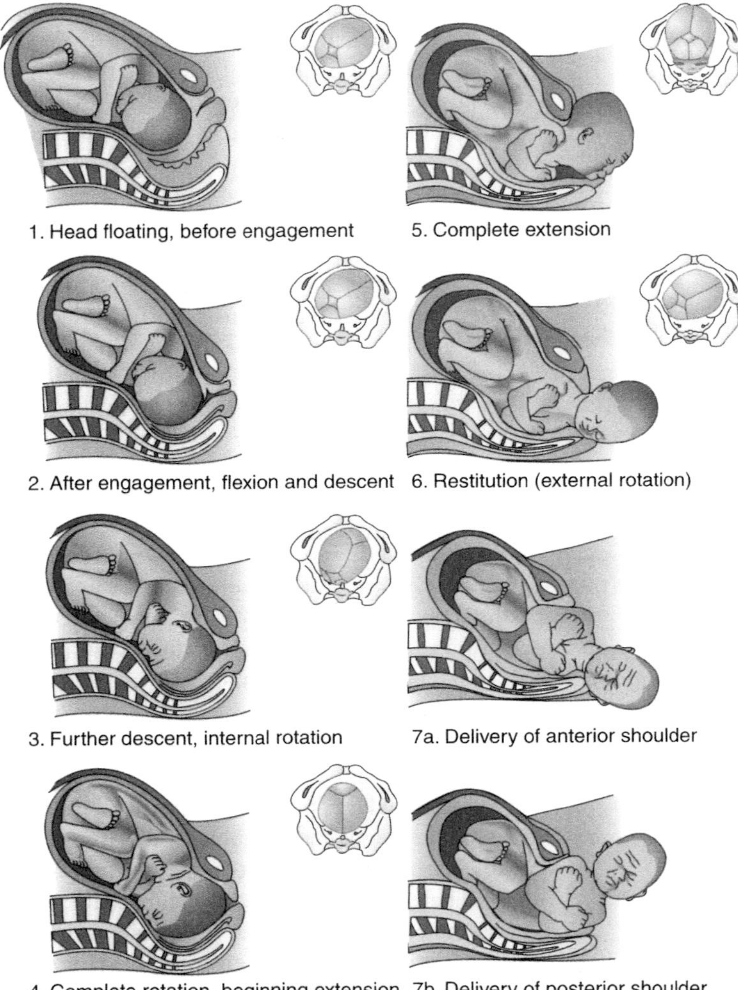

FIGURE 11.2 The seven cardinal movements of labor.

Fortunately, we don't remember our own birth. It must have been a traumatic experience! Despite the relatively short gestational period of humans, a baby's head has quite a tight squeeze. This is made possible by unclosed **sutures** and **fontanelles** (spaces), which make the newborn's head flexible (see Chapter 14, "Skeleton and muscles"). After the emergence of the shoulders, the rest of the baby rapidly emerges from the vaginal opening. The expulsion stage lasts for about 30–90 minutes.

Stage 3: placental stage

The placental stage lasts for 5–15 minutes, extending from the birth of the baby to the birth of the placenta. The placenta is no longer needed and must separate from the uterine lining to prevent future problems. As the placenta separates, the raw surface of the uterine cavity is exposed with some normal, minimal bleeding. Continued uterine contractions, which expelled the baby, functionally clamp off uterine blood vessels and normally prevent postpartum (after delivery) hemorrhage.

To help stop uterine bleeding, the new mother is encouraged to gently massage her nipples in order to stimulate release of oxytocin, the posterior pituitary hormone that stimulates contraction of the smooth muscle of the uterus.

The new mom must be carefully followed to make sure none of the placenta is left in the uterus and to ensure that postpartum hemorrhage does not go unnoticed. If part of the placenta remains in the uterus, it can cause subsequent problems, such as bleeding. In addition, placental tissue left behind may cause choriocarcinoma, a very malignant cancer. Fortunately, treatment of this cancer is one of the success stories of modern oncology, and it is 100% curable with chemotherapy.

Finally, the obstetric team must evaluate the newborn, but that is the beginning of another story.

Multiple pregnancies

Humans most often give birth to a single child at a time, but occasionally **twins** and rarely **triplets** are born. Higher numbers of offspring resulting from a multiple pregnancy are so rare as to be considered a dramatic occurrence. On the other hand, multiple pregnancies are common for many other mammals, such as cats and dogs. Fish, reptiles, and amphibians generally have large and possibly huge numbers of offspring. Also, in recent years, the use of drugs to stimulate ovaries to yield abnormally large numbers of eggs as a treatment for infertility has made multiple pregnancies and births much more common. This has even led to a procedure called **pregnancy reduction** (see Chapter 21).

There are two categories of multiple births, based on origin. For example, if twins arise from a single fertilized egg, they are referred to as **identical** or **monozygotic twins**; if twins originate from two eggs, they are referred to as **fraternal** or **dizygotic twins**. It is interesting that although the incidence of dizygotic twinning in humans varies widely among different populations (from 2.7 per 1,000 births in Japan to 42 per 1,000 births in Nigeria), the incidence of monozygotic twinning is much lower and varies little among different populations (from 2.9 per 1,000 births in Japan to 4.0 per 1,000 births in Nigeria).

Monozygotic twins

In animal experiments, when early blastomeres (from two- or four-cell embryos) are separated from each other, each blastomere may give rise to a normal animal. We also know from experience that human embryos can naturally divide into smaller units, with each giving rise to a separate individual if the separation is complete.

If the separation is not complete, a condition of **conjoined twins** (**"Siamese twins"**) results. The conjoining may be trivial and easily corrected surgically, or it may be a profound joining that cannot be safely severed by the surgeon's knife. There might even be a situation known as **fetus in fetu**, in which one twin is rather normal except that a tiny second

FIGURE 11.3 Siamese twins, Chang and Eng, the original Siamese twins. (Visuals Unlimited.)

twin is attached as an appendage. The story of the original "Siamese twins," Chang and Eng, is fascinating reading (Figure 11.3).

Because both monozygotic twins are derived from the same zygote, they are genetically identical and always of the same sex. Such twins are remarkably alike in appearance and may be difficult to distinguish from each other by someone who is not a member of the family. This similarity between identical twins may even extend to behavioral characteristics.

Dizygotic twins

Ordinary brothers and sisters share the experience of having come from the same uterus, though at different times. Like ordinary brothers and sisters, dizygotic twins come from different fertilized eggs and develop in the same uterus, but the twins happen to occupy the uterus at the same time. Although sharing a birthday, dizygotic twins are no more similar genetically than ordinary brothers and sisters. They may or may not be of the same sex and do not resemble each other (physically or behaviorally) any more than ordinary brothers and sisters do.

Triplets

Triplets may be monozygotic, dizygotic, or **trizygotic**. Like twins, if triplets are monozygotic, we expect them to be similar physically and behaviorally. If the triplets are dizygotic, two are "identical," and the third is an ordinary brother or sister. If the triplets are trizygotic, we expect them all to be as different as ordinary brothers and sisters. The degree of intimacy of their amnions, chorionic sacs, and placentas are affected by the same factors as for twins. Understandably, we expect the birth weight of individual triplets to be even lower than that of individual twins.

Multiple births

In recent years, fertility drugs have made multiple births more common. However, among women not taking these drugs, it can be predicted how likely a woman is to have a multiple pregnancy. According to Eastman and Hellman, a mathematical relationship (referred to as **Hellin's Law**) between the various orders of multiple births was first stated by Hellin, who claimed that twins occurred in 1 in 89 births, triplets in 1 in 89^2 (7,921) births, and quadruplets in 1 in 89^3 (704,969) births. More recently, Kay Cassill, in her engaging book, *Twins*, states that Hellin's Law still provides a generally accepted prediction about the frequency of all multiple births.

Superfecundation Twins with two different fathers: the term *fecund* means the ability to produce offspring. The term *superfecundation* refers to fertilization of two or more eggs during the same uterine cycle by sperm from separate acts of sexual intercourse; this can lead to twins from two separate biological fathers. Heteropaternal superfecundation refers to fertilization of two separate eggs by two different fathers. Homopaternal superfecundation refers to the fertilization of two separate eggs by the same father, resulting in fraternal twins. Although superfecundation is rare, it can occur through either separate occurrences of sexual intercourse or through artificial insemination.

Study questions

The pregnant woman

1. The life of a pregnant woman is to a great extent affected by what two factors?
2. Name three common gastrointestinal effects of pregnancy. What are their causes?
3. What are three common changes involving the circulatory, urinary, and musculoskeletal systems during pregnancy?

Childbirth

4. Childbirth is part of what larger process, and what are its three stages? What is generally true of these stages in multigravidas compared to primigravidas?
5. What begins and ends the expulsion stage of labor?
6. What are the four aspects of the fetus's alignment in the uterus that influence how the baby is born?

7. In order, what are the seven cardinal movements of labor?
8. What will normally prevent postpartum hemorrhage?

Multiple pregnancies

9. What are the two categories of twins? What are alternative names for each category?
10. If, during twinning, separation is not complete, what results?
11. Regarding the uterus, how do fraternal twins differ from ordinary siblings?
12. Distinguish between heteropaternal superfecundation and homopaternal superfecundation.

Critical thinking

1. The pregnant woman may experience frequent urination, backache, and shortness of breath. All of these are related through what common occurrence?
2. In what sense is the invasiveness of implantation a double-edged sword?
3. Taking the phenomenon of sex determination into account, why are identical twins, but not fraternal twins, always the same sex?
4. Why is it not surprising that twins have lower birth weights than singletons?

PART II

Some details of human development

Part I has provided an overview of human development, beginning with cells and finishing with childbirth. Part II presents some of the details of development.

After consideration of the skin (Chapter 12), the outermost part of which derives from the surface ectoderm, we take up development of the nervous system (Chapter 13), which early on leaves the surface of the embryo as neural tube and neural crest. In recent years, the neural crest has been referred to as the fourth germ layer because it gives rise to such important and far-flung parts of the embryo.

The skeleton and muscles (Chapter 14) are both derived almost totally from mesoderm, and both play important roles in posture and movement. Perhaps unexpectedly, endocrine glands (which produce hormones, in contrast to the exocrine glands, such as salivary glands and part of the pancreas, which do not release their secretions into the bloodstream) are considered together with the circulatory system (Chapter 15). This is not because these systems have germ layer origins (endocrine glands develop from all three germ layers) in common, but because hormones are transported to their target cells by the circulatory system.

The digestive and respiratory systems (Chapter 16) both arise with linings of endoderm. In fact, the respiratory system originates as an outgrowth of the digestive system. With such an intimate early relationship, it is reasonable to consider these two systems together.

Even without the brain, the head is the most complex part of the body. Although the throat just begins in the head (extending into the neck) and the tactile sense, unlike the other four senses, is not confined to the head, the developments of the mouth, throat, face, and the five senses are considered together (Chapter 17).

We finish Part II in consideration of the urinary and reproductive systems (Chapter 18), which are so inextricably intertwined that their development cannot be considered independently. Closely allied to the development of these systems is the development of the external genitalia, and so we wind up on the surface of the body, which was where we began our consideration of some of the details of human development.

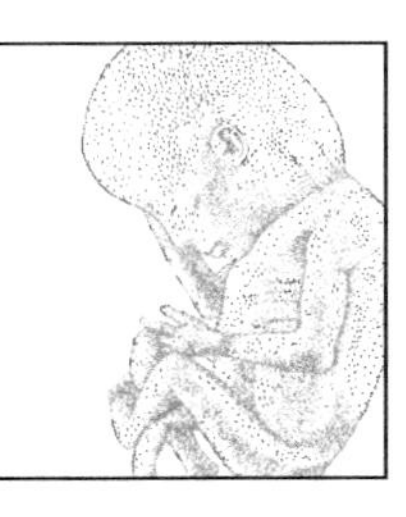

CHAPTER TWELVE

The skin

> **CHAPTER OBJECTIVES**
>
> After studying this chapter, you should be able to:
>
> 1. Discuss the two general parts of the skin and their origins.
> 2. List the appendages of the skin and discuss the development of hair and glands.
> 3. Discuss the developmental interaction between the epidermis and dermis.

Through our skin, we are able to touch our environment. But the skin has functions that go far beyond sensory perception. It prevents us from drying out, protects us against abrasion, and imposes a physical barrier against disease-causing agents. In addition, **Langerhans cells**, members of the immune system, survey our skin and play an active role in cell-mediated immunity, protecting us from invasion by foreign organisms. By the time we are adults, our skin is made up of two principal components, the **epidermis** and the **dermis** (Figure 12.1).

We begin our consideration of the skin with the surface, the epidermis. Then we go deeper, into the dermis. Specialized parts of the skin—hair, glands, and nails—arise from interactions between the epidermis and dermis, and we then consider these. One of the most striking characteristics of skin is its color, and this chapter ends with a consideration of skin pigmentation.

Epidermis

The epidermis consists of several **strata** (layers). From the inside out, they are **stratum germinativum, stratum spinosum, stratum granulosum, stratum lucidum**, and **stratum corneum** (Figure 12.2). Epidermal cells originate in the germinativum layer and move progressively outward, differentiating and finally dying. This is normal cell death under genetic

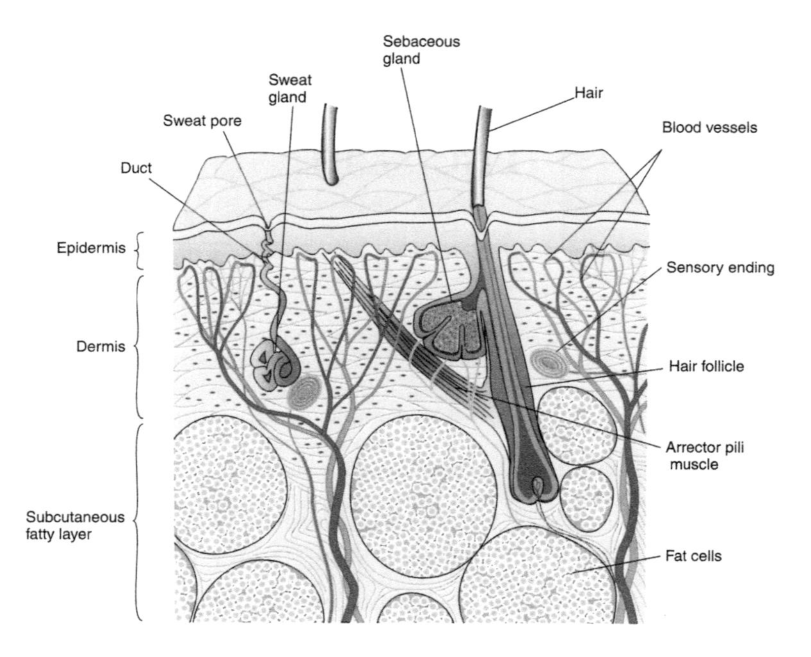

FIGURE 12.1 A section through adult human skin. Notice the relatively thin outer epidermis and the relatively thick inner dermis.

regulation, and the corneum layer is composed naturally of dead cells. As the cells differentiate toward death, they synthesize and accumulate the fibrous protein **keratin**. Organized into **intermediate filaments**, keratin allows skin cells to function so admirably in preventing dehydration and limiting the effects of abrasion.

Earlier in development, humans do not have such a multilayered surface. At 4 weeks, the original skin of the embryo consists of a single layer of cells, the **ectoderm**. By the end of the embryonic period (8 weeks), the skin develops into two layers: the outer **periderm** and the inner **basal layer** (Figure 12.3). Before the end of the first trimester, an intermediate layer (produced by the basal layer) appears between the basal layer and the periderm. By the sixth month, the periderm disappears, and by birth, the layers typical of adult skin are present. Some of the epidermis's cells originate in the neural crest.

In addition to Langerhans cells derived from the bone marrow, two additional cell types are not of epidermal origin. **Melanocytes** (pigment cells) and **Merkel cells** (sensory cells) arise from the **neural crest** (see Figure 8.9) and subsequently migrate into the epidermis. The melanocytes provide pigmentation to the epidermis, and Merkel cells function as skin **mechanoreceptors**, which sensitize the skin to touch.

Dermis

The dermis of the skin is primarily derived from the mesoderm. The dermis of the dorsal part of the body develops from the dermatomal portions

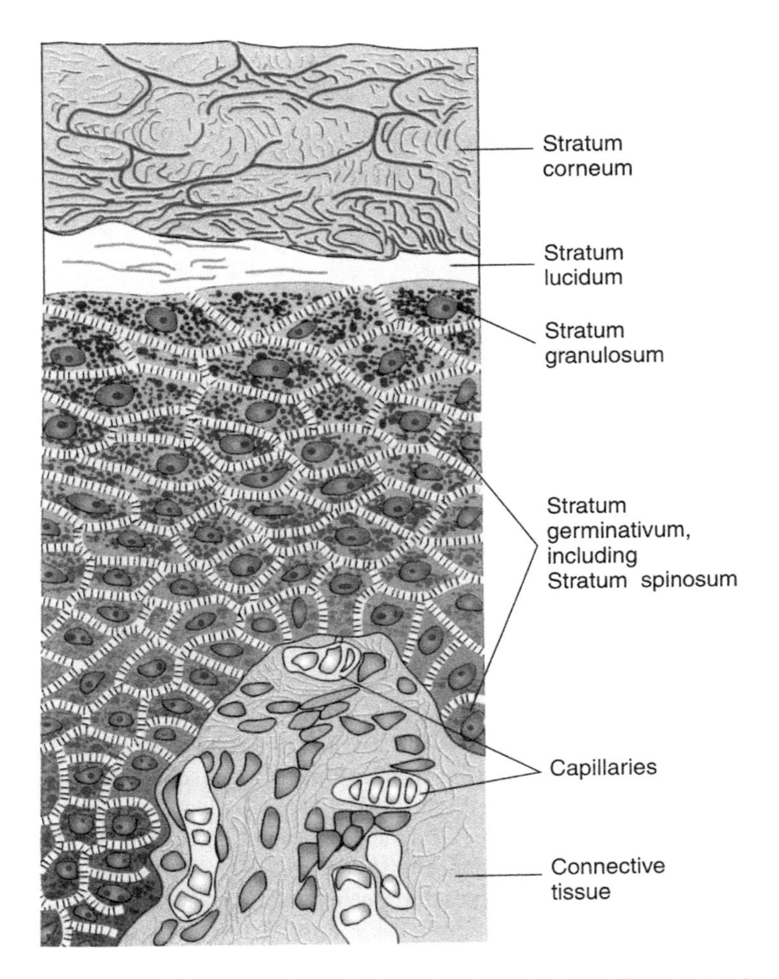

FIGURE 12.2 Diagrammatic vertical section through the epidermis. The five layers of the skin from the lowermost layer to the most superficial layer are shown. New cells are continuously generated in the stratum germinativum, and dead cells are continuously lost from the stratum corneum.

of somites, and the ventral dermis develops from somatic mesoderm. However, a portion of the dermis of the head comes from the neural crest.

Both components of the skin, epidermis and dermis, depend on each other for their normal development. This can be demonstrated experimentally by separating the components and growing them in isolation from each other. In such circumstances, neither epidermis nor dermis develops normally. Many of the structures produced by the skin (such as hair and nails) depend on interaction between epidermis and dermis.

Hair

Although hair begins to develop in the 12th week of development, it is not until the 20th week that it becomes readily visible. Hair first appears

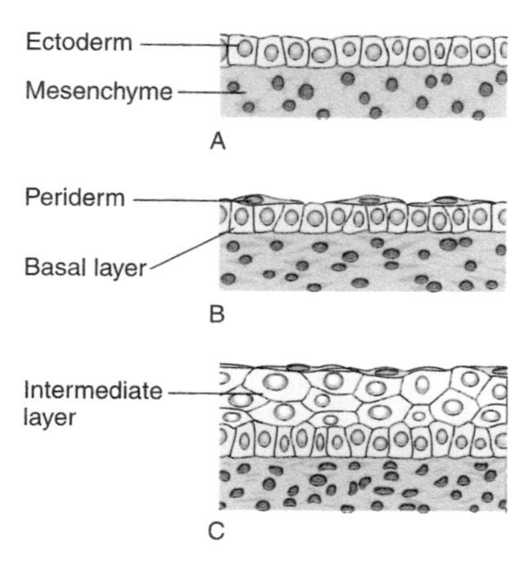

FIGURE 12.3 Development of the epidermis. Diagrammatic vertical sections are shown through the skin at different stages of development. (A) Embryonic skin at 4 weeks consists of a single layer, the ectoderm. (B) At 8 weeks, the end of the embryonic period, the skin consists of two layers—the outer periderm and the inner basal layer. (C) At 4 months, during the fetal period, the skin consists of three layers, with the addition of the intermediate layer between the two layers of embryonic skin. By birth, the layers typical of adult skin are present (see Figure 12.2).

on the head and then progresses toward the tail end of the fetus. The first hair to appear is **lanugo**, a fine, downlike hair confined to the fetal period (see Figure 9.7). Gradually, lanugo is replaced by hair more typical of babies. As the hair develops, **sebaceous glands** appear near the bases of the hair shafts (see Figure 12.1). The secretions of these glands (**sebum**), the periderm cells, and the lanugo together make up a whitish, cheesy coat on the skin of the fetus, called **vernix caseosa** (Figure 12.4). This protective layer prevents what would otherwise be harmful effects of

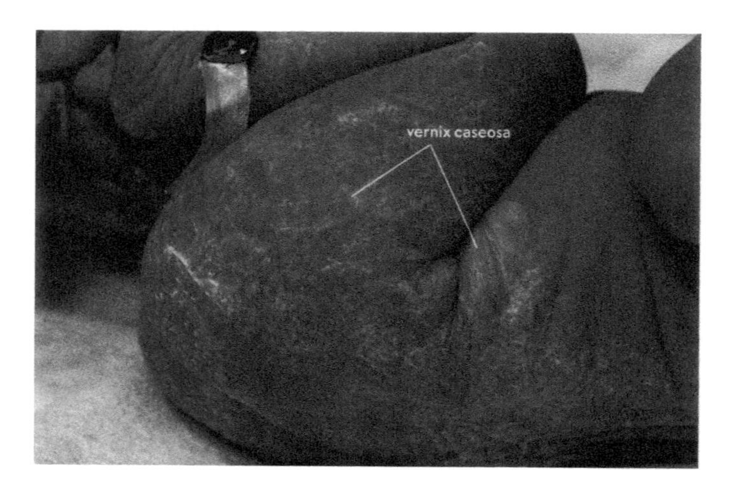

FIGURE 12.4 Vernix caseosa. Photograph of a fetus during the sixth month shows the vernix caseosa. (Keith/Custom Medical Stock Photo.)

amniotic fluid on the delicate skin of the fetus. Each hair has an associated smooth muscle fiber, which, unlike the ectodermally derived hair shaft and sebaceous gland, is derived from the dermis. These **arrector pili muscles** cause us to have a "hair-raising" experience under certain stressful conditions (see Figure 12.1).

Glands

In addition to the sebaceous glands, which we considered in conjunction with hair development, other important glands originate in the ectoderm. As do many other derivatives of the skin, **sweat glands** first appear as epidermal down-growths into the underlying dermis (Figure 12.5). This solid core of epidermal cells becomes highly contorted at its distal end within the dermis, but it remains rather straight closer to the surface of the skin. The straight portion hollows out and becomes the duct, and the contorted distal portion becomes the secretory part of the gland.

Development of **mammary glands** (the milk-producing glands) is foreshadowed by the appearance of bilateral **mammary ridges** extending from the armpits to the groin of the 4-week embryo (Figure 12.6). By the sixth week, these ridges normally disappear, except in the pectoral (chest) region, where they play a role in breast development. As with the sweat glands, a solid core of epidermal cells grows down into the underlying dermis. However, unlike the contortion of the distal region of the developing sweat gland, the distal portion of the developing mammary gland becomes highly branched, with portions proximal to the surface of the skin giving rise to **lactiferous ducts** (the milk-carrying ducts) and the distal regions giving rise to the secretory portions of the mammary glands (Figure 12.7).

If the mammary ridges do not regress properly, the embryo may develop more than the two breasts standard for humans. Such extra breasts, which may occur anywhere along the original mammary ridges, are referred to as **supernumerary breasts** (Figure 12.8).

Nails

The fetus develops nails near the tips of the **digits** (fingers and toes) (Figure 12.9). This development begins early in the fetal period (at 10 weeks). Despite their seeming similarity, the fingernails and toenails develop at different paces, with the fingernails reaching the tips of the digits at 32 weeks and the toenails at 36 weeks.

Pigmentation

In addition to the other functions already listed, the skin protects us from solar radiation. This protection is afforded by melanin pigment provided

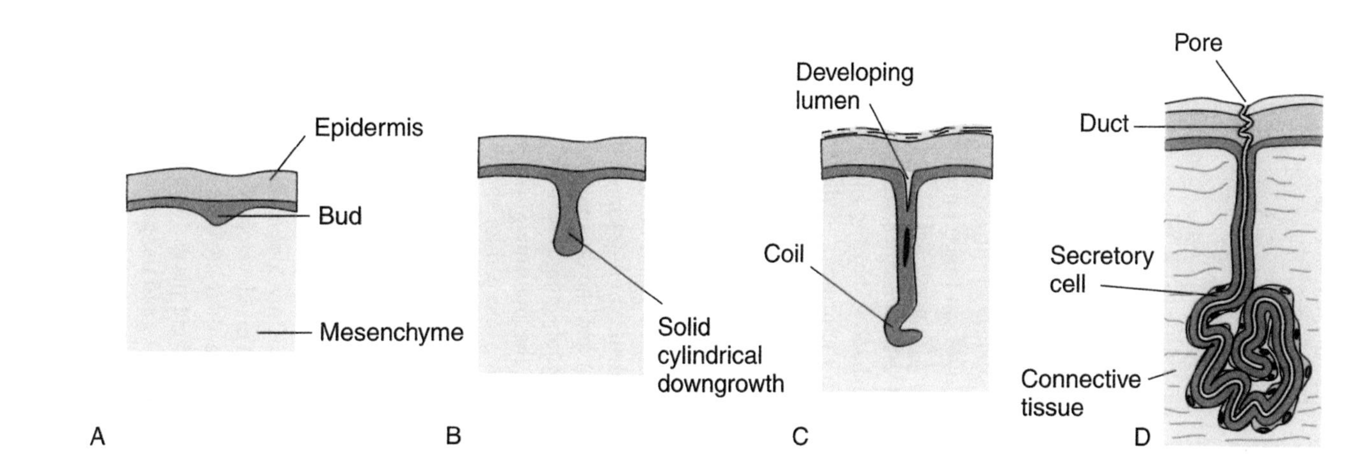

FIGURE 12.5 Development of a sweat gland. Diagrammatic vertical sections through the skin show the origin of a sweat gland as a bud of epidermis (at about 20 weeks) (A), which grows down into the dermis (B and C), and its subsequent development into a sweat gland (D). Not visible are the interactions between the epidermis and dermis that underlie this morphologic development.

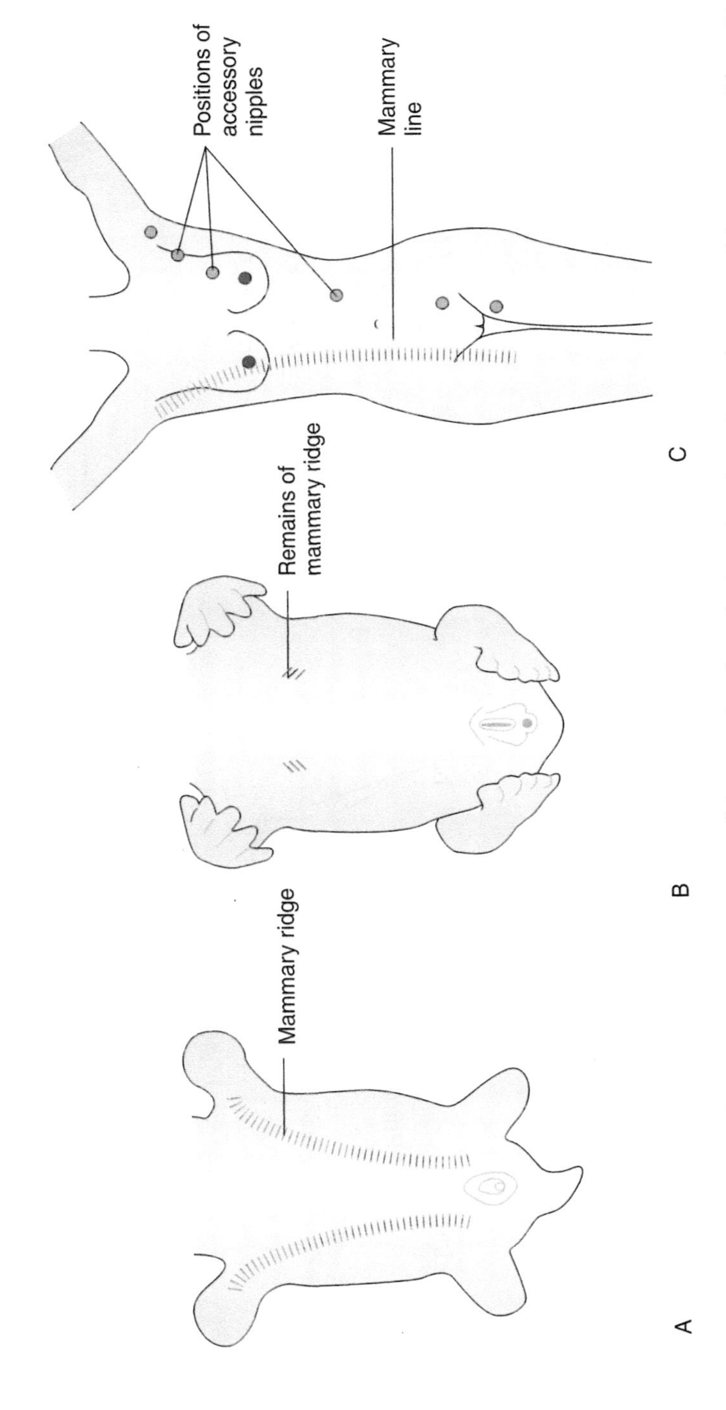

FIGURE 12.6 The mammary ridges. (A) Diagrammatic view of the ventral surface of a 4-week embryo shows the extent of the mammary ridges. (B) Similar view of a 6-week embryo shows the remains of the mammary ridges, where the mammary glands normally form. (C) Diagrammatic view of the ventral surface of an adult woman shows the relative positions of the original mammary ridges and also where accessory nipples (and mammary glands) may form.

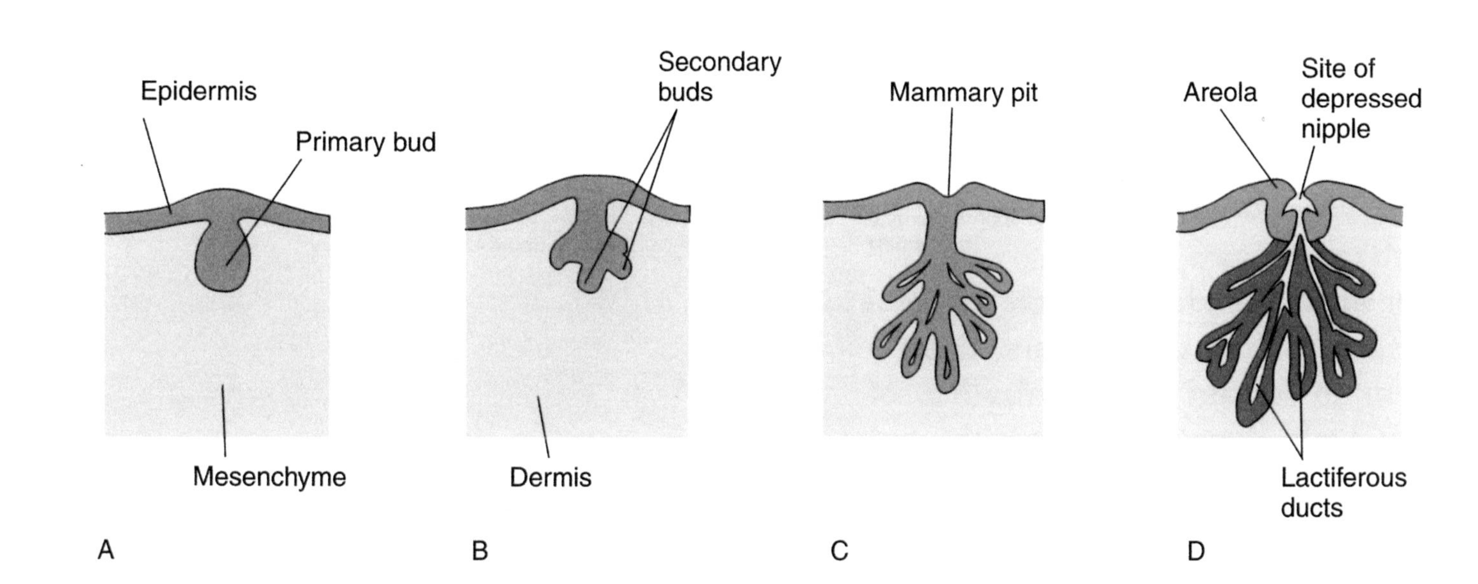

FIGURE 12.7 Development of a mammary gland. Diagrammatic vertical sections through the skin in the region of a mammary ridge show the origin of a mammary gland as a bud of epidermis (A), which grows down into the dermis (B and C), and its subsequent development into a mammary gland (D) between 6 weeks and birth. Not visible are the interactions between the epidermis and dermis that underlie this morphologic development.

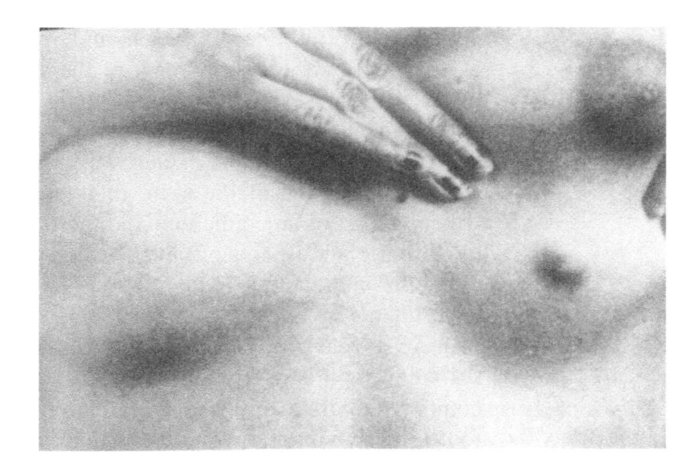

FIGURE 12.8 Supernumerary breast. Photograph of a woman shows a supernumerary breast, in this instance, below her normal left breast. This abnormality results from improper regression of the mammary ridges. (From Moore, K.L. and Persaud, T.V.N. *Before We Are Born*, 4th edition. WB Saunders, Philadelphia, 1993, Figure 19.6, p. 319.)

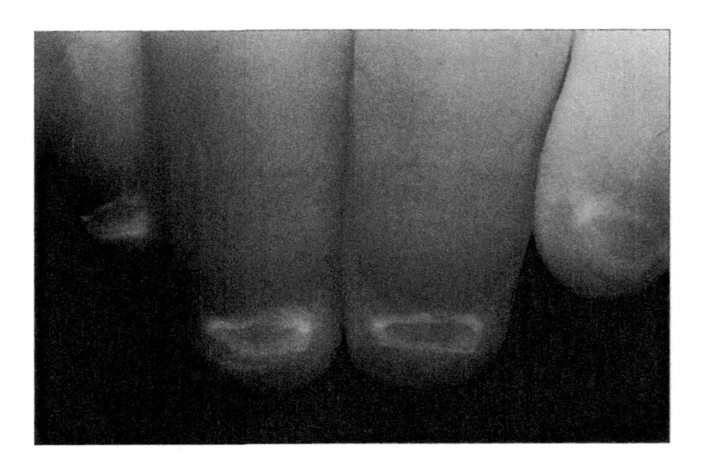

FIGURE 12.9 Fingernail development. Photograph of the fingertips of a 15-week fetus shows the developing fingernails. (England, M.A. *Color Atlas of Life Before Birth*, Year Book Medical Publishers, Chicago, 1983, Figure E, p. 173.)

for the skin by the melanocytes. Different levels of pigmentation lead to differences in protection from the sun. Albinos, persons who lack pigmentation, are especially sensitive to sunlight, whereas ethnic groups with dark pigmentation have a lower risk of developing skin cancer than other groups. Melanocytes also provide pigmentation for hair; when our pigment disappears, we develop gray hair.

Epithelial–mesenchymal interactions

The role of epithelial–mesenchymal interactions in development is exemplified by the development of the skin and its derivatives. These interactions are examples of embryonic inductions, which include the development of the skin, in which epithelium, which gives rise to the epidermis of the skin and its derivatives, and mesenchyme, which gives rise to the dermis of the skin, interact. The development of organs, which consist of an epithelium and mesenchyme, are clearly among the most important aspects of development. See Chapter 17 for the role of epithelial–mesenchymal interactions in tooth development. Classical experiments by Saunders on chick skin development demonstrated the instructive role of the mesenchyme in the regional specificity of chick skin development. Specifically, if wing epidermal epithelium was combined alternatively with wing mesenchyme, thigh mesenchyme, or foot mesenchyme, the wing epithelium would respectively give rise to wing feathers, thigh feathers, or scales.

Study questions

1. What are the two principal components of adult skin? By when during development are the layers typical of the adult skin present?
2. Langerhans cells play a role in what type of immunity? From what do these cells arise?
3. Melanocytes provide the epidermis with what? From what do these cells originate?
4. Which cells in the epidermis, derived from the neural crest, function as skin mechanoreceptors?
5. Which part of the skin is primarily derived from mesoderm? Specifically, from what are dorsal dermis, ventral dermis, and head dermis derived?
6. Development of skin structures—for example, hair and nails—depends on interactions between what?
7. When does hair first begin to develop? What is lanugo and to what period of development is it confined?
8. By means of what does the skin protect us from sunlight? This is provided by which cells?
9. Name three types of glands that originate from the epidermis.
10. When do nails begin to develop? Which nails reach the tips of the digits first?
11. The development of the skin, as well as the development of teeth and other organs, involves interactions between what two tissues?

Critical thinking

1. The outer cells of the epidermis and its derivatives (e.g., hair and nails) are dead. Apoptosis (genetically programmed cell death), initially a developmental phenomenon, plays a role in skin function. What might be the consequences of deficient apoptosis?
2. Because so many skin cells die to become functional, they must be replaced by stem cell populations. In what sense does the skin never stop developing? (Hint: Think about the three basic processes of development.)

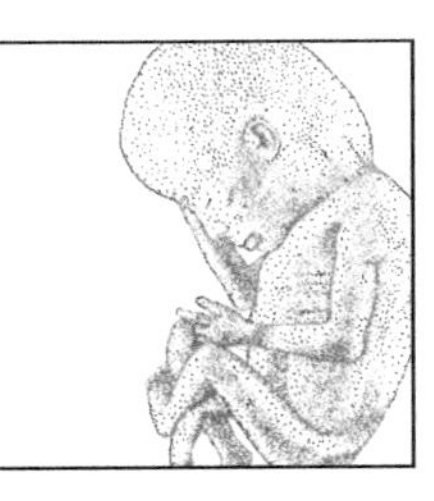

CHAPTER THIRTEEN

The nervous system

CHAPTER OBJECTIVES

After studying this chapter, you should be able to:

1. Discuss the origin of the spinal cord, including the spinal canal, the three layers of the wall of the spinal cord, and its white and gray matter.
2. Discuss the origin of the spinal nerves.
3. Discuss the origin of (1) the three-vesicle and (2) the five-vesicle brain, including the ventricles and cerebrospinal fluid.
4. List some of the parts of the adult brain derived from the parts of the five-vesicle brain.
5. Discuss the origin of spinal and cranial ganglia and nerves.

Our nervous system, among other things, provides us with memory, anticipation, sentience, emotion, and the abilities to move, to fear, and to think. Not only does it give us an impression of what is out there—however subjective that may be—but it also allows us to imagine, design, and build extensions of our abilities. We have telescopes to see farther, microscopes to see the invisible, and spectrophotometers to analyze the chemical composition of distant stars. But even without the aid of such instruments, our natural senses are truly impressive. In an age in which computer technology can provide us with tremendous potential for data retrieval, we can only marvel at the power of our nervous system.

The capabilities of the most sophisticated camera are dwarfed by the combination of the human eye and brain. We can see well in a wide range of light, faster, in three dimensions, and with an enormous depth of field. Moreover, the "development of the film" takes place instantly. True, we do not get a "hard copy," but we do document what we see in our memory.

We owe our abilities to the cooperative activity of many different kinds of cells, but they all are derived from a single cell, the zygote. As the embryo develops from the zygote, the embryo's neural tube develops into the spinal cord toward its caudal end and into the brain at its cephalic end. These parts make up our central nervous system. After considering the development of the CNS, we will take a look at the peripheral nervous system.

Central nervous system

The CNS is made up of the spinal cord and the brain.

Spinal cord

The fluid-filled lumen that runs through the **neural tube** (structure that gives rise to the CNS) is called the **neural canal** or **neurocoele**. This relatively simple structure gives rise to the **spinal canal** of the spinal cord, and its contained fluid becomes the **cerebrospinal fluid**. The walls of the neural tube are quite thin, *originally* organized into epithelium that is the thickness of one cell layer. This **neuroepithelium** gives rise to three kinds of cells:

- *Ependymal cells*: Make up the inside lining of the neural tube
- *Neuroblasts*: Give rise to the various kinds of neurons (nerve cells) found in the brain and spinal cord (or CNS)
- *Glioblasts*: From which two types of CNS glial cells arise—**astrocytes** and **oligodendrocytes** (Figure 13.1)

As the spinal cord develops, it becomes much thicker in comparison to the *surrounded* fluid than the neural tube was. The cells of the wall

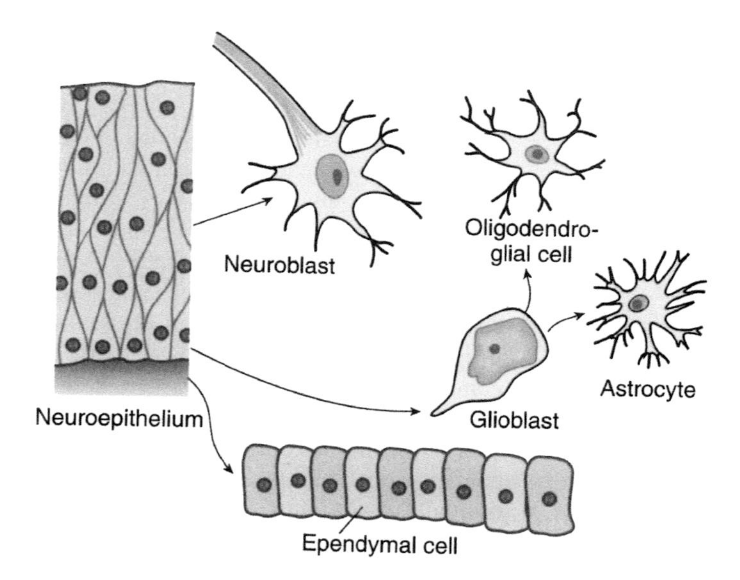

FIGURE 13.1 Cell types derived from the neuroepithelium. Ependymal cells, neuroblasts, and glioblasts. Ependymal cells line the central canal of the brain and spinal cord, neuroblasts give rise to various kinds of neurons of the central nervous system, and glioblasts give rise to glial cells of the CNS.

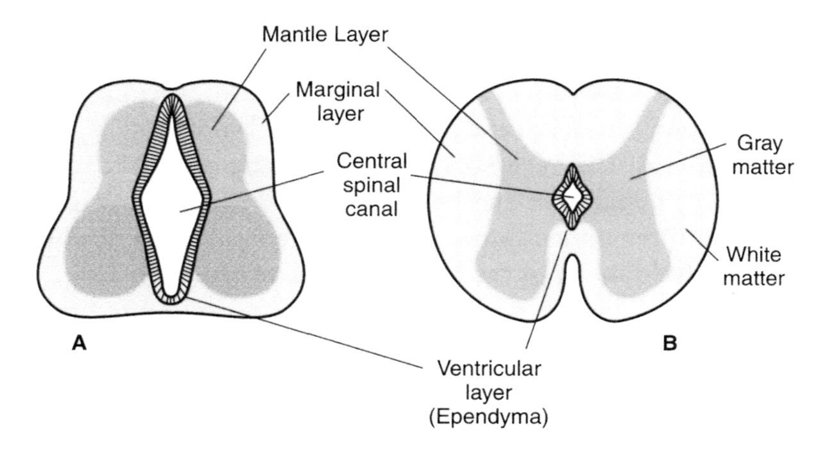

FIGURE 13.2 Spinal cord development. Two successive stages, earlier (A) and later (B) in spinal cord development. The relative size of the central canal decreases compared with the thickness of the spinal cord wall. The ventricular layer is made up of the ependymal cells. Cell bodies of the neuroblasts make up the mantle (intermediate) layer, which gives rise to the gray matter. The marginal layer is made up of processes of the developing neurons and gives rise to the white matter.

of the spinal cord are arranged into three layers (from the spinal canal outward):

- *Ventricular layer*: Contains the ependymal cells
- *Mantle (intermediate) layer*: Contains the cell bodies of the neuroblasts, which in turn contain the nuclei of the nerve cells
- *Marginal layer*: Is made up of the processes (**neurites: axons** and **dendrites**) of the developing neurons

The mantle layer has a gray appearance and is therefore called **gray matter**; the marginal layer is similarly referred to as **white matter** because of its appearance. From the center outward, we have spinal canal, ependyma, gray matter, and white matter (Figure 13.2).

It is through the gray matter that the spinal cord communicates with the body. Nerve cell bodies in the lower (ventral) half of the gray matter send axons out of the spinal cord to become the **ventral (motor) roots of spinal nerves**, so called because these nerves will **innervate** (supply with nerves) skeletal muscle and so control our motor activity. Nerve cell bodies in the upper (dorsal) half of the spinal cord receive axons of nerve cell bodies found in **spinal ganglia** (aggregates or masses of nerve cell bodies derived from the neural crest) outside the spinal cord. These axons form the **dorsal (sensory) roots of spinal nerves** and bring sensory input from peripheral parts of the body, such as skin. Beyond the spinal ganglia, the spinal nerves have both sensory and motor components and therefore make up what are called **mixed nerves** (Figure 13.3).

Brain

The brain has its origins early in human development. It originates as the anterior region of the neural tube, which begins to form during the third

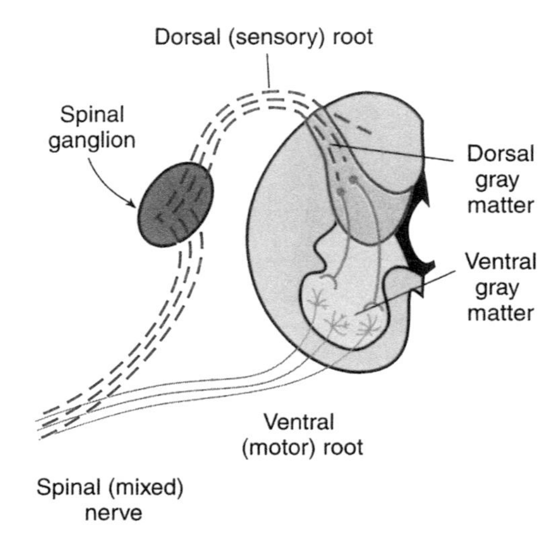

FIGURE 13.3 Spinal nerves. A diagrammatic representation of a spinal ganglion and spinal nerve on one side of the spinal cord (the other member of the bilaterally symmetric pair is not shown in this figure). The spinal ganglion is on the sensory dorsal root of the spinal nerve, near where the sensory dorsal root joins the motor ventral root. Spinal nerves are mixed nerves with both sensory and motor components.

week of development. During the fourth week, the brain portion of the neural tube, which is wider than the spinal cord portion, will develop the three **primary brain vesicles: forebrain** (**prosencephalon**), **midbrain** (**mesencephalon**), and **hindbrain** (**rhombencephalon**) (Figure 13.4).

Initially, the cells are arranged into the same three layers (ventricular, mantle, and marginal) as found in the developing spinal cord. However, in some parts of the brain, such as the developing **cerebrum** (the largest portion of the brain), some of the gray matter of the mantle layer migrates to the surface of the brain outside of the white matter. When we "use our gray matter," the reference is to the **cortex** (outer portion) of our cerebrum.

As the neural tube forms during the fourth week, it bends along with the head end of the embryo. This bending is referred to as **flexion**. For the sake of simplification, we can think of the early brain as being a straight tube. However, it does not remain so for long. By the end of the fifth week, it exhibits **cephalic, pontine**, and **cervical flexures** (Figure 13.5A). In addition, by the seventh week, cerebral hemispheres are growing out of the walls of the telencephalon (see following text); by 11 weeks, they will have overgrown the rest of the forebrain. By 30 weeks, only two large regional divisions of the brain—cerebrum and cerebellum—are visible from a surface view of the brain (see Figure 13.5B).

Vesicles and ventricles

During the fifth week of development, the three **primary vesicles** (small fluid-filled sacs) give rise to a brain made up of five **vesicles**. The **telencephalon** and **diencephalon** arise from the forebrain, the midbrain persists as the **mesencephalon**, and the hindbrain gives rise to the **metencephalon** and the **myelencephalon** (see Figure 13.5A).

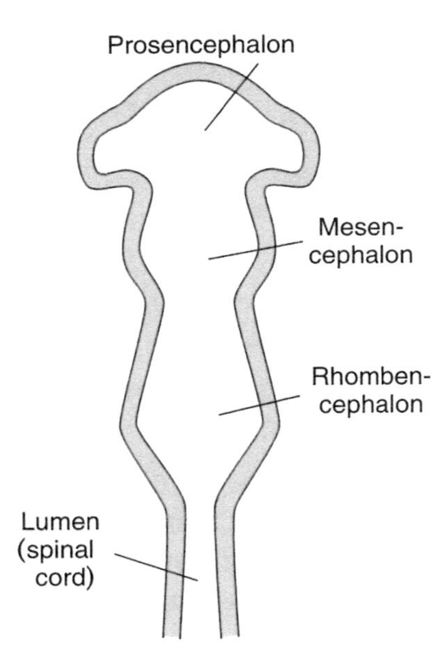

FIGURE 13.4 The three-vesicle brain. A highly diagrammatic representation of the three primary vesicles of the early brain (at 4 weeks): forebrain (prosencephalon), midbrain (mesencephalon), and hindbrain (rhombencephalon).

Some of the better-known structures of the brain have their origins in these vesicles. The **retinas** and **optic nerves** of the eyes are really extensions of the brain wall and arise as paired outgrowths (optic vesicles) of the walls of the diencephalon. Glands are also derived from the vesicles. The **pineal gland** arises as an outgrowth from the roof of the diencephalon, and a portion of the **pituitary gland** (**neurohypophysis**) arises as an outgrowth (**infundibulum**) of the floor of the diencephalon. The roof of the metencephalon gives rise to the **cerebellum**, and its floor gives rise to the **pons**. The **medulla** of the brain arises from the myelencephalon.

Like the spinal cord, the brain contains a cerebrospinal fluid-filled lumen. The lumen makes up the four **ventricles** (small cavities) and the **aqueduct of Sylvius** of the brain. Two outgrowths of the telencephalon—the **cerebral vesicles**—give rise to the **cerebral hemispheres**, which contain ventricles I and II. These ventricles communicate with ventricle III, making up the remainder of the telencephalon and diencephalon by means of the **foramina of Monro** (Monro's foramen). Ventricle III in turn communicates—through the aqueduct of Sylvius (lumen of the mesencephalon)—with ventricle IV, which is the lumen of the metencephalon and myelencephalon (Figure 13.6). From the walls of the lateral ventricles (I and II) and roofs of ventricles III and IV, vascular **choroid plexuses** extend into the ventricles and secrete cerebrospinal fluid.

Cranial ganglia

Just as the spinal cord has spinal ganglia associated with it, the brain has **cranial ganglia** associated with it (Figure 13.7). There are 12 pairs of

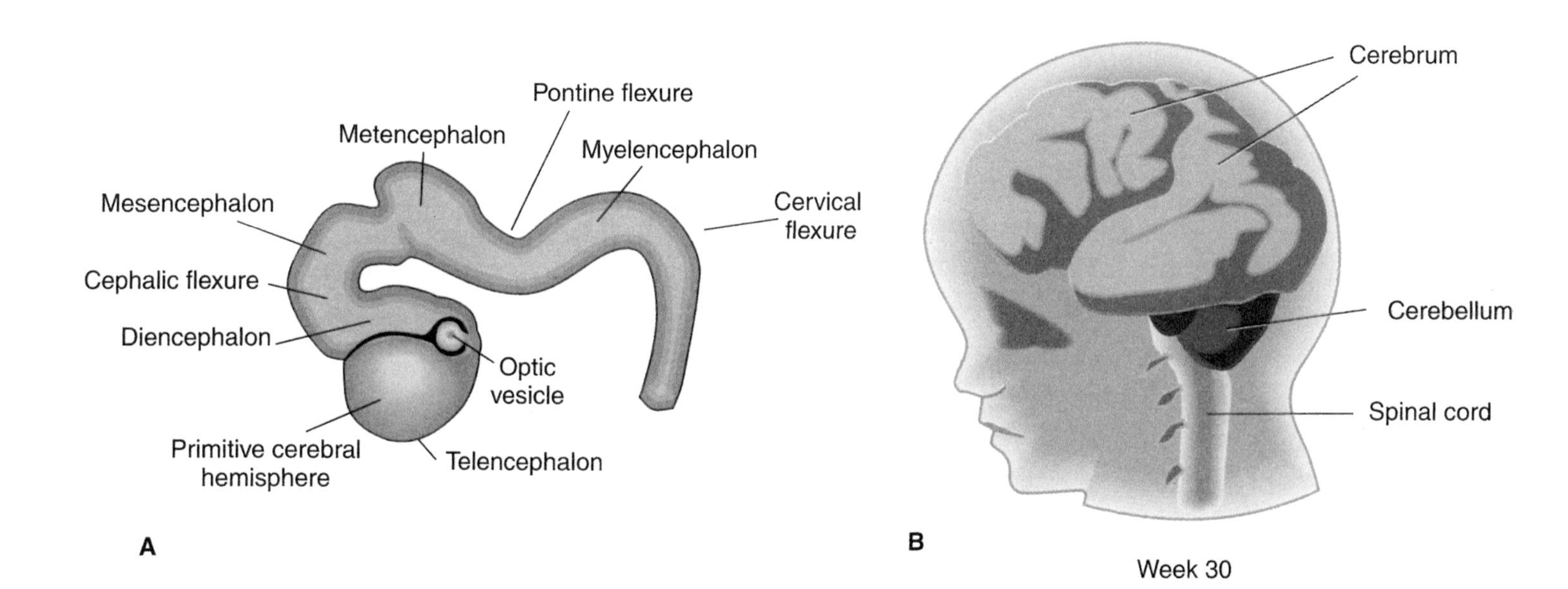

FIGURE 13.5 Flexures and vesicles of the brain. (A) As the brain forms, it bends to give rise to three flexures: cephalic, pontine, and cervical flexures. The five-vesicle brain—a highly diagrammatic representation of the five vesicles of the early brain (at 5 weeks). The prosencephalon gives rise to the telencephalon and the diencephalon, the mesencephalon persists, and the rhombencephalon gives rise to the metencephalon and myelencephalon. (B) The surface of the brain at 30 weeks.

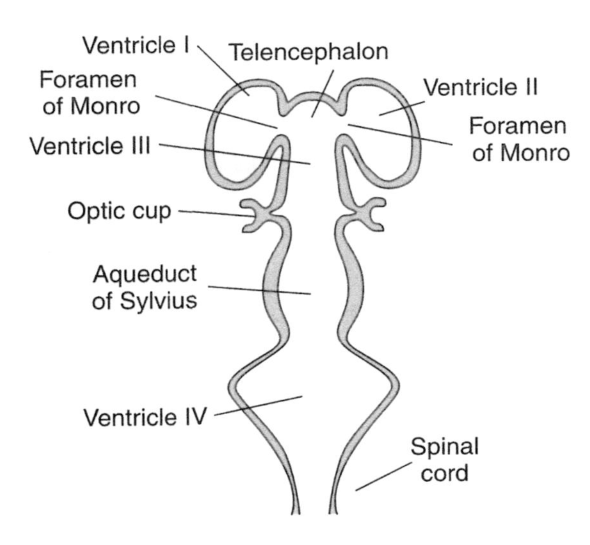

FIGURE 13.6 Lumen of the brain. Four ventricles (I–IV) and the aqueduct of Sylvius make up the cerebrospinal fluid–filled lumen of the brain.

cranial nerves, which arise from three sources: the brain itself, the neural crest, and thickenings of the ectoderm called **ectodermal placodes**. Cranial nerves may be classified according to their composition into one of three categories: sensory, motor, or mixed. Two of the three sensory nerves, the **olfactory nerve** (concerned with smell) and the optic nerve, are considered parts of the brain. Like the five purely motor cranial nerves, these two nerves have no associated ganglia. The third sensory nerve, the auditory nerve, and the four mixed (i.e., having sensory and motor components) cranial nerves have associated ganglia derived from neural crest or ectodermal placodes.

Peripheral nervous system

The **peripheral nervous system** (PNS) is the portion of the nervous system *outside* the brain and spinal cord. Outside the CNS, an aggregate of nerve cell bodies is called a **ganglion**, and a bundle of cell processes (primarily axons) is called a nerve. (Inside the CNS, the equivalent structures are called **nuclei** and **nerve tracts**, respectively.) We have already made reference to the origin of spinal ganglia and spinal nerves. The PNS has another component composed of autonomic nerves ("autonomic" means independent of the will to act) and ganglia, which arise from neural crest cells. The sensory components of the cranial nerves (except for the olfactory and optic nerves) arise from cranial ganglia in a manner analogous to the formation of the sensory components of spinal nerves from spinal ganglia. Likewise, the motor components of cranial nerves arise from the brain in the same way motor components of spinal ganglia arise from the spinal cord.

Most axons, which make up nerves in the PNS and nerve tracts in the CNS, become coated with a fatlike substance called **myelin**. The process is called **myelination**. **Glial cells** myelinate nerves. Specifically, **oligodendrocytes** perform this function in the CNS, and **Schwann**

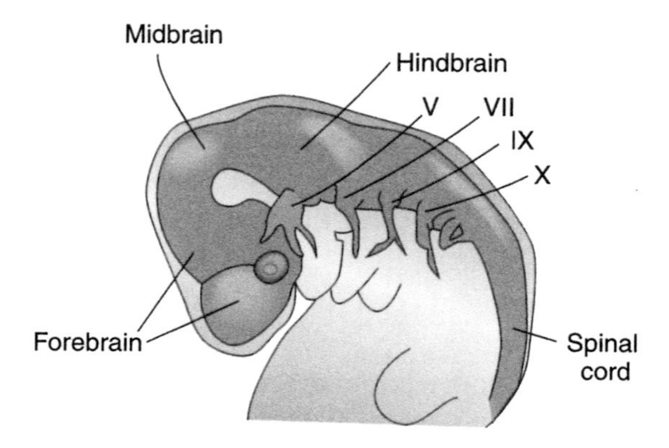

FIGURE 13.7 Cranial ganglia. Lateral view of the embryonic brain (during the sixth week). Note that four of the cranial ganglia (one member of each bilateral pair) are visible: V–semilunar ganglion, VII–geniculate ganglion, IX–superior ganglion of the glossopharyngeal nerve, and X–superior ganglion of the vagus nerve.

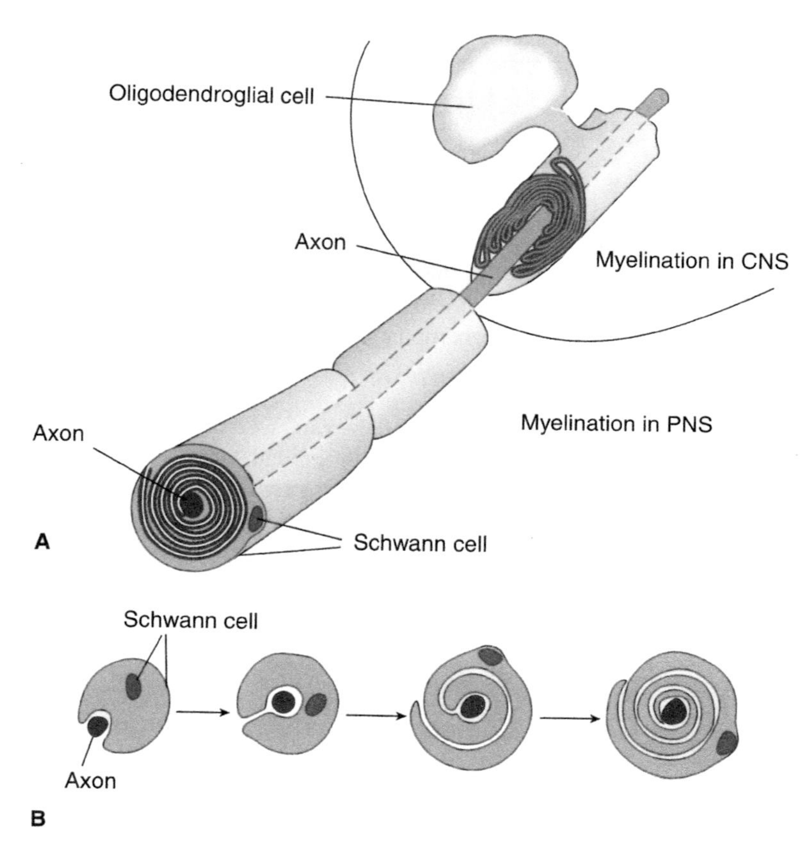

FIGURE 13.8 Myelination. (A) In the central nervous system, axons are myelinated by oligodendroglial cells. In the peripheral nervous system, axons are myelinated by Schwann cells. (B) Cross-sectional views of a Schwann cell myelinating an axon.

cells do so in the PNS (Figure 13.8). As nerves and nerve tracts become myelinated, they become functional.

In addition to being a factor in the regulation of cell number in the development of many tissues, **programmed cell death** (**apoptosis**) also plays such a role in the development of the nervous system.

Paradigm lost

At the beginning of the twentieth century, it was thought that neurons (nerve cells) were not formed in the mammalian brain after birth. Toward the end of the twentieth century, evidence accumulated that indeed the postnatal mammalian brain does produce new neurons. The process by which neurons are generated from neural stem cells is called neurogenesis. Consequently, we now talk of two general types of neurogenesis, developmental neurogenesis and adult neurogenesis. In the brains of studied mammals, adult neurogenesis occurs in two regions of the adult brain: the subgranular zone (SGZ) of the dentate gyrus of the hippocampus and the subventricular zone (SVZ) of the striatum.

Study questions

1. What is the neurocoele and to what does it give rise? What fluid does it come to contain?
2. The neuroepithelium gives rise to what three kinds of cells? To what does each of these kinds of cells give rise?
3. What are the three layers of the spinal cord (from the spinal canal outward) and what does each contain?
4. Distinguish between gray matter and white matter.
5. Why are spinal nerves called mixed nerves?
6. From what does the human brain originate?
7. Our "gray matter" refers to what?
8. What are the five brain vesicles that arise during the fifth week and from what do they arise?
9. Relate the parts of the fluid-filled lumen of the brain to the five vesicles of the brain.
10. From what three sources do the 12 pairs of cranial nerves arise?
11. What is the central nervous system equivalent of ganglia in the peripheral nervous system?
12. What is the central nervous system equivalent of nerves in the peripheral nervous system?
13. From what do the autonomic nerves and ganglia arise?
14. What is myelination?
15. Distinguish between developmental neurogenesis and adult neurogenesis.

Critical thinking

1. Give the general origin of the five-vesicle brain from the three-vesicle brain, and give the general derivatives of the five-vesicle brain.
2. Give the relationship between the five-vesicle brain and the fluid-filled cavities of the brain.

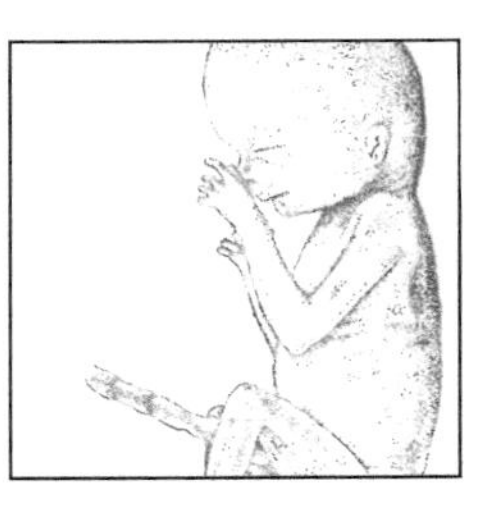

CHAPTER FOURTEEN

The skeleton and muscles

CHAPTER OBJECTIVES

After studying this chapter, you should be able to:

1. Discuss the difference between the two types of osteogenesis.
2. Explain the origin of bone marrow.
3. Discuss the development of somites, including the origin of the vertebrae, intervertebral discs, and the nucleus pulposus of the discs.
4. Explain the development of the skull.
5. Explain: what limb buds are, the development of long bones, and how this development is related to bone growth.
6. Distinguish between osteoblasts and osteoclasts.
7. Discuss the origins of myotomes and lateral mesoderm and their contributions to muscle development.
8. Compare the three major types of muscles in the human body, their origins, and their differentiation.

Without our skeleton and muscles, we would not be able to move about or maintain our posture. Moreover, our ability to hear also depends on three small bones—the incus, malleus, and stapes—in our middle ear and the muscles attached to these bones. Like other vertebrates, we have an **endoskeleton** (internal skeleton), which provides protection for our soft parts. We do not have an **exoskeleton** (shell) like some animals, but if you have ever banged your head against something hard, you can appreciate the skull's protection of your brain! Our skeleton is divided into two general parts (Figure 14.1): the **axial skeleton** is composed of the skull, vertebral column (backbone), sternum, and ribs, and the **appendicular skeleton** is composed of limbs, pelvic girdle, and pectoral girdle.

In addition to allowing us to move and maintain our posture, our muscles perform many of the functions necessary for life. They allow us to chew our food, process it through the digestive system, and excrete the

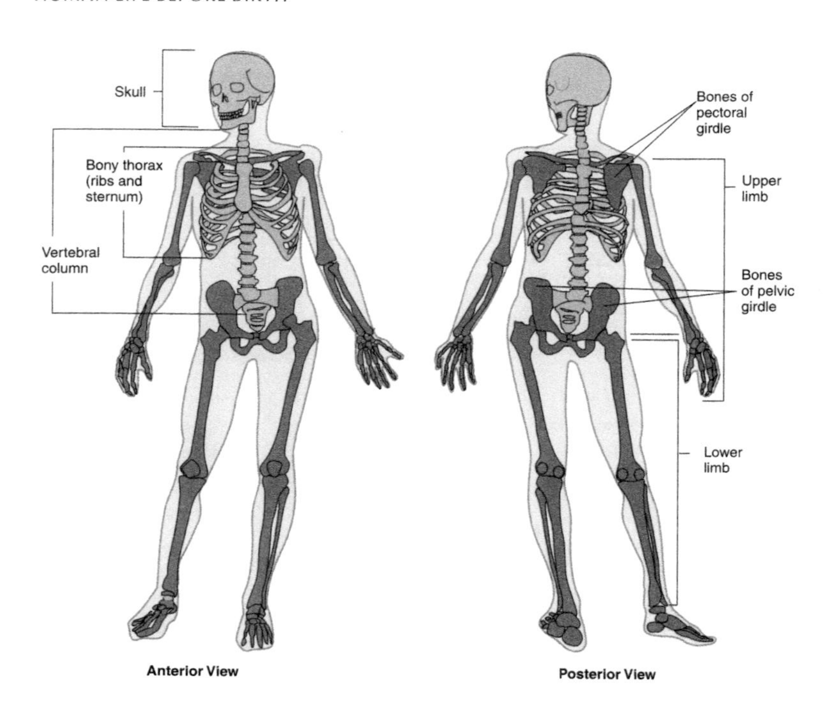

FIGURE 14.1 The skeleton is divided into two general parts: the axial skeleton and the appendicular skeleton.

remains. They help us see by regulating the amount of light entering our eyes and focusing our vision. Not only do muscles pump our blood, but they also contribute to the vessels through which the blood moves. Most of our muscles originate in the mesoderm, and the three major types of muscle are skeletal muscle, cardiac muscle, and smooth muscle.

The skeleton

Origin of bones

Two general types of **osteogenesis** (bone formation) occur: **intramembranous ossification** and **endochondral ossification** (ossification is the formation of bone substance). Both types begin with **mesenchymal cells** (loose connective tissue), which are usually of mesodermal origin. However, the **cranial neural crest** contributes to bone formation in the head.

The initiation of **intramembranous ossification** occurs when specific mesenchymal cells called **osteoblasts** begin to secrete a specific **extracellular matrix**. This matrix, together with an enzyme, alkaline phosphatase, causes deposits of calcium phosphate crystals to form, resulting in the formation of bone.

Endochondral ossification occurs when mesenchymal cells called **chondrocytes** aggregate and begin to secrete proteins. **Cartilage** is a type of connective tissue composed of these chondrocytes, their secreted extracellular matrix, and the surrounding layer of **perichondrium** (the

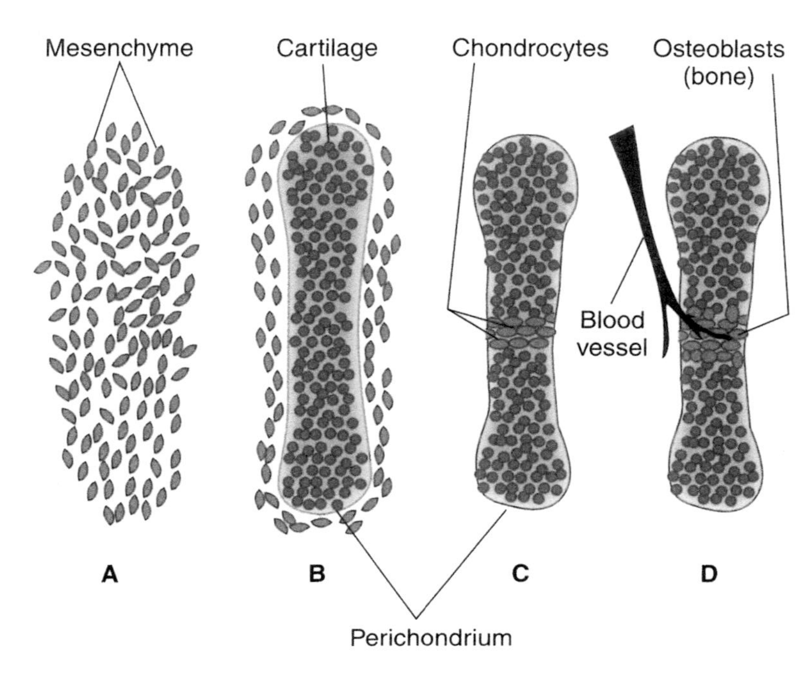

FIGURE 14.2 Endochondral ossification. Begins when mesenchymal cells aggregate and begin to secrete proteins (A). This activity initiates cartilage formation (B). The cartilage has the general shape of the bone to be formed (C) and is gradually replaced by bone (D).

fibrous connective tissue covering cartilage). This cartilage has the general shape of the bone that is to be formed (Figure 14.2).

Gradually, the cartilage is replaced by bone. This process involves **cell hypertrophy**, deposition of calcium phosphate crystals, cell death, and erosion of the calcified matrix. The cartilage takes on a honeycombed appearance and is infiltrated by extensions of the covering perichondrium called **periosteal buds**, which carry blood vessels and mesenchymal cells into the eroded region of the cartilage. Cells brought in by this means replace the cartilage with bone and provide the **bone marrow** (the blood-forming tissue found within some bones) that occupies the cavities of some bones. As cartilage is replaced by bone, the perichondrium becomes the **periosteum**.

The axial skeleton

The vertebrae forming the vertebral column are derived from the somites. Soon after its formation (during gastrulation), the mesoderm is organized into four regions: **head mesoderm, paraxial mesoderm, intermediate mesoderm**, and **lateral mesoderm**. Each of these is bilaterally arrayed across the midline of the body (Figure 14.3A).

Paraxial mesoderm becomes segmented into paired aggregates of tissue called **somites** (see Figure 14.3B). The first somites to appear (early in the embryonic period) are in the head region. Subsequently, additional pairs of somites form in a progressively caudal direction. Each somite is organized into two regions: sclerotome and dermatomyotome (see Figure 14.3C). Cells of the sclerotome leave the somites and aggregate

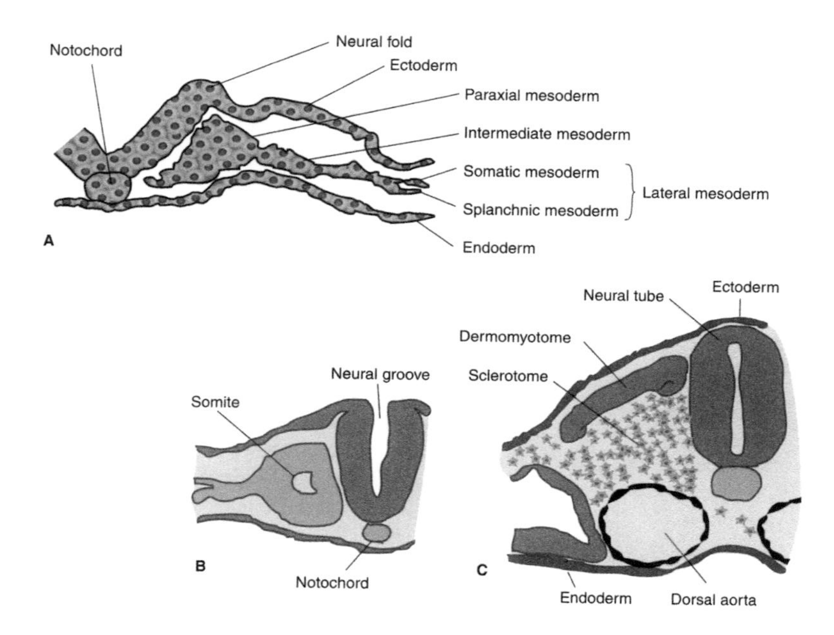

FIGURE 14.3 Somite development. Earlier (A) and later (B and C) diagrammatic transverse sections showing somite development. In each case, only one somite of each bilaterally symmetric pair is shown. Each somite becomes organized into two regions: a dermomyotome and a sclerotome.

around the notochord (the primitive axial skeleton of the embryo) and developing spinal cord. These cells give rise to the vertebrae (by the process of endochondral ossification) and to the **intervertebral discs** (masses of fibrocartilage between vertebrae). Despite the importance of the notochord to the developing embryo, the notochord's only contribution to the adult is the **nucleus pulposus**, which occupies the core of each disc. Ossification (bone formation) replaces the original cartilaginous vertebral column with a bony one. This process begins during the embryonic period and continues into about the 25th year of life.

The **skull** consists of two general parts: the **neurocranium**, which provides a container for the brain, and the **viscerocranium**, from which the face arises (Figure 14.4). Both parts of the skull are formed partially from intramembranous ossification and partially from endochondral ossification. The neurocranium consists of the **calvarium** (cranial vault), derived from intramembranous ossification, and the **base of the skull**, derived from endochondral ossification. The viscerocranium, which also results from both intramembranous and endochondral ossification, finds its origin primarily in mesenchymal cells originating in the neural crest.

The plates of membranous bone making up the calvarium of the skull are each derived from a **primary ossification center**, from which bone formation spreads outward (Figure 14.5). However, the individual plates do not fuse with each other during prenatal development. As a consequence, newborn babies have unclosed **sutures** (in which two such

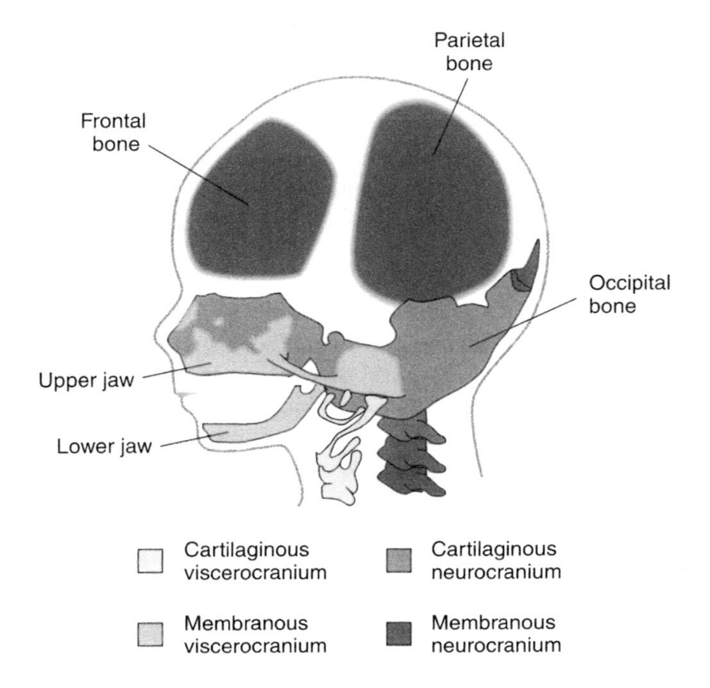

FIGURE 14.4 Diagram of the skull of a 3-month-old fetus. The skull consists of two general parts: the neurocranium and the viscerocranium. As the shading key indicates, each of the two parts of the skull is partially formed from intramembranous ossification and partially formed from endochondral ossification.

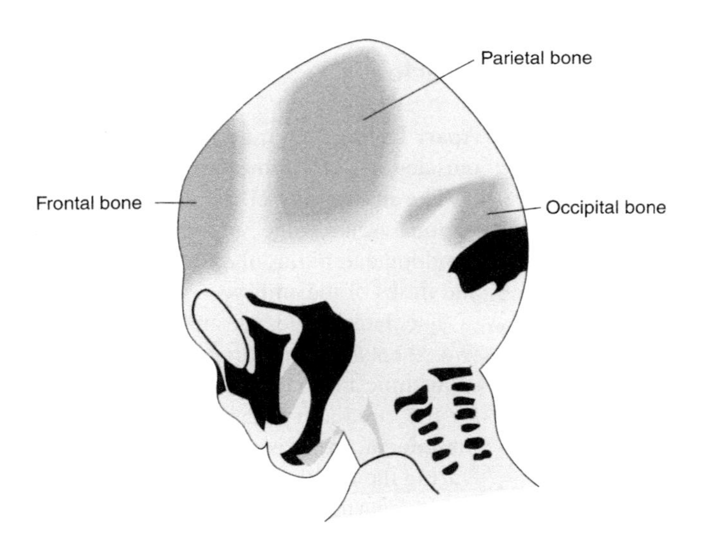

FIGURE 14.5 Fetal skull. Diagram of the skull of the 3-month fetus shows primary ossification centers of the developing neurocranium. Note the primary ossification centers for frontal, parietal, and occipital bones.

bones come close together) and **fontanelles** (spaces in which more than two bones come together). These temporary discontinuities between the bones of the calvarium aid passage of the head through the birth canal at childbirth and permit an increase in the size of the skull to match brain growth after birth. The smaller fontanelles close during the first year, and the larger anterior fontanelle closes during the second year after birth. However, some of the sutures actually remain open until adulthood.

The appendicular skeleton

The limbs first appear as buds on the sides of the embryo near the end of the fourth week of development and undergo dramatic morphogenetic changes by the end of the embryonic period (Figure 14.6). Originally, each **limb bud** consists of a core of **mesenchyme** covered by a layer of ectoderm. The bones of the arms and legs develop (by endochondral ossification) from this mesenchyme.

By the end of the embryonic period, primary ossification begins in the **diaphyses** (shafts) of the long bones. After birth, secondary ossification centers appear in the **epiphyses** (ends) of these bones. As long as cartilage (in the form of **epiphyseal plates**) persists in the ends of the long bones between the ossification centers, growth continues. The persistence of epiphyseal plates provide for **interstitial growth** in the length of the long bones, whereas the periosteum provides for **appositional growth** in the girth of these bones. With adolescence, elevated sex steroid levels cause both **epiphyseal fusion** of the ossification centers and cessation of the long bones' growth.

As the bones grow in response to the activity of the **osteoblasts** (bone-building cells), the interiors of long bones are hollowed out by the activity of the bone-degrading **osteoclasts**. This provides chambers for bone marrow produced by blood-forming cells supplied by blood vessels.

Muscles

Mesoderm

Apart from a few muscles in the eyes and certain glands, almost all muscles arise from **mesoderm**.

The **myotomes** of the somites (each somite is organized into two regions, sclerotome and dermomyotome) play a significant role in development, aiding the formation of muscle in the throat, neck, trunk, and limbs of the embryo.

The lateral mesoderm on each side of the body separates into two sheets: dorsal (upper) somatic mesoderm and ventral (lower) splanchnic mesoderm. **Splanchnic mesoderm** provides muscle tissue for the circulatory, gastrointestinal, and respiratory systems. **Somatic mesoderm** gives rise to the appendicular skeleton (Figure 14.7A and B). During the fourth week of development, the spaces between these layers of mesoderm become a single cavity—the **coelom** or general body cavity (see Figure 14.7C–E) During the second month of development, the coelom is partitioned into the peritoneal cavity, which contains the viscera; the pericardial cavity, which contains the heart; and the pleural cavity, which contains the lungs.

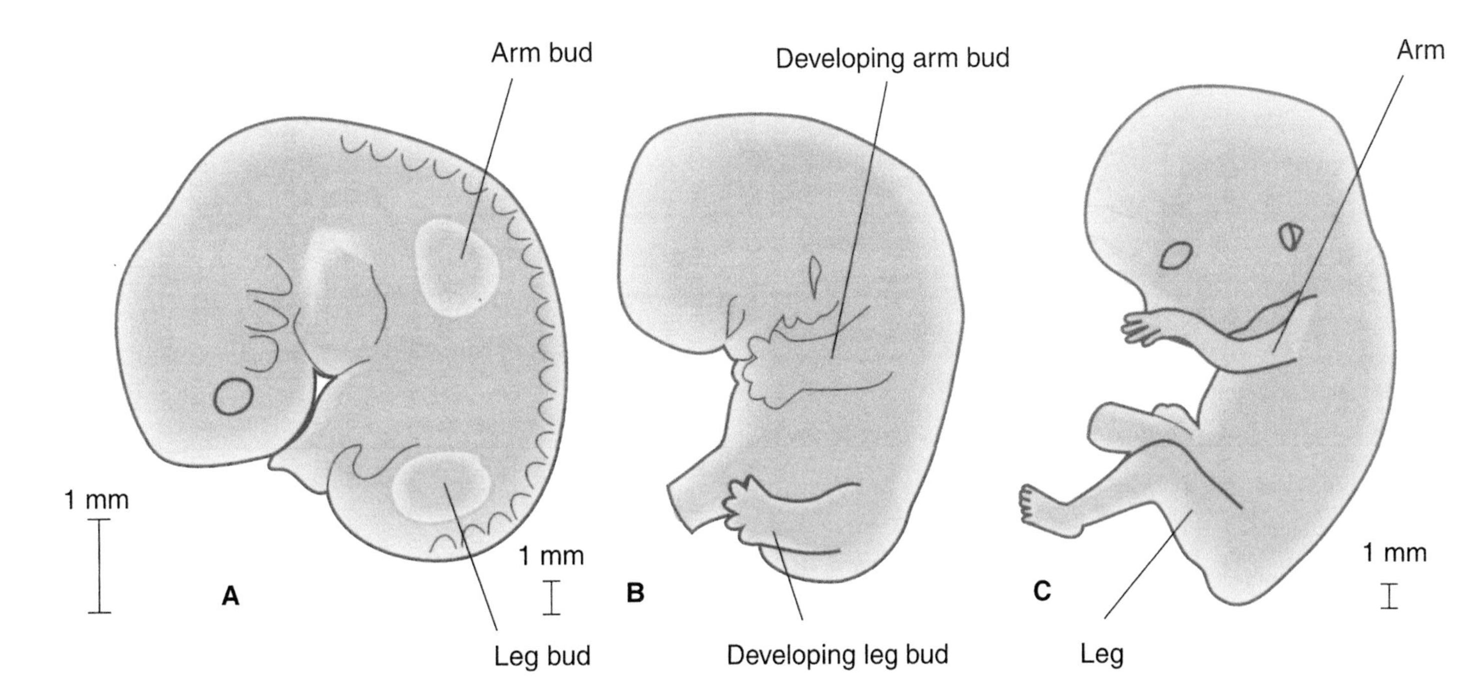

FIGURE 14.6 Development of the limb buds. Drawings are of human embryos at 5, 6, and 8 weeks of development (A, B, and C, respectively). (A) The limb buds arise as flattened-outgrowths of the sides of the embryonic body. (B) Arm bud development is more advanced than leg bud development, and cells are dying in columns on the hand plate, which sculpts the fingers out of the hand plate. (C) At 8 weeks, the fully developed embryo, on the threshold of the fetal period, has recognizable human arms and legs.

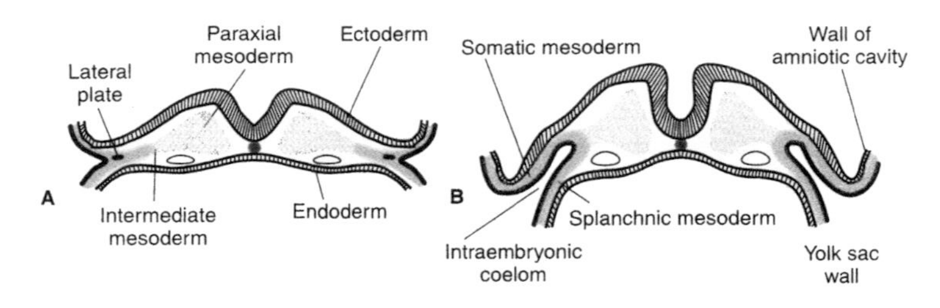

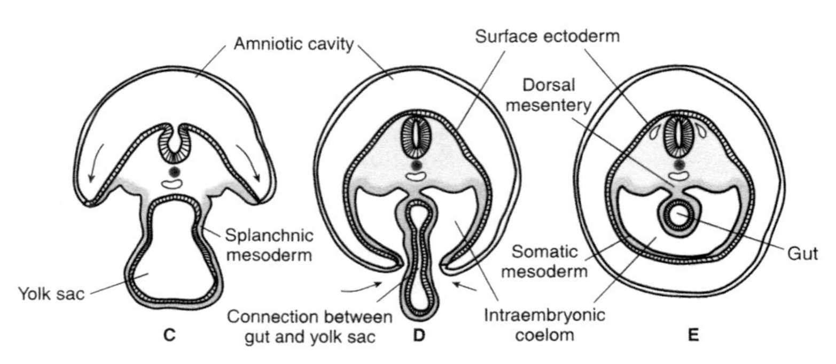

FIGURE 14.7 Development of the lateral mesoderm and coelom is seen in transverse sections through embryos at various stages of development. (A) At 19 days, the three general parts of the bilateral mesoderm are visible—paraxial, intermediate, and lateral. (B) At 20 days, the lateral mesoderm is split bilaterally into an upper somatic layer and a lower splanchnic layer. The spaces between these two layers will give rise to the coelom. (C–E) Development of the single coelom (body cavity) occurs by the end of the fourth week. Note that the somatic mesoderm and splanchnic mesoderm are continuous and make up the mesodermal lining of the coelom.

Muscle types and proteins

With minor exceptions (such as the Purkinje fibers of the heart and muscle spindles found elsewhere), three major kinds of muscle are found in the human body. Our arms and legs are powered by **skeletal muscle**, our stomachs and bladders are able to contract because of **smooth muscle**, and our blood is pumped by the **cardiac muscle** of the heart.

All these muscle types are derived from mesenchymal cells, versatile cells that also give rise to fibroblasts, chondrocytes (cartilage cells), and osteocytes (bone cells). What makes muscle cells different from other cells during the cell differentiation process is that muscle cells produce and accumulate high levels of **actin** and **myosin**, the so-called **contractile proteins**.

Skeletal muscle

Mesenchymal cells destined to become skeletal muscle first give rise to single cells called **myoblasts**. These cells fuse together to form multinucleated cells (cells with many nuclei) called **myotubes** (Figure 14.8). As the myotube begins to accumulate the contractile proteins, its major organelles—nuclei and mitochondria—are displaced to the periphery of the myotube. Actin, myosin, and other contractile proteins are organized into subcellular structures called **sarcomeres**.

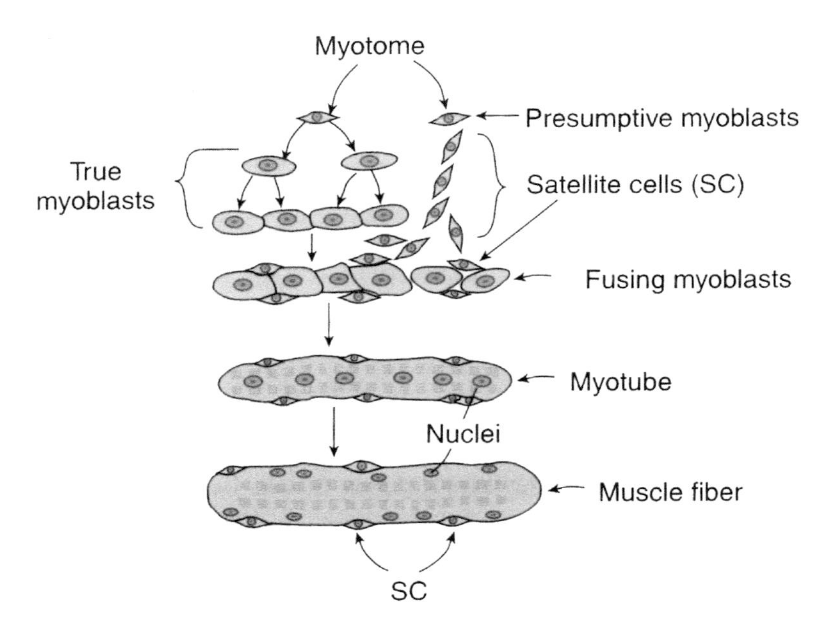

FIGURE 14.8 Histogenesis of the skeletal muscle fiber. The diagram shows the formation of a skeletal muscle fiber, beginning with the fusion of individual myoblasts. Note that since each myoblast brings a nucleus to the fusion, the muscle fiber is a multinucleated entity (a syncytium). Furthermore, the high degree of organization of its contractile proteins, that is, actin and myosin (see Figure 14.9), gives the muscle fiber its characteristic striated (striped) appearance.

The sarcomeres are responsible for the skeletal muscle's **striated** (striped) appearance, which visibly differentiates it from other types of muscle. Sarcomeres are lined up in rows called **myofibrils**. When the myofibrils align in register (like freight trains parked side by side), the sarcomeres (the "boxcars") are also in register, thus imparting the striated appearance to the skeletal muscle fiber (Figure 14.9). The muscle fibers also contain **satellite cells**, which may serve as a source of muscle fiber regeneration in postnatal life (see Figure 14.8).

Cardiac muscle

The mesenchymal cells that give rise to cardiac muscle originate from splanchnic mesoderm, which forms the heart tube. This tube, through a complex and fascinating morphogenesis, gives rise to the beating, four-chambered heart. The mesenchymal cells produce the very versatile **cardiac myoblasts**, which can accumulate contractile proteins, beat, and divide at the same time. This versatility is necessary because the cardiovascular system is the first organ system to "come online." Just 21 days after fertilization, the embryo already has a beating heart pumping blood. Not endowed with a large amount of yolk, the rapidly growing embryo must develop a functional circulatory system to obtain food and eliminate wastes.

Unlike skeletal muscle myoblasts, cardiac myoblasts do not fuse with each other to form a syncytium (multinucleated mass of cytoplasm).

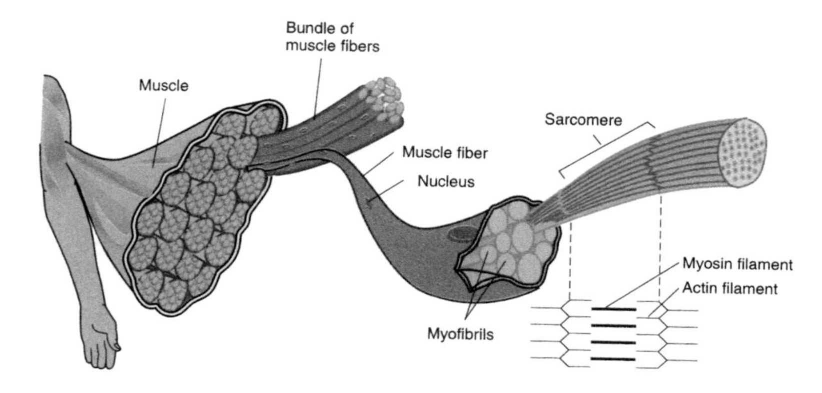

FIGURE 14.9 Diagrammatic representation of the hierarchy of organization in skeletal muscle. The sarcomere is the functional unit of skeletal muscle and the building block of the myofibrils. The highly organized contractile proteins in the myofibrils give them their striated (striped) appearance. Muscle fibers contain numerous myofibrils and nuclei. Bundles of muscle fibers make up skeletal muscles.

Rather, they develop unique cell junctions that become the **intercalated discs** of the heart. Some cardiac myoblasts give rise to **Purkinje fibers** (modified heart muscle fibers), which make up the electrical conducting system of the heart.

Smooth muscle

Although frequently relegated to "housekeeping chores," such as moving food along the digestive system by **peristalsis** (progressive waves of contraction along the tubular gut), smooth muscle also plays important roles at the beginning and end of prenatal development. Even before conception, contraction of smooth muscle in the mother's fallopian tubes aids the upward passage of sperm. After conception, smooth muscle aids the downward passage of the early embryo. During labor and childbirth, smooth muscle contractions expel the fetus from the mother's womb. Contractions of both the fallopian tubes and the uterus are regulated by hormones. Even with induced labor, doctors use the hormone **oxytocin** to stimulate the uterine smooth muscle.

As expected, the development of smooth muscle differs from that of the other two types. Myoblasts destined to become smooth muscle neither accumulate contractile proteins to the extent that skeletal muscle myoblasts do, nor do they spontaneously beat, as in the case of cardiac myoblasts.

A dynamic tissue

During the second half of the embryonic period (which, recall, makes up the first 8 weeks of development), bone formation begins. Although the skeleton has a structural role, as well as a role in movement, unlike other structural elements, such as the steel girders that support a building, bones are dynamic entities. The dynamic nature of bone is realized from a consideration of the types of cells in bones and their functions. There are four types of cells associated with bones: osteoblasts, osteoclasts, osteocytes, and osteogenic cells. The osteoblast is the bone cell responsible for forming new bones; they do not divide, but synthesize and secrete the collagen matrix and calcium salts that make up bones. As the osteoblast becomes surrounded by calcified matrix, it becomes trapped within it, changes its shape, and becomes an osteocyte. The osteocytes, the primary cells of mature bone and the most common type of bone cell, maintain the mineral concentration of the matrix. Like osteoblasts, osteocytes also do not divide; both cell types are replenished by osteogenic cells, which are the only bone cells that divide. The dynamic nature of bone depends on the creation of new bone by osteoblasts and osteocytes and by the resorption (breakdown) of bone by the fourth type of bone cell, osteoclasts. Unlike osteoblasts and osteocytes, which arise from osteogenic cells, osteoclasts originate from monocytes and macrophages (two types of white blood cells). The continual balance between osteoblasts and osteoclasts is responsible for the constant, albeit subtle, reshaping of bone.

https://courses.lumenlearning.com/boundless-biology/chapter/bone/

Stem cells in muscle

Skeletal muscle function is required for movement, breathing, maintenance and alteration of posture, and generation of heat (e.g., shivering). Multinucleated fibers of skeletal muscle do not proliferate. Maintenance of muscle tissue depends on satellite cells, found in close proximity to the muscle fibers. These satellite cells are a heterogeneous population with a small subset of muscle stem cells, termed satellite stem cells. These satellite stem cells are prepared for mobilization by stimuli, including physical trauma and growth signals. Once mobilized, satellite stem cells undergo symmetric divisions to expand their number or asymmetric divisions to give rise to committed satellite cells and thus progenitors. Myogenic progenitors proliferate and eventually differentiate through fusion with each other or to damaged fibers to reconstitute fiber integrity and function.

https://www.ncbi.nlm.nih.gov/pubmed/26140708

Study questions

Skeleton

1. What are the two general parts of our skeleton and what makes up each part?
2. What is osteogenesis and what are the two general types?
3. From what are the vertebrae of the vertebral column derived?
4. What is the notochord's only contribution to the adult?
5. What are the two parts of the skull and what is each derived from?
6. The newborn baby's calvarium has sutures and fontanelles. What are they and what is their significance at birth?
7. What provides for interstitial growth and appositional growth of the long bones of the limbs?
8. What causes an end to the growth of the long bones?

Muscles

9. From what do almost all muscles arise?
10. What two sheets of mesoderm are derived from lateral mesoderm? To what does each give rise?
11. What is another name for the general body cavity and what three cavities does it give rise to during the second month of development?
12. What three major kinds of muscles are found in the body, and what are some examples of the functions of each?
13. The contractile proteins of skeletal muscle fibers are organized into what subcellular structures?
14. Which cells serve as a source of muscle fiber regeneration in postnatal life?
15. Why must the rapidly growing embryo develop a functional circulatory system early in development?
16. What are the intercalated discs of the heart?

Critical thinking

1. Muscle cell differentiation is characterized by the accumulation of the proteins actin and myosin. How are these proteins differently organized in the three different types of muscles?
2. What are important roles played by smooth muscle at the beginning and end of prenatal development?

CHAPTER FIFTEEN

The circulatory system and hormones

CHAPTER OBJECTIVES

After studying this chapter, you should be able to:

1. Beginning with splanchnic mesoderm, explain the development of the valved, four-chambered beating heart.
2. Discuss the changes that must take place at the foramen ovale and ductus arteriosus at birth and why they must occur.
3. Discuss the three circulatory arcs of the human body and the changes that occur in them at birth.
4. Explain the development of blood islands and their significance.
5. Explain the relationship of the endocrine glands to the germ layers.
6. Explain the relationship between hormones and the circulatory system.
7. Discuss the role of hormones in development, giving at least five examples.

Just 21 days after conception, the circulatory system begins to permeate the body of the fetus with blood carrying food, oxygen, hormones, and waste products under the propulsive force of the heart. Imagine—one cell in this short time will develop a pump, a system of tubes, and the transported fluid. The hormones carried by the circulatory system are the products of endocrine glands, which, by definition, secrete their hormones directly into the bloodstream.

This elaborate system is possible only when the original cell, destined to give rise to the circulatory system, becomes millions of cells through **cell proliferation**. All of these cells are derived from mesoderm, but they are of many kinds: cardiac cells that contract, endothelial cells that line the blood vessels, and red blood cells that flow in an endless circuit.

Cell differentiation also plays a major role in the development of the circulatory system. The cardiac cells produce appreciable amounts of the contractile proteins actin and myosin, whereas the red blood cells are packed with a different protein (hemoglobin) and have even lost their nuclei! The formation of the circulatory system also involves a good deal of morphogenesis (changes in shape).

Heart

Morphogenesis is a dramatic aspect of heart development: The heart starts out as a pair of tubes and develops into a complex four-chambered pump with valves.

Splanchnic mesoderm

The sophisticated adult heart is derived from a portion of the lateral mesoderm called **splanchnic mesoderm**, which is located in the head end of the embryo near the developing foregut and the hindbrain. The heart actually begins its development outside of the body in a region called the **cardiogenic area**! Not until the fourth week of development does the head fold bring the heart into its proper position inside the embryonic body (see Figure 8.12).

Early in development, the heart consists of tubes derived from splanchnic mesoderm. Two portions of thickened splanchnic mesoderm are separated across the midline of the embryo. From the medial face of each layer of mesoderm, a group of mesenchymal cells (loose connective tissue) arises. Each group organizes itself into a hollow tube. The two tubes are separated from each other across the midline.

The two tubes then fuse together at the midline into a single tube (the **endocardial tube**) (Figure 15.1). The splanchnic mesoderm fuses above and below the endocardial tube, forming an outer tube (the **epimyocardial tube**). At this point, the heart is a tube within a tube, with the outer epimyocardial tube separated from the inner endocardial tube by a type of connective tissue—**cardiac jelly**. The inner tube gives rise to the lining of the heart (the **endocardium**), whereas the outer tube gives rise to the muscular wall of the heart (the **myocardium**) and its surface covering (the **epicardium**) (see Figure 15.1).

The splanchnic mesoderm also plays a role in the suspension of the heart within the pericardial cavity. The splanchnic mesoderm's formation of the epimyocardial tube also gives rise to a short-lived membrane called the **ventral mesocardium**. Its disappearance results in the formation of a single **pericardial cavity** from two heretofore separate regions of the coelom—the amniocardiac vesicles. Above the epimyocardial tube, fusion of splanchnic mesoderm gives rise to the more persistent **dorsal mesocardium**, which suspends the tubular heart in the pericardial cavity (see Figure 15.1).

Regions of the tubular heart

The tubular heart undergoes considerable changes as it bends, forming the basic functional parts of the heart: **sinus venosus, atrium, ventricle**, and **truncus arteriosus** (blood enters the tubular heart at the sinus venosus end and is discharged into blood vessels at the truncus arteriosus end; Figure 15.2). As the initial pair of tubes fuses into the single heart

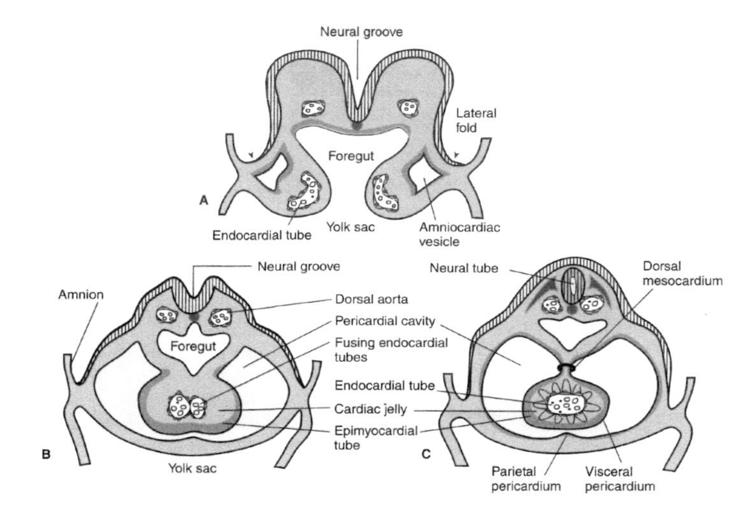

FIGURE 15.1 Formation of the heart (endocardial) tube is illustrated in diagrams of transverse sections through the heart-forming regions of the early embryo. (A) At 20 days, note the paired endocardial heart tubes and the paired precursors of the pericardial cavity separated across the midline of the embryo. (B) At 21 days, the lateral body folds have brought the pair of endocardial tubes together in the midline, where they will fuse into a single heart tube. Simultaneously, the precursors of the pericardial cavity (amniocardiac vesicles) have been brought together to form the single pericardial cavity. (C) At 22 days, note that the single heart tube—with an inner endocardial lining, intermediate cardiac jelly, and outer epimyocardial layer—is suspended by the dorsal mesocardium in the pericardial cavity.

tube, the direction of fusion is truncoventricular region, atrial region, and region of the sinus venosus.

Experiments with chick embryos have demonstrated that as the cardiac tubes fuse, successive regions of the tubular heart take over the function of **pacemaker**. First the ventricle, then the atrium, and finally the sinus venosus function as pacemaker. When the tubular chick heart is cut into these three regions, each beats at its original characteristic rate—with the ventricle beating slowest (50 beats per minute) and the sinus venosus beating fastest (140 beats per minute).

The four chambers

The four chambers of the heart (two atria and two ventricles) result from processes involving dramatic external morphogenetic movements and internal partitioning. The internal partitioning is accomplished by mesenchymal tissue called **endocardial cushion tissue**.

Division into ventricles With the folding of the cardiac tube into a loop, the truncoventricular region comes to press on the ventral surface of the atrial region. This gives the atrial region a two-lobed configuration, which is an external indication of the two atria that are to form. The ventricular loop, formed by bending of the cardiac tube, develops a depression over the apex of its outer surface, the **interventricular sulcus** (Figure 15.3). Opposite this sulcus, strands of heart muscle (**trabeculae**

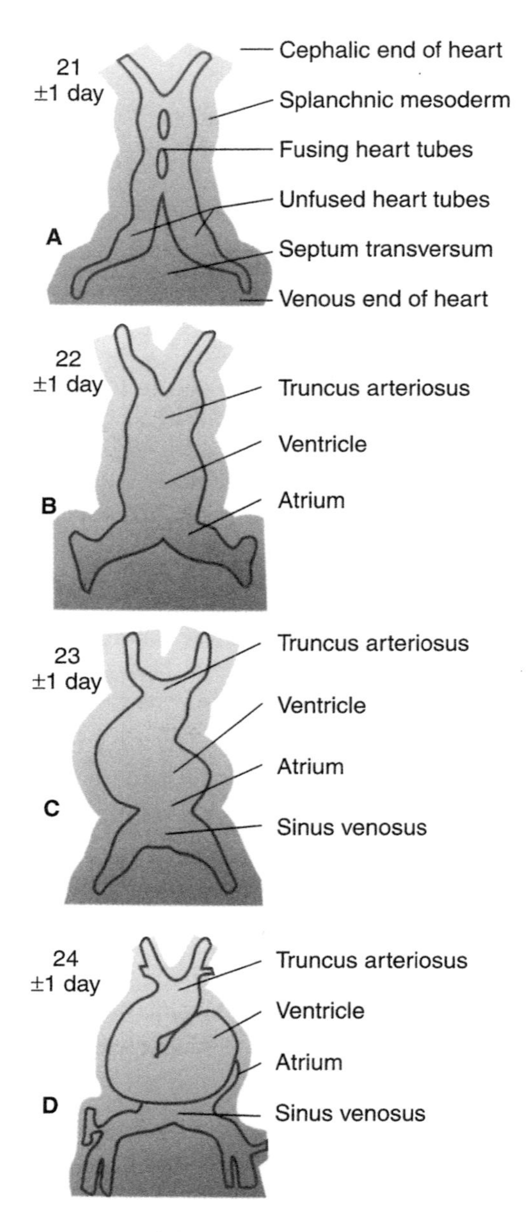

FIGURE 15.2 Regions of the early heart are shown in diagrams of the ventral side of the heart during the fourth week. The head (cephalic) end of the embryo is at the top of the figure. The direction of blood flow is sinus venosus, atrium, ventricle, and truncus arteriosus. The order of fusion of the two heart tubes is truncoventricular region, atrial region, and region of the sinus venosus. (A) At 21 days, the two heart tubes are just beginning to fuse in the midline. (B) At 22 days, the truncoventricular regions have fused. (C) At 23 days, the sinus venosus region has just fused. (D) At 24 days, the, heretofore straight heart tube is forming a loop.

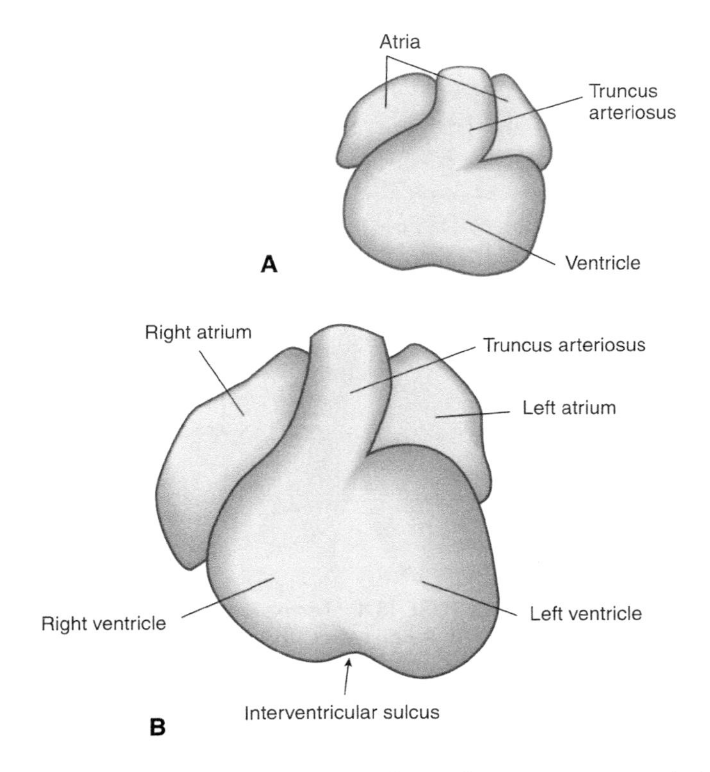

FIGURE 15.3 Formation of the ventricles is shown in two ventral views of the developing embryonic heart. Bending of the original straight heart tube causes the truncus arteriosus to press against the ventral side of the atrial region and the ventricular region of the heart to form a loop. If one compares (A), earlier, with (B), later, it is apparent that a surface groove, the interventricular sulcus, appears over the apex of the ventricular region. Opposite this groove, an internal interventricular septum appears on the inner wall of the heart. This septum contributes to the partitioning of the ventricular regions into left and right ventricles.

carneae) coalesce to form the **interventricular septum** on the interior of the ventricular wall. As its name indicates, this septum contributes to the partitioning of the ventricular region of the cardiac loop into right and left ventricles.

Foramen ovale Meanwhile, a partition called the **septum primum** begins to form from the cephalic wall of the atrial region at the level of its compression by the truncoventricular region. This crescent-shaped tissue has its limbs directed over the dorsal and ventral walls of the atrial region toward a constriction in the heart tube (the **atrioventricular canal**), between the atrial and ventricular regions of the cardiac tube. As the septum primum develops and its opening (the **foramen primum**) is about to close, a new opening appears in its cephalic region, the **foramen secundum**. This opening becomes the **foramen ovale**, an opening that persists between the right and left atria until the time of childbirth (Figure 15.4A).

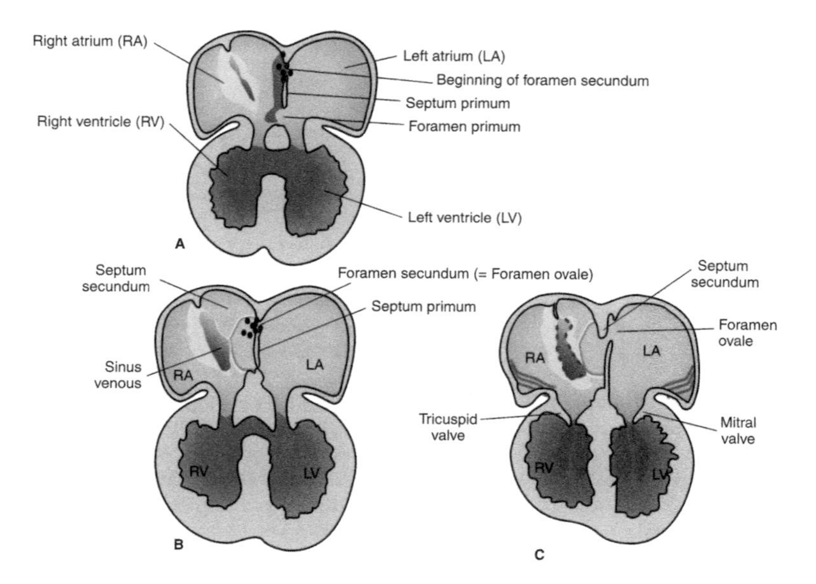

FIGURE 15.4 Formation of the atria and the foramen ovale (ventral views). In these diagrams, the cephalic end of the heart is toward the top, and the heart appears as if it has been cut in half and you are looking into the interior of the heart. (A) In the embryonic heart during the sixth week, note that the septum primum is about to partition off the left and right atria. Furthermore, as the foramen primum is about to close, the foramen secundum is appearing as a group of perforations in the cephalic part of the septum primum; the foramen secundum becomes the foramen ovale. (B) Subsequently, a new partition, septum secundum, arises from the cephalic wall of the heart to the right of the septum primum. Although the opening of the septum secundum diminishes in size, it does not close. Rather, its edges provide a valvular mechanism for the foramen ovale. (C) The fetal heart during the third month illustrates that the tricuspid valve between the right atrium and right ventricle and the mitral valve between the left atrium and the left ventricle are visible. Endocardial cushion tissue provides these valves even as it partitions the atrioventricular canal into left and right channels.

In addition, a second partition arises from the wall of the atrial region of the heart, the **septum secundum**. This crescent-shaped partition has its limbs directed toward where the sinus venosus enters the atrial region of the developing heart. The initial opening of the septum secundum does not close. Rather, the edges of its opening provide a valvular mechanism for the foramen ovale (see Figure 15.4B). This valve is designed so that blood returning from the rest of the body will supply both atria. The necessity for this valve has to do with the flow of blood. In the adult, blood returns to the left atrium from the lungs. However, because of the underdeveloped nature of the lungs during prenatal existence, the lungs supply essentially no blood to the left atrium. Instead, blood is supplied to the left atrium from the right atrium via the foramen ovale. This provides the left atrium with a load to pump, which is necessary for its proper development.

Valve development Endocardial cushion tissue provides the valves between the left atrium and left ventricle (**mitral valve**) and between the right atrium and right ventricle (**tricuspid valve**), even as it partitions the atrioventricular canal into left and right channels (Figure 15.4C).

The interventricular septum, the endocardial tissue of the atrioventricular canal, and additional tissue from the outlet of the heart (truncus arteriosus) cooperate to partition the ventricular region of the heart into left and right ventricles (see Figure 15.4C). The right ventricle is faced with a problem not unlike that of the left atrium. In the adult, the right ventricle supplies blood to the lungs. Again, prenatal lungs are underdeveloped and do not require a large supply of blood. Two major arteries—the **aorta** and the **pulmonary trunk**—leave the heart. In the adult, blood is pumped by the right ventricle through the pulmonary trunk into the pulmonary arteries and out to the lungs. Because of complicated reconstruction of a series of blood vessels called **aortic arches**, a shunt, the **ductus arteriosus**, is provided between the pulmonary trunk, leaving the right ventricle, and the aorta, leaving the left ventricle. This way, the right ventricle has somewhere to pump its load of blood and, like the right atrium, can exercise and develop normally.

Both the foramen ovale and the ductus arteriosus are adaptations to intrauterine life. Both openings functionally close at childbirth with the expansion of the lungs at first breath (Figure 15.5). As a result, the blood is oxygenated in the lungs and no longer by the placenta. How does the organism "know" enough to prepare during its development for the transition from aqueous to terrestrial existence? The organism's genes program the necessary information.

Blood vessels

We have three basic kinds of blood vessels: arteries, veins, and capillaries. **Arteries** carry oxygen-rich blood *away* from the heart, **veins** carry oxygen-poor blood *to* the heart, and **capillaries** supply oxygen *through* their thin walls to the cells of the body. The capillaries also supply the cells with food and remove carbon dioxide and other wastes.

Circulatory arcs

In our adult lives, we possess two kinds of circulatory arcs: pulmonary and systemic. In **pulmonary circulation**, blood is carried from the right ventricle, through the pulmonary arteries, through the capillaries of the lungs (where the blood is oxygenated), and then through the pulmonary veins back to the left atrium of the heart (see Figure 15.5).

Systemic circulation carries blood away from the left ventricle and out to the body through the aorta and its various arterial branches (see Figure 15.5). Oxygen and nutrients pass across the capillary walls to the cells of the body. Similarly, the cells pass carbon dioxide and other wastes back across the capillary walls. The blood containing these wastes returns to the right atrium of the heart by means of the venous blood

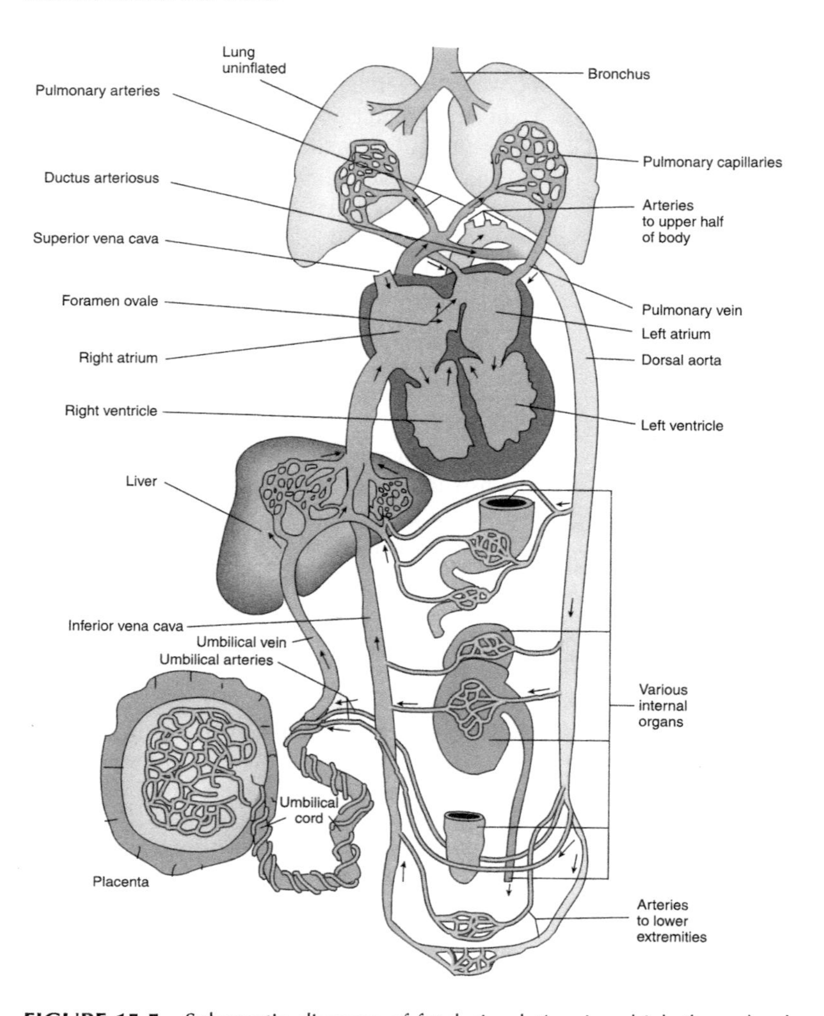

FIGURE 15.5 Schematic diagram of fetal circulation in which the valved foramen ovale is visible between the two atria, allowing blood to flow directly from the right atrium into the left atrium. The ductus arteriosus, shunting blood from the pulmonary trunk into the aorta, is also visible. At this time, the lungs are not expanded and are not receiving a full complement of blood. Note also that two umbilical arteries carry blood from the fetus to the placenta, and one umbilical vein carries blood back from the placenta to the fetus.

vessels: the **superior vena cava** returns blood from the head, and the **inferior vena cava** returns blood from the rest of the body.

Before our birth, we had a third circulatory arc, **placental circulation** (see Figure 15.5). The fetus attaches to the placenta by means of the umbilical cord and the two umbilical arteries run (in a spiral course around the single umbilical vein) through the umbilical cord. The **umbilical arteries** carry blood that is deficient in oxygen and nutrients out to the placenta, and the **umbilical vein** carries replenished blood back into the body of the fetus.

Blood

The circulatory system is the stream of life. Fish use a stream of water to get their food and oxygen and to eliminate wastes. Our bloodstream does that and much more. It carries regulatory molecules (such as hormones), which allow our various systems to operate in harmony. For example, the brain and the gonads constantly interact by means of these chemical messengers.

This process begins early in development, *outside* of the embryonic body! Aggregations of mesenchyme cells, derived from extraembryonic mesoderm, appear during the third week on the yolk sac. Called **blood islands**, these clusters of mesenchyme cells begin to differentiate even as the endocardial tubes of the heart begin to form within the embryo (Figure 15.6). The outer cells of the blood islands flatten out, join together, and give rise to the endothelial linings of the blood vessels, whereas the central cells separate from each other, round up, and give rise to **primitive blood cells**. The fluid within the blood islands becomes the first **blood plasma**.

As blood vessels form in this manner, they coalesce with neighboring blood vessels and create networks of blood vessels. Like the first blood cells, the first blood vessels appear on the yolk sac. Subsequently, additional blood vessels form (from mesenchymal cells) inside the embryo.

The formation of blood cells is called **hematopoiesis**. As we have just learned, the first hematopoietic site in human development is the yolk sac. The embryo and fetus produce blood cells in the liver, spleen, and bone marrow, but the adult produces blood cells only in the bone marrow and spleen.

Hormones

Without some mechanism to orchestrate the body's numerous activities, utter chaos would result. Two of the systems of the body are particularly

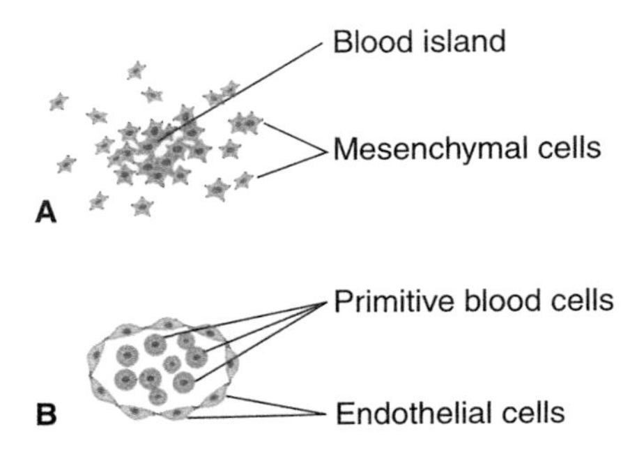

FIGURE 15.6 Blood vessel formation. (A) Aggregating mesenchymal cells on the yolk sac forming a blood island. (B) The outer cells flatten out and join together to form the endothelial lining of the blood vessel, while the central cells round up and give rise to primitive blood cells.

concerned with such coordination: the **nervous system** and the **endocrine system**. The nervous system generally integrates systems rapidly through **nerve impulses**, which are propagated almost instantaneously along nerves. Recall the last time you inadvertently touched something hot! That's your nerve impulse at work. On the other hand, hormones, which are carried to distant sites of activity, tend to exert their effects more slowly. This generalization notwithstanding, adrenaline (**epinephrine**) can have rapid effects.

Endocrine glands

We have already considered development of the nervous system (Chapter 13). Here, we consider the developmental origin of the various endocrine glands.

At the outset, it is important to make a general point about the endocrine glands. Unlike the nervous system, skeleton, or gut lining, each of which is derived from a single germ layer (ectoderm, mesoderm, and endoderm, respectively), the endocrine glands, collectively, derive from all three germ layers. We have already considered two glands derived from ectoderm: the pineal gland, from the roof of the diencephalon, and the pituitary gland, from the floor of the diencephalon and the roof of the mouth (**Rathke's pocket**). Similarly, we will consider the origin of the **thyroid gland, thymus gland**, and **parathyroid glands** (Chapter 17) from the endoderm lining the foregut and the development of the gonads (Chapter 18) from mesoderm.

Some glands are derived from more than one germ layer. The **adrenal glands** have a dual structure reflecting a dual origin. The outer cortical regions of these glands are derived from mesoderm. However, the inner medullary regions are derived from **neural crest** cells, which are derived from ectoderm.

Hormones have profound effects on the developing organism. They play a role in the onset of **puberty**, and a surge of luteinizing hormone (which regulates the menstrual cycle) triggers ovulation in potential new mothers. Estrogen rebuilds the lining of the uterus after menstruation, and progesterone stimulates the sweeping of the end of the fallopian tube over the surface of the ovary at the time of ovulation (in addition to maintaining pregnancy). During pregnancy, chorionic gonadotropin rescues the corpus luteum and saves the early pregnancy, and oxytocin stimulates the smooth muscle contraction initiating labor and childbirth. In anticipation of the new baby, placental lactogen prepares the breasts for postnatal **lactation** (breastfeeding).

Cardiac organoids: During an ischemic (oxygen-deprived) event, although the *adult* human heart exhibits limited regenerative potential, it undergoes pathological changes in response to such injury. However, it is unknown if the *immature* human heart can undergo complete regeneration, even though cardiac regeneration has been documented in zebrafish and neonatal mouse hearts. Recently, researchers at the University of Queensland and the University of Melbourne, in Australia, using advances in pluripotent stem cell differentiation and tissue engineering, developed human cardiac organoids (hCOs). These organoids resemble fetal heart tissue and were used to ascertain the regenerative capacity of immature human heart tissue in response to injury, specifically cryoinjury with a dry ice probe. They discovered that the hCOs exhibited a regenerative response, with full recovery 2 weeks after such injury. This recovery occurred without pathological fibrosis or cardiomyocyte hypertrophy. It was noted that cardiomyocyte proliferation may have been responsible for the regenerative capacity of the hCOs. The researchers suggested that immature human heart tissue has an intrinsic capacity to regenerate.

Study questions

1. What brings the developing heart into its proper relative position during the fourth week of development?
2. What is cardiac jelly?
3. The disappearance of the ventral mesocardium gives rise to which single cavity? This cavity arises from what two preexisting cavities?
4. What suspends the tubular heart in the pericardial cavity?
5. From blood intake end to blood discharge end, what are the four basic functional parts of the heart?
6. What are the interventricular sulcus and the interventricular septum? The interventricular septum arises from what, and what is its significance?
7. The foramen ovale is found between which chambers of the heart? To which chamber does it provide blood?
8. Where is the ductus arteriosus found and what does it provide for the developing heart?
9. What are the three basic kinds of blood vessels in the body? From which germ layer do they all arise?
10. What blood vessels pass through the umbilical cord and in which direction do they carry blood?
11. What are blood islands, where do they form, and to what do they give rise?
12. What is hematopoiesis?
13. To carry out its regulatory responsibilities, what do the cells of the endocrine system produce?
14. List endocrine glands derived from each of the germ layers.
15. Give seven examples of important roles of hormones in development; list these in chronological order.
16. Human cardiac organoids (hCOs) resemble what tissue?
17. hCOs have been used to ascertain what?
18. Researchers suggest that immature human heart tissue has an intrinsic capacity to do what?

Critical thinking

1. Beginning with lateral plate mesoderm during the third week of development, describe the development of the four-chambered, beating human heart.
2. Regarding the circulatory system, at the time of birth, two openings must functionally close. Explain why each must close at the time of birth.
3. Discuss the role of cell proliferation, cell differentiation, and morphogenesis in the development of the circulatory system.

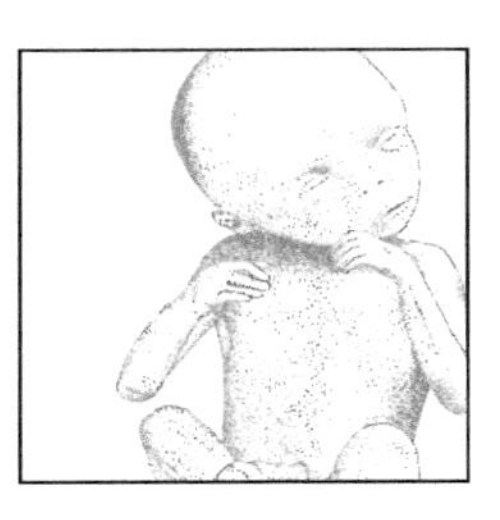

CHAPTER SIXTEEN

The digestive system and the respiratory system

CHAPTER OBJECTIVES

After studying this chapter, you should be able to:

1. Explain the origin of the digestive system and its three general parts.
2. Explain the contributions of endoderm and splanchnic mesoderm to the development of the digestive system.
3. Discuss the relationship between the foregut, midgut, and hindgut of the embryo and the adult parts of the digestive system.
4. Discuss digestion before and after birth.
5. Explain the contributions of endoderm and mesenchyme to the development of the respiratory system.
6. Explain the origin of the respiratory system.
7. Discuss the role of surfactant in lung development and its relationship to respiratory distress syndrome.

The digestive and respiratory systems allow us to subsist on the food and oxygen in our environment. Our digestive system serves us in two obvious but important ways. The height (and weight) we gain by eating properly shows us that the digestive system provides the building blocks of life. The relief we feel from our first meal after exercise (or just after a long time without eating) is a demonstration of how the digestive system provides us with energy. We cannot live for more than a few minutes without oxygen. Our cells have a constant supply of oxygen because our respiratory system oxygenates our blood, which, in turn, carries oxygen to our cells.

Both the digestive and respiratory systems are lined with endoderm. The respiratory system arises as an outpocketing of endoderm—the **laryngotracheal groove**—from the floor of the pharynx (part of the digestive system located between the mouth cavity and esophagus). Thus, in this chapter, we first consider development of the digestive system and then development of the respiratory system.

Digestive system

Something of what we eat—water, salts, vitamins, and other small entities—readily enters our circulatory system without modification. However, other substances, such as polysaccharides (carbohydrates), fats, and proteins, are large molecules and must be broken down into smaller molecules before they gain access to blood and the rest of the body. The digestive system provides enzymes to digest these large molecules and provides bile to emulsify (suspend in a liquid) fats so that such nutritive substances can enter our body. Topologically speaking, the lumen (space inside of a tube) of the digestive system is part of the outside environment.

Once digested, these substances provide metabolic energy and supply the building blocks for making our own large molecules. For example, our immune system would reject the cow protein in beef if it were injected into our circulatory system. However, our digestive system breaks cow protein into its amino acids, which our circulatory system distributes to the cells of our body. These cells use the "cow" amino acids to make human proteins, which our immune system does not reject as foreign.

Origin of our digestive system

Although the embryo relies on the placenta for nourishment, the digestive system develops very early. During the third week of development, we undergo gastrulation (the process in which three germ layers are produced) and are transformed from a two-layered embryo into a three-layered embryo. The bottom-most of these three flat layers is the endodermal layer, which will give rise to the lining of the mostly tubular digestive system.

It is not until the fourth week of development that the embryo undergoes truly widespread morphogenetic changes. As a result of bending and folding of the embryo, the flat endoderm is transformed into the tubular **digestive tube** (see Figure 8.12). Initially, the endoderm is the roof of the **yolk sac**, located outside the embryonic body, but, as a result of the transformation during the fourth week, the developing digestive tube acquires a floor of endoderm in two regions. The region of the tube in the forming head of the embryo is the **foregut**, and the region in the tail is the **hindgut** (Figure 16.1). Between the cephalic foregut and the caudal hindgut is the **midgut**, the portion of the gut that is still open to the yolk sac. Each of these three regions gives rise to specific parts of the adult digestive system.

Other important constituents of the digestive system also begin to develop. As the embryonic body is formed, so, too, are the coelom and mesenteries. The **coelom** is the body cavity in which much of the digestive system is suspended, and the **mesenteries** are the membranes that suspend the digestive organs (Figure 16.2).

All of those epithelial cells that uniquely belong to the digestive system develop from the **endodermal** lining of the gut. Examples are gastric cells, which secrete pepsin; liver cells, which make glycogen;

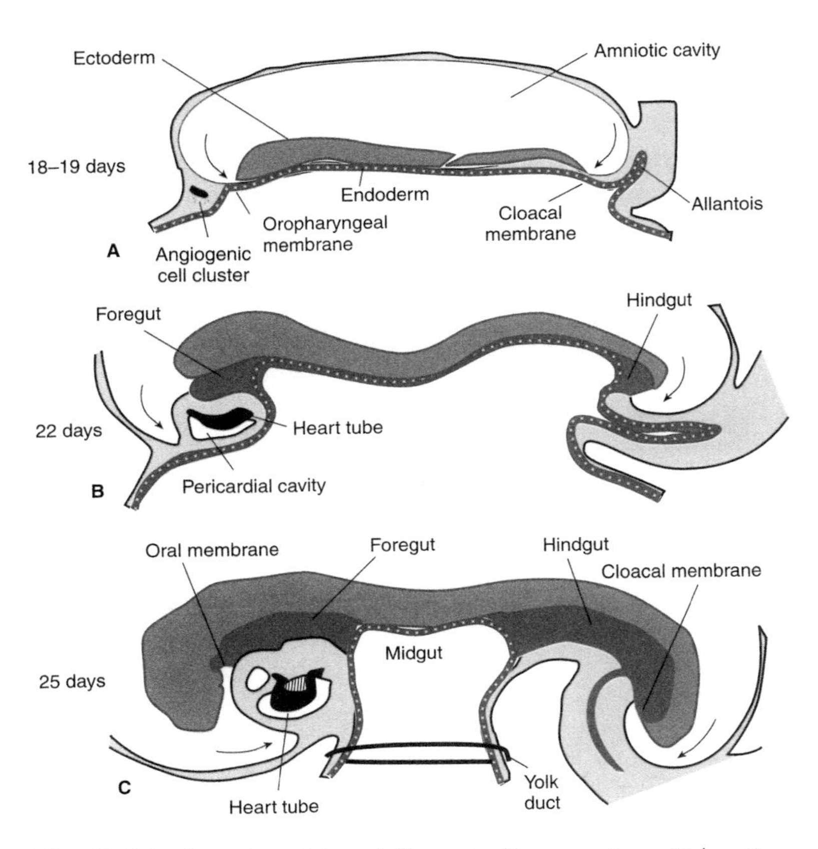

FIGURE 16.1 Formation of the gut. Shown in diagrammatic sagittal sections of the embryo from the late third week (A) into the fourth week (B, C). As the body folds form the head and tail of the embryo, endoderm is caught up in the process of forming the foregut and hindgut. Note that the foregut and hindgut develop and lengthen with the acquisition of a floor of endoderm. The portion of the gut that persists without a floor of endoderm is the midgut, which becomes proportionally smaller with development. Food does not enter the embryonic body through the yolk (vitelline) duct, but through the circulatory system.

and pancreatic cells, which secrete digestive enzymes. Nonepithelial components of the digestive system, such as muscles and connective tissue, are provided by splanchnic mesoderm.

Major parts of the digestive system

Foregut The foregut gives rise to many important structures: the **pharynx, esophagus, stomach, liver, biliary apparatus, pancreas**, and a portion of the **small intestine** (the cephalic portion of the duodenum) (Figure 16.3). Initially, the foregut is a blind tube at its cephalic (head) end, but during the fourth week, with the rupture of the oropharyngeal membrane, the mouth opens to the amniotic cavity.

The **esophagus** is a tube that runs from the end of the pharynx to the stomach. Although it grows during development, it does not undergo the dramatic alterations in shape undergone by some of the other components

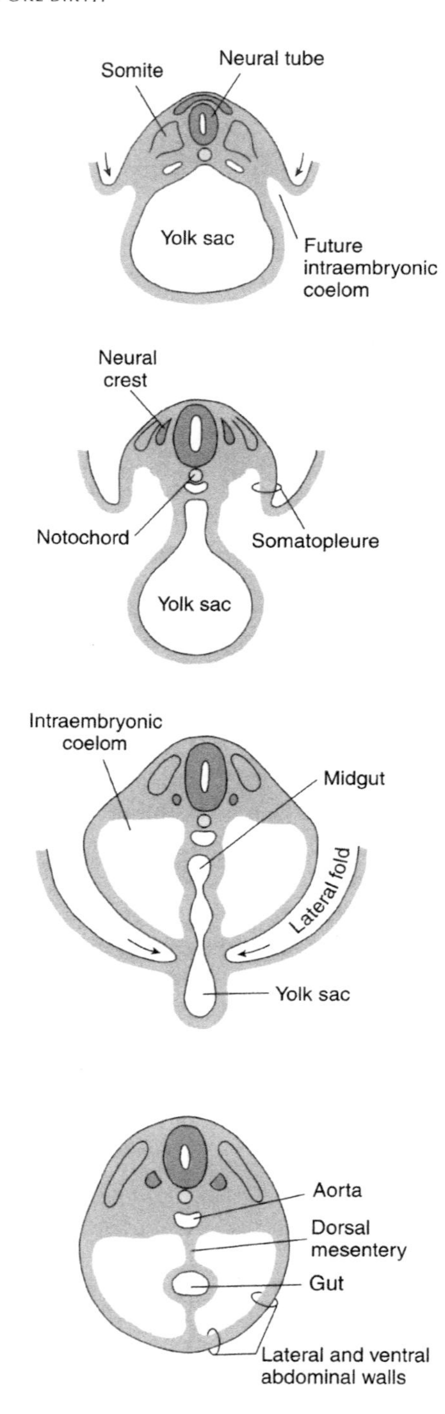

FIGURE 16.2 Coelom and mesentery development. Shown in diagrammatic transverse sections of the embryo during the fourth week. As the lateral body folds form the sides of the embryonic body and the gut is separated from the yolk sac, the intraembryonic coelom is formed and the gut comes to be suspended in the coelom by mesenteries.

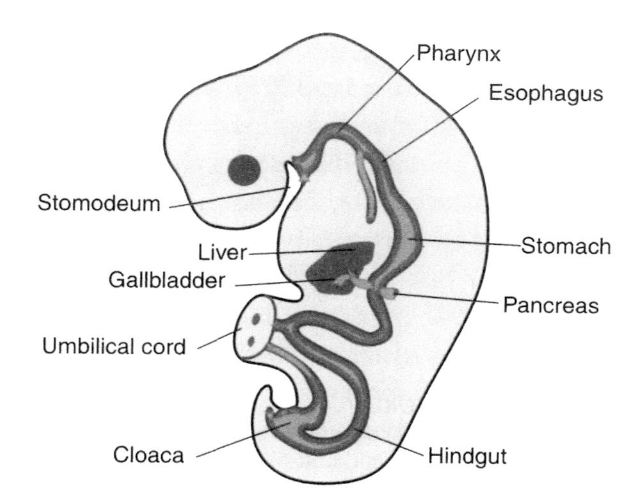

FIGURE 16.3 Development of the foregut. Seen in a diagrammatic lateral view of the developing gut of the embryo. The pharynx, esophagus, stomach, liver, pancreas, and a portion of the small intestine are derived from the foregut.

of the digestive system. For a time during late embryonic and early fetal life, the esophagus may become occluded (blocked) by an overproduction of epithelial cells. If such blockage does occur, it is removed during about the tenth week of development, reestablishing a pathway between pharynx and stomach.

Unlike the esophagus, the **stomach** undergoes marked changes in shape and position during its early development. Initially just a dilated region of the foregut behind the level of the esophagus, its growth pattern causes it to undergo metamorphosis into its characteristic adult shape, with its greater and lesser curvatures. A clockwise rotation of the stomach brings it to adult position, carrying its supporting membrane (the **mesogastrium**) with it, thus creating the **omental bursa**, an outpocketing of the general peritoneal cavity.

Caudal to the level of the stomach, there appears the **hepatic diverticulum**, a ventral outpocketing of the foregut. This diverticulum gives rise to the characteristic **parenchymal cells** of the liver, but not to its hematopoietic cells or **Kupffer's cells**. The **gallbladder**, **cystic duct**, **hepatic duct**, common **bile duct**, and ventral pancreatic bud also originate in the hepatic diverticulum (Figure 16.4).

The ventral pancreatic bud is one of the two rudiments that give rise to the **pancreas**. At about the level of the hepatic diverticulum, the dorsal pancreatic bud grows out of the foregut. The ventral pancreatic bud and the dorsal pancreatic bud fuse when the duodenal loop of the foregut and midgut rotates in a clockwise direction. The dorsal **pancreatic duct** regresses and the pancreas opens (together with the common bile duct) into the duodenum through the ventral pancreatic duct (see Figure 16.4).

Midgut The midgut gives rise to much of the **colon** and **small intestine**, as well as the **cecum** (the blind pouch at the junction of the small and

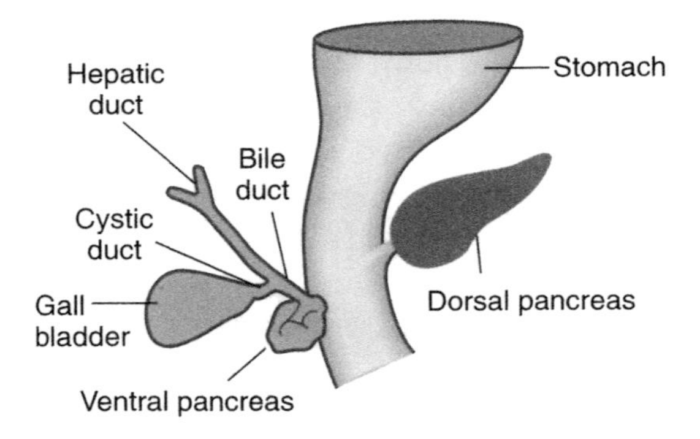

FIGURE 16.4 The pancreas arises from two initially separate structures, the ventral pancreatic bud and the dorsal pancreatic bud. By 35 days, the ventral pancreatic bud migrates around the foregut in the direction of the dorsal pancreatic bud. The dorsal and ventral pancreatic buds fuse.

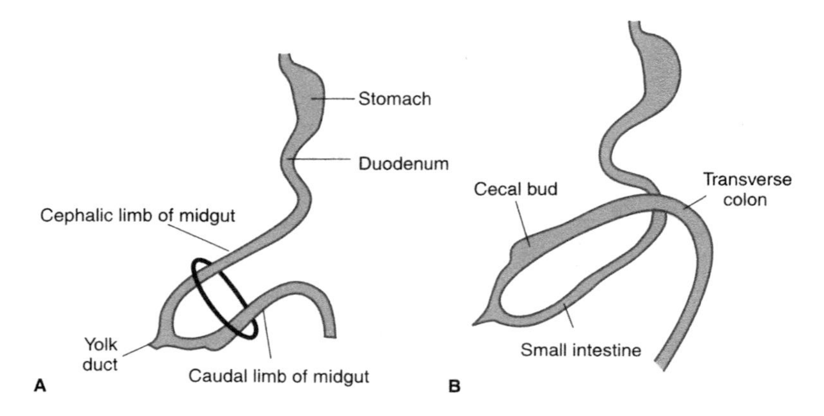

FIGURE 16.5 The midgut (A) gives rise to much of the colon and small intestine, as well as to the cecum (B) (the appendix arises from the end of the cecum).

large intestine) and the **appendix** (Figure 16.5). The development of the intestines is marked by a substantial increase in length and considerable changes in position. During the latter part of the embryonic period and the early part of the fetal period, the developing intestines extend into the umbilical cord, making up the "umbilical hernia," which provides the intestines with space for development. The **midgut loop** extending into the umbilical cord twists and elongates, forming multiple loops of small intestine, as well as the cecal diverticulum (forerunner of the cecum and appendix) and the proximal part of the colon.

As the yolk stalk (to which the original midgut was attached) degenerates and the intestines are withdrawn into the body cavity, further twisting of the gut puts the small intestine, ascending colon, transverse colon, and descending colon into their proper adult positions. Despite the intestines' tangled appearance, their positions are predetermined. As a consequence, **appendicitis** (inflammation of the appendix, which is a medical emergency) normally causes a sharp pain in the lower right region of the abdomen. Even though the intestines *appear* to have a tangled ("random") appearance, they really do not—usually, the appendix is in a specific location.

Hindgut A large part of the colon and the rectum have their origins in the hindgut (Figure 16.6). The caudal end of the hindgut is a blind tube ending at the **cloacal membrane**, the boundary between the rectum and the proctodeum (an inpocketing of ectoderm on the ventral side of the tail

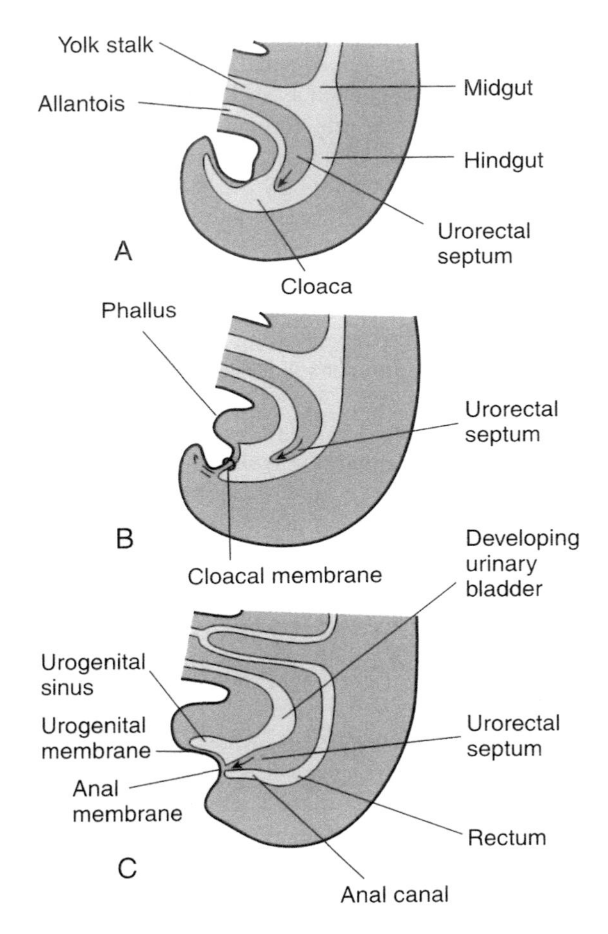

FIGURE 16.6 Hindgut development. Shown in diagrams of sagittal sections of the caudal portions of embryos, showing the origin of part of the colon and the rectum from the hindgut. The most caudal portion of the hindgut is the cloaca. The urorectal septum divides the cloaca into the urogenital sinus and the rectum. (A) At 4 weeks, the cloaca is the caudal-most portion of the hindgut. (B) At 6 weeks, the urogenital septum grows toward the cloacal membrane. (C) At 7 weeks, the urogenital septum has divided the cloaca into the urogenital sinus and the rectum.

end of the embryo). The portion of the hindgut near the cloacal membrane is called the **cloaca**. From the cloaca's ventral surface, the **allantoic diverticulum** extends for a short distance into the umbilical cord. The **urorectal septum** grows between the hindgut and allantoic diverticulum toward the cloacal membrane, separating the cloacal membrane into an **anal membrane** and a **urogenital membrane** (see Figure 16.6). When the anal membrane ruptures during the seventh week, the **anus** opens to provide a passageway from the digestive tube into the amniotic fluid.

Digestion before birth

Before the end of the embryonic period, the digestive tube is complete in the sense that it is open at both ends. However, although the cells of the digestive system differentiate during the fetal period, and even secrete small amounts of digestive enzymes near term, the digestive system does not function as such before birth. Rather, it accumulates contents called **meconium**, composed of cells and hair from the amniotic fluid, cells from the lining of the digestive tube, and other matter.

Even though the fetus normally drinks, as well as urinates into, amniotic fluid, it does not normally defecate into it. In fact, *meconium-stained amniotic fluid found during amniocentesis is a sign of fetal stress.* During prenatal life, the fetal circulatory system is providing nutrients from the mother by way of the placenta. There is no need for the fetus to eliminate anything from its digestive system, as indigestible residue of the mother's diet is eliminated by her.

Respiratory system

The respiratory system arises from the early digestive system and is thus lined by endoderm. Like the digestive system, development of the respiratory system depends on the mesenchyme (loose connective tissue) surrounding the endoderm. Particular types of mesenchyme act in association with respiratory endoderm to develop particular regions of the respiratory system, such as the trachea or bronchi.

Origin of the respiratory system

The respiratory system appears as the **laryngotracheal groove** in the floor of the back of the pharynx (Figure 16.7). As the groove deepens, it gives rise to the **trachea**, a tube that runs parallel with the upper portion of the esophagus. Two **lung buds** arise as the distal portion of the trachea bifurcates (splits into two). Growth and branching of these buds result in the **bronchial trees** of the lungs (Figure 16.8).

Lung development can be divided into four phases: the **pseudoglandular period**, the **canalicular period**, the **terminal sac period**, and the **alveolar period** (Table 16.1). These periods overlap because of the cephalic-to-caudal progression of lung development. For example, if the "top" of the lung is in the canalicular period, the "bottom" of the lung might still be in the pseudoglandular period. Complete lung development is vital. Only from the beginning of the terminal sac period is birth likely to result in a viable baby. Lung development does not stop with birth. The alveolar period extends into childhood.

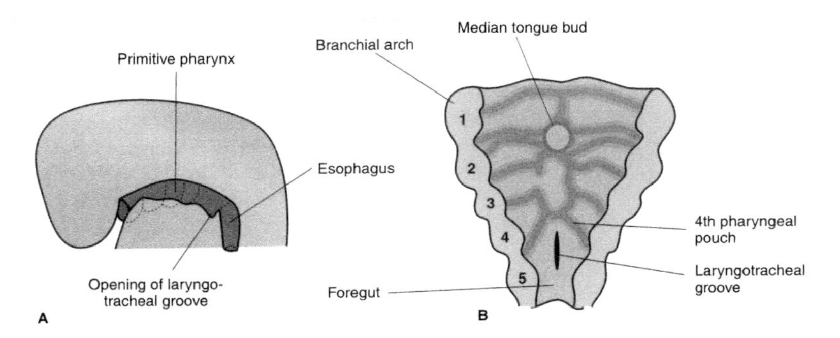

FIGURE 16.7 Origin of the respiratory system. (A) Diagram of a sagittal section through the cephalic end of a 5-week embryo. The respiratory system arises as an outgrowth (diverticulum) of the pharynx, which is part of the foregut. This diverticulum first appears as a groove, the laryngotracheal groove. Note that since the respiratory system is an outgrowth of the gut, which is lined with endoderm, the respiratory system is also lined with endoderm. (B) A diagrammatic view, looking at the floor of the pharynx of the embryo. Note the laryngotracheal groove in the floor of the pharynx, which gives rise to the respiratory system.

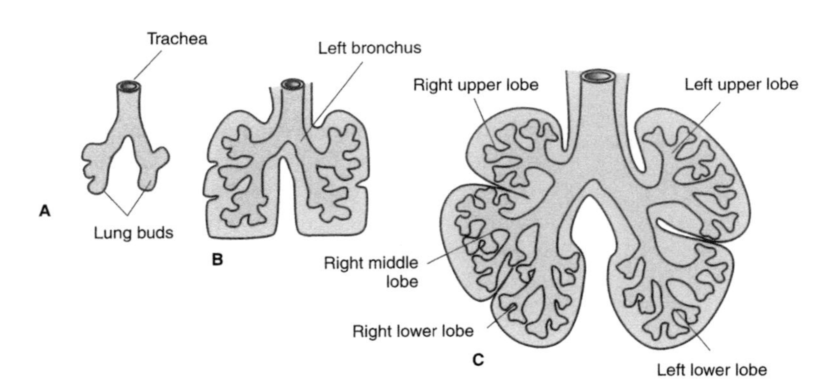

FIGURE 16.8 Development of the respiratory system. The laryngotracheal groove deepens to give rise to the trachea, which in turn bifurcates to give rise to the lung buds. (A) By 5 weeks, the trachea has bifurcated (divided) to give rise to the two lung buds. (B) At 6 weeks, the two bronchi have developed from the lung buds. (C) At 8 weeks, the two lobes of the left lung and the three lobes of the right lung are developing.

The first breath

As fetuses, our blood was oxygenated by maternal blood as both fetal and maternal blood circulated through the placenta. Of course, oxygenation of maternal blood is accomplished by the mother's respiratory system. Maternal oxygenation of the fetal blood means that the fetus's respiratory system need not begin to function until birth. Nevertheless, the respiratory system begins to develop during the fourth week, in preparation for birth.

For breathing to occur normally, two important conditions must be present. First, the central nervous system's respiratory centers must

Table 16.1 Phases of lung development

Name of period of development (in order of appearance)	Begins	Ends
Pseudoglandular	5th week	17th week
Canalicular	16th week	25th week
Terminal sac	24th week	Birth
Alveolar	Late fetal period	8 years after birth

Source: Adapted from Moore, K.L. *Before We Are Born*, 3rd edition. WB Saunders, Philadelphia, 1989.

be sufficiently developed. Second, the developing lungs must have synthesized and accumulated a sufficient amount of **surfactant**, a lipid (fatty material) mixture that prevents the lungs from collapsing with each breath. It is during the terminal sac period that surfactant begins to be produced, so that before birth the lungs are "waterlogged." At birth, the first few breaths inflate the lungs and require some effort on the part of the infant. The presence of surfactant prevents the lungs from collapsing between breaths and prevents the infant from developing respiratory distress syndrome.

Lung organoids: Researchers recently reported the generation of lung bud organoids (LBOs) from hPSCs that contained mesoderm and pulmonary endoderm and developed into branching airway and early alveolar structures in Matrigel 3D culture. Analysis indicated that the branching structures reached the second trimester of human gestation. Further analysis, using viral infections (to cause a condition resembling bronchiolitis in infants) and mutations (to cause a condition resembling intractable pulmonary fibrosis) suggested that these lung bud organoids, in addition to recapitulating lung development, may also provide a useful tool to model lung disease.

https://www.nature.com/articles/ncb3510#auth-14

Study questions

Digestive system

1. What gives rise to the lining of the digestive system?
2. Which two parts of the forming digestive system are the first to acquire floors of endoderm? What is the portion of the digestive system without a floor called?
3. What are mesenteries?
4. What parts of the digestive system are derived from the foregut?
5. What may happen to the esophagus during late embryonic and early fetal life?
6. How many rudiments give rise to the pancreas? What are they?
7. What parts of the digestive system are derived from the midgut?
8. What does the umbilical hernia provide for the developing digestive system?

9. What parts of the digestive system are derived from the hindgut?
10. When does the anal membrane rupture and open the anus?
11. What substance does the digestive system accumulate and what is its composition?
12. What does the fetus normally not do into the amniotic fluid? What is a sign of fetal distress involving the digestive system?

Respiratory system

13. From what does the respiratory system arise? Therefore, the respiratory system is lined with what?
14. Lung buds arise from the bifurcation of what? What results in the bronchial trees of the lungs?
15. What are the four phases of lung development? Why do these phases overlap?
16. From when in lung development is birth likely to result in a viable baby?
17. What is surfactant and its significance? When does surfactant begin to be developed?
18. Lung bud organoids (LBOs) appear to recapitulate what in culture?
19. Lung bud organoids may also provide a useful tool to model what?

Critical thinking

1. Why is it so that if cow protein is injected into the circulatory system, the body will reject it, but if cow protein is eaten, it is used as a nutrient?
2. What pain indicates that, despite the intestines' tangled appearance, their positions are predetermined?
3. Why is it not necessary for the digestive system to function before birth?

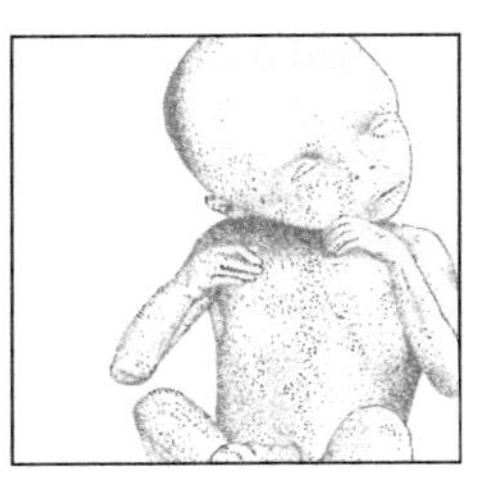

CHAPTER SEVENTEEN

Mouth and throat, face, and the five senses

CHAPTER OBJECTIVES

After studying this chapter, you should be able to:

1. Explain the origin of the mouth.
2. Discuss the development of the pharynx.
3. Beginning with the mouth and pharyngeal arches, discuss the development of the face and its features.
4. Explain the origins of the hard palate, soft palate, and cleft palate.
5. Beginning with the forebrain, discuss the origin of the eyes, including the role of embryonic induction.
6. Discuss the origin of the ears, including inner, middle, and outer ears.
7. Explain the origins of the senses of smell, taste, and touch.

On the face of it, the title of this chapter suggests a rather dissimilar group of topics. However, there is a concept that holds these topics together—the head. This most complex part of the body possesses the mouth and part of the throat, it carries the face, and it is the exclusive realm of four of the five senses—sight, hearing, smell, and taste. The sense of touch is dispersed widely throughout the body.

After considering the development of the mouth and throat, we will look at the development of the face. A study of the face is a good foundation for consideration of the five senses because the eyes, ears, nose, and tongue are features associated with the face.

Mouth and throat

The mouth (or oral cavity) arises from the **stomodeum**, an inpocketing of ectoderm on the ventral side of the embryo's head. The stomodeum is separated from the end of the foregut by the **oropharyngeal membrane**. When the oropharyngeal membrane breaks down during the 24th day of development, the fetus's mouth opens for the first time. This breakdown of the oropharyngeal membrane occurs in the boundary between the oral cavity and the throat or pharynx, which becomes the **tonsillar region** (where the tonsils are located) of the adult.

Pharynx

The pharynx is the anterior-most portion of the foregut and is therefore lined by endoderm.

Thyroid gland From the anterior portion of the floor of the pharynx, there arises a ventral outpocketing (pouch) of endoderm—the **thyroglossal duct**—which gives rise to the **thyroid gland** (Figure 17.1). As previously noted, a ventral outpocketing from the posterior floor of the pharynx gives rise to the respiratory system.

Branchial arches The walls of the pharynx have a series of paired, lateral outpocketings or pouches. These pharyngeal pouches of endoderm come into close proximity to a corresponding series of **branchial grooves**, which are inpocketings of ectoderm on the sides of the embryo's developing head. Between the pharyngeal pouches and the ectodermal branchial grooves, the walls of the pharynx are subdivided into a series of **branchial arches**, which are also called "gill arches" because in lower vertebrates, they give rise to actual gills. The branchial arches are only transitory structures in human development. Nevertheless, they are conspicuous structures on the surface of the embryo during the middle of the embryonic period, and they give rise to a number of very important

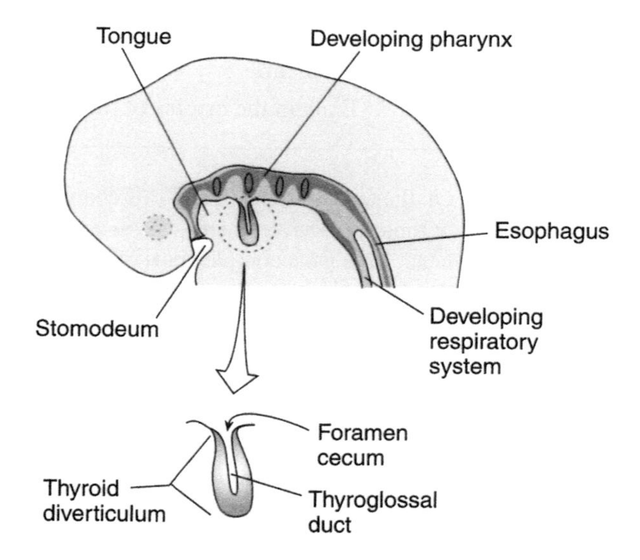

FIGURE 17.1 Thyroid gland. Diagram of the cephalic end of a 5-week embryo, showing part of the foregut in lateral view. Note the thyroglossal duct and thyroid diverticulum arising from the floor of the pharynx.

permanent structures. Furthermore, the **aortic arches** (blood vessels) running through the branchial arches give rise to a number of other important blood vessels. Examples include the arch of the **aorta**, the **pulmonary trunk**, and the **carotid arteries**, all important adult blood vessels, as well as the **ductus arteriosus**, a prenatal blood vessel shunt (bypass), which plays an important role in the development of the heart.

The first pair of branchial arches, the **mandibular arches**, play an important role in development of the face, forming much of the jaws (Figure 17.2). Between the first and second pairs of branchial arches (the **hyoid arches**) are the paired first branchial furrows called **hyomandibular furrows** (also called "gill clefts," even though they are really depressions and not clefts). These furrows give rise to the **external ear canals**. Mesenchyme of the first and second pairs of branchial arches also gives rise to the three **bony ossicles** (small bones) of the middle ear, which transmit sound from the eardrum to the inner ear (Figure 17.3). Two kinds of endocrine (hormone-secreting) glands arise from the third and fourth branchial pouches: the **thymus gland** and the **parathyroid glands**.

The latter is not an exhaustive list of structures derived from the branchial arches, but it is sufficient to demonstrate the important roles played by these **transitory structures** in human development. You may recall that other transitory structures play an important role in development. For example, the yolk sac gives rise to germ cells and the first blood cells, and the placenta plays an indispensable role in many aspects of development.

The face

It is interesting to reflect on how much importance is placed on physical appearance in our society from something that results from no design or effort on the part of the developing individual, but rather from dramatic morphogenetic movements that sculpt the face. On this depends the difference between beautiful and not so beautiful. In fact, so much value is placed on facial beauty that one would think the possessor had something to do with it, other than the inheritance of a random collection of genes.

The human face and its origin

With its diverse array of expressions, the face provides insight into much of what makes us human. A person trying with difficulty to remember something may look lost or confused, or a scared person may look as if he has seen a ghost. A mother's smile expresses love, an employer's stern gaze expresses displeasure, and a wrinkled brow may denote an unanswered question.

Anatomists define the face as the anterior part of the head and its features. In his recent book, *Making Faces* (2017), Adam Wilkins demonstrates "how the physical evolution of the human face has been inextricably intertwined with our species' growing social complexity." Wilkins argues that "it was both the product and enabler of human sociality." From a developmental point of view, it is known that a part of the neural crest makes a significant contribution to the substance of the developing head.

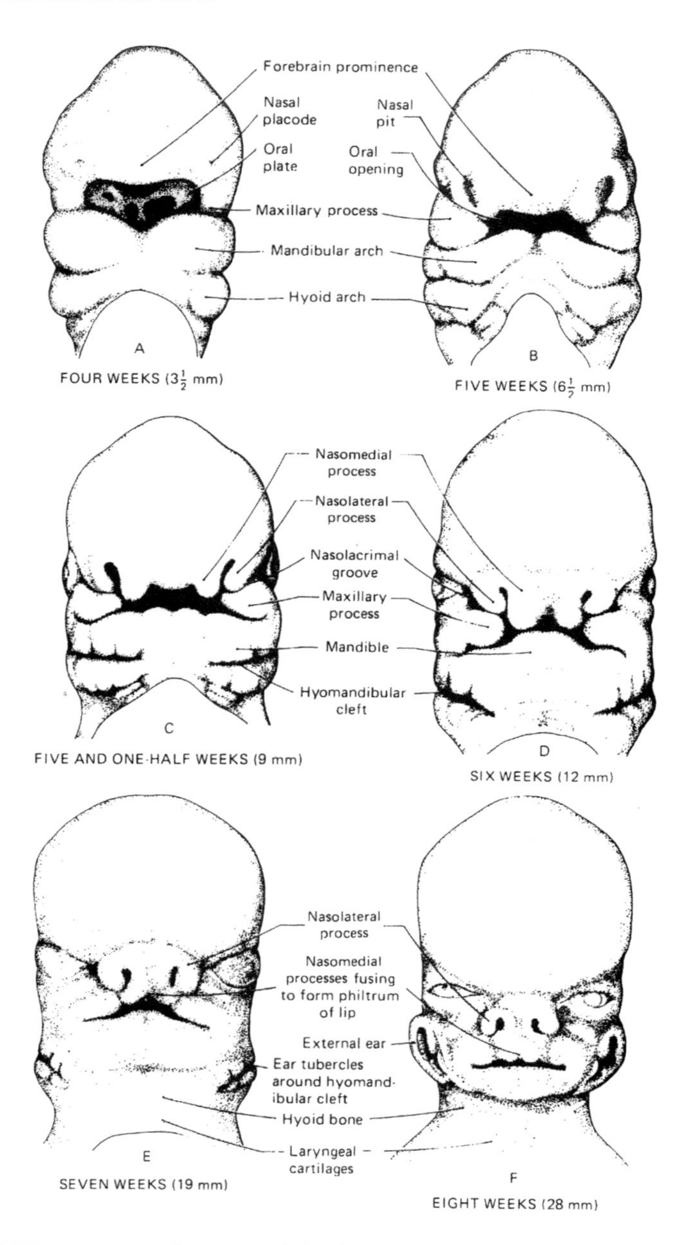

FIGURE 17.2 Development of the face. Diagrams of frontal views of the developing face of the embryo from 4 to 8 weeks. (A) At 4 weeks, note the mandibular arches with their maxillary processes and the oropharyngeal membrane (oral plate), which is disappearing. (B) At 5 weeks, note the nasal pits and the growth of the maxillary processes. (C) At 5 1/2 weeks, note the hyomandibular cleft between the mandibular and hyoid arches. (D) At 6 weeks, note the developing jaw (mandible). (E) At 7 weeks, note the naso-medial processes fusing with the maxillary processes. (F) By the end of the embryonic period (here, at 8 weeks), the face looks like a human face. (From Carlson, B. *Patten's Foundations of Embryology*, 5th edition. 1988. Reproduced with permission of the McGraw-Hill Companies.)

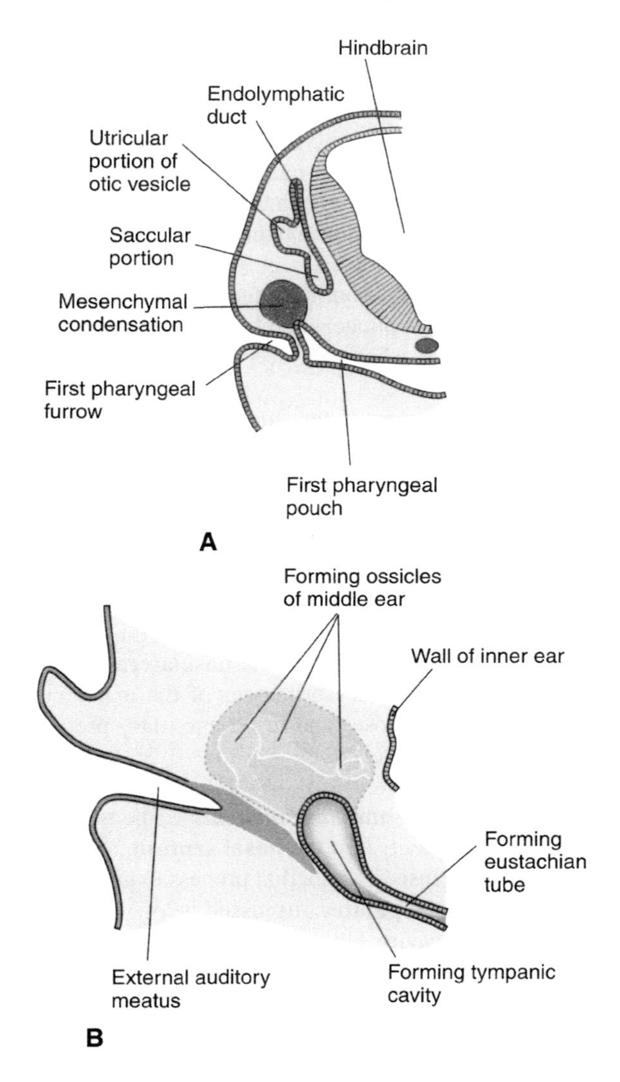

FIGURE 17.3 Diagrammatic representation of ear development during the fetal stage. The first pharyngeal (hyomandibular) furrow (A) gives rise to the external ear canal (B) (external auditory meatus). Note the three bony auditory ossicles of the middle ear that arise from mesenchyme of the first and second branchial arches.

By the end of the fourth week of development, the developing head is flexed down onto the pharyngeal region with its branchial arches. The oropharyngeal membrane, which marks the approximate site of the future mouth, is now surrounded by a number of structures that play important roles in the development of the face (see Figure 17.2).

On the surface of the head, rostral to the mouth region, are paired, bilaterally symmetric depressions, called **nasal pits**, which are surrounded by horseshoe-shaped elevations. Each horseshoe has two limbs (**nasomedial and nasolateral processes**) directed back toward the mouth region. Between the nasomedial processes is a conspicuous portion of the head referred to—appropriately—as the **frontonasal prominence**. The caudal and lateral borders of the mouth region are provided by the **mandibular arches**. Each of these mandibular arches develops an outgrowth called the **maxillary process**, positioned so as to make up a lateral border of the mouth region. The fusing mandibular arches make up the posterior border of the mouth region (see Figure 17.2).

Nose

This prominent feature of the human face, the nose, undergoes great morphogenetic changes throughout life. The nose of an elderly person on a baby, or vice versa, would be so out of place as to be truly startling.

External developments Despite its later prominence, the nose initially arises as a pair of thickenings (**nasal placodes**) on the ventral surface of the embryonic head. Because of both a sinking in of the nasal placodes and an outgrowth of surrounding tissue, depressions (nasal pits) are formed. The openings into these nasal pits are the future **nostrils**. On the medial side of each nasal pit, the elevated tissue is called the **nasomedial process**. These two processes eventually merge to form the roof and parts of the sides of the nose. On the lateral side of each nasal pit, the elevated tissue is called the **nasolateral process**. These processes fuse with the maxillary processes of the upper jaw to form the rest of the sides of the nose. The fused maxillary processes form the floor of the nose (see Figure 17.2).

Internal developments The nostrils initially open into a single nasal cavity, but the **nasal septum**, a vertical partition, grows down from the fused nasomedial process to give two nasal cavities. The development of the **palates** (discussed below) separates the nasal cavity from the mouth cavity. Ultimately, the nasal cavities open into the **nasopharynx** (throat) as well as to the outside. Inside, on the roof of each nasal cavity, the specialized **olfactory epithelium** arises. It contains sensory **olfactory cells** that extend odor receptors into the nasal cavities to detect odors and send nerve fibers back toward the brain. These nerve fibers collectively make up the **olfactory nerves** (for a more detailed discussion, see "Smell," **below**).

Development of the eyes and ears

In considering the development of the head, let's look at a few structural notes regarding the eyes and ears.

The eyes originate on the sides of the head and are gradually brought to the front of the head, a necessity for **binocular vision** (seeing a single field of view with two eyes). The **eyelids** develop as folds of skin that grow back over the corneas. They fuse by the 10th week and reopen by about the 26th week (Figure 17.4A and B).

Like the eyes, the ears migrate during development. The ears originate on the sides of the neck and are gradually brought onto the sides of the head. The **pinnae** (flaps) of the ears arise from tissue elevations flanking the first branchial furrows as the furrows themselves give rise to the **external ear canals** (Figure 17.5).

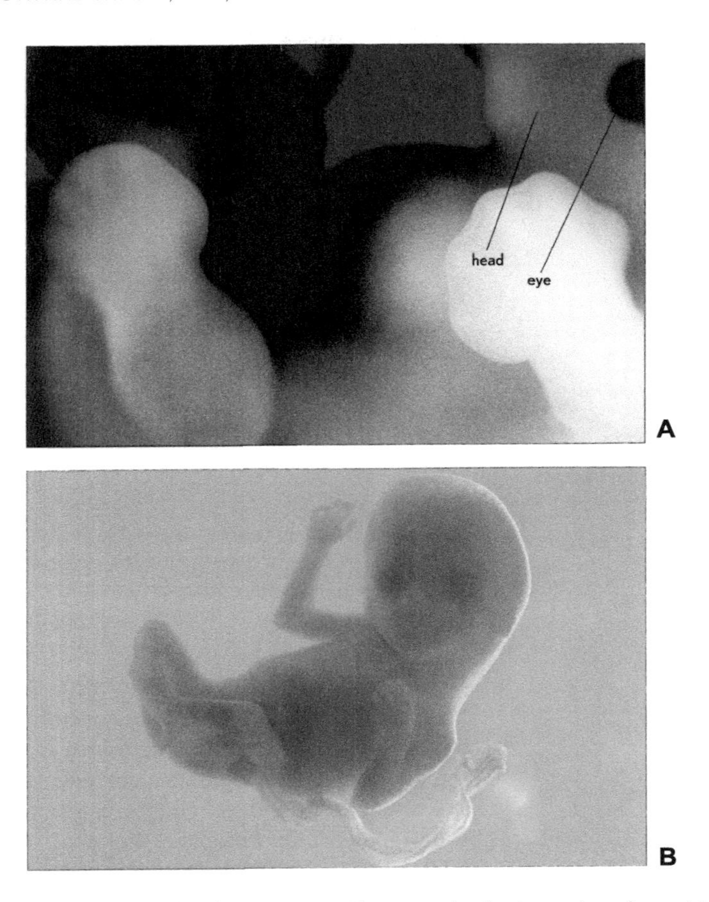

FIGURE 17.4 Eye development. (A) Photograph of a 6-week embryo. Note that the eyes are open. (B) Photograph of a 12-week embryo. Note that the eyelids have fused, to reopen by about the 26th week. (A: Cabisco/Visuals Unlimited; B: Vincent Zuber/Custom Medical Stock Photo.)

Jaws, lips, and teeth

In the morphogenesis of the face, the nasomedial processes and the maxillary processes come together and fuse to form the upper jaw (**maxilla**), while the mandibular processes come together and fuse in the midline to form the lower jaw (**mandible**) (see Figure 17.2). More specifically, the fused nasomedial processes give rise to the center (premaxillary) portion of the maxilla, and the maxillary processes give rise to the balance of the upper jaw. Subsequently, the premaxillary portion gives rise to the external **philtrum** (the depression on the surface of the upper lip immediately beneath the nose) of the lip, the **incisor teeth**, and the **median palatine process**. The balance of the maxilla gives rise to the rest of the upper lip, the rest of the upper teeth, and the **paired lateral palatine processes**. The mandible gives rise to the external lower lip and the lower teeth. In both the upper and lower jaws, a thickening of ectoderm (the **labiogingival lamina**) grows down into the underlying mesenchyme, then degenerates, forming the deep labiogingival groove that separates the lips from the gums.

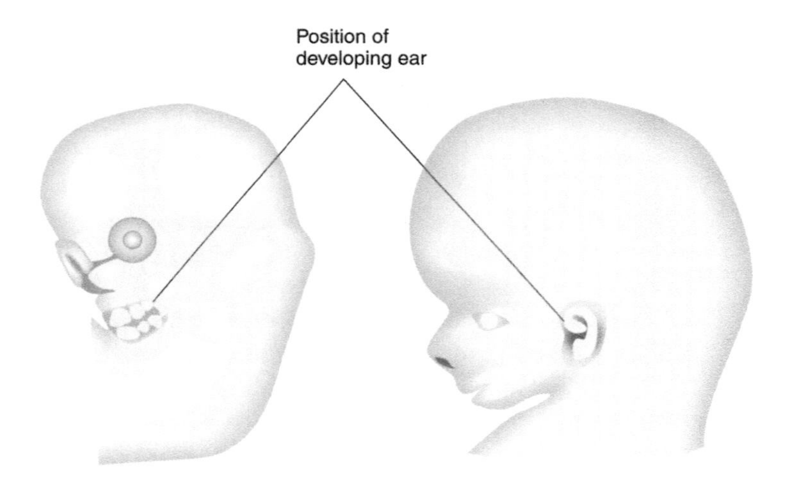

FIGURE 17.5 Migration of the ears. Drawings of the left side of the head and neck regions of two embryos. Six-week embryo (left) and 8-week embryo (right). Compare the position of the ears on the two embryos. Note that the ears have moved from the sides of the neck to the sides of the head.

Some of these morphogenetic processes are readily apparent by observation of the exterior surfaces of embryos during the fourth through eighth weeks. At the end of the fourth week, the human embryo looks more like a fish than a human (see Figure 8.13). By the end of the eighth week, however, the developing organism looks human enough to now be designated as a **fetus** (see Figure 8.14). Perhaps the most human characteristic of the early fetus is its face.

Palates

At the same time that the externally visible changes are occurring in the development of the face, changes just as dramatic are taking place internally. From the inner, medial borders of the maxillary processes, additional outgrowths (**lateral palatine processes**) arise, change positions from nearly vertical to horizontal, grow toward each other, and fuse with each other as well as with a median process (**median palatine process**) that grows from the innermost part of the **premaxilla** (Figure 17.6). As a result, the **hard palate** (roof of the mouth) is formed, dividing the original oral cavity into an oral cavity and a nasal cavity. Simultaneously, the development of a vertical nasal septum subdivides the nasal cavity into a pair of **nasal passages**, which connect the external openings of the nose with the throat. From the posterior margins of the hard palate, a pair of tissue folds grow backward (caudad) and by their fusion give rise to the **soft palate** and the **uvula**.

In normal development of the palate, the palatal shelves must fuse with each other, a process that depends on genetically programmed cell death. If this does not happen, a **cleft palate** results. Thus, we again have an example of normal cell death playing an important role in development, just as it shapes the fingers and toes, decreases the number of female germ cells, and decreases the number of neurons in both sexes.

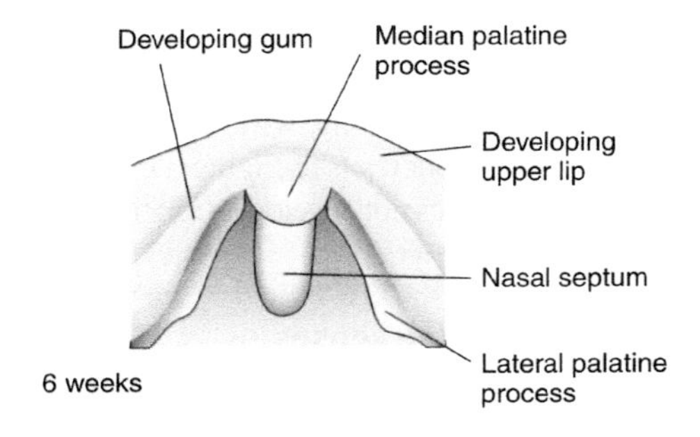

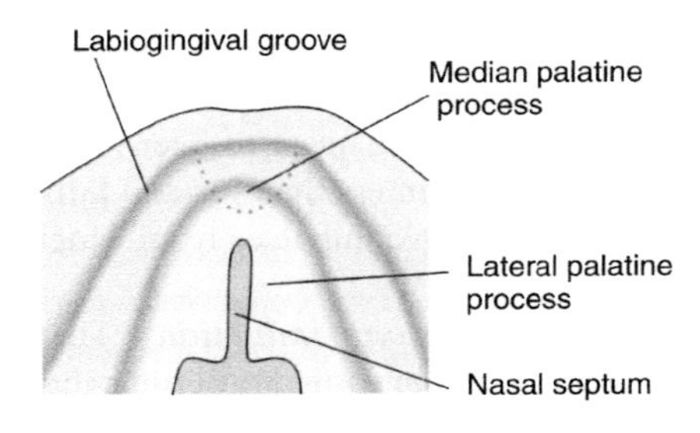

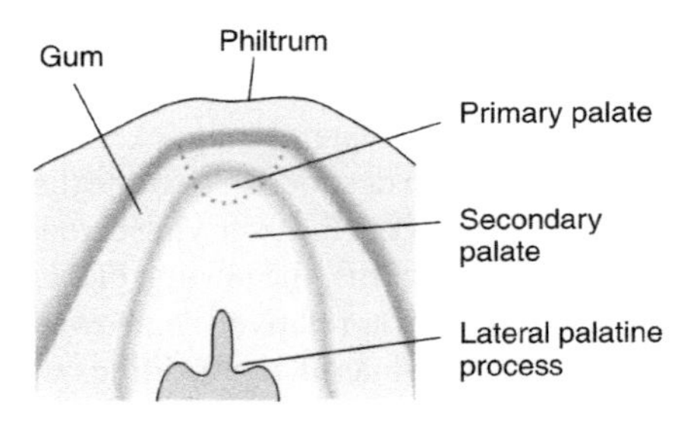

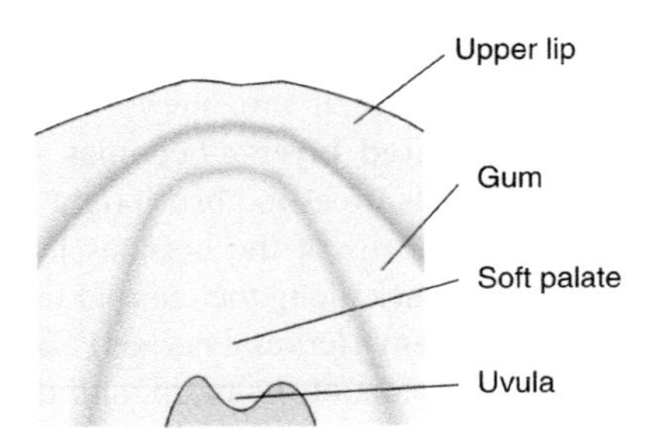

12 weeks

FIGURE 17.6 Development of the palates. Diagrams of the roof of the mouth from 6 to 12 weeks. The hard palate is formed by the growth toward each other and the fusion of the median palatine process, an outgrowth of the premaxilla, and the two lateral palatine processes, outgrowths of the medial borders of the maxillary processes.

Forehead and chin

Unlike the features in between, the top and bottom of the face do not go through particularly complicated developmental stages. The forehead develops along with the underlying forebrain, especially the cerebral hemispheres (see Figure 17.2). The chin is found at the bottom of the face and does not develop its relative prominence until the approach of adulthood.

The five senses

Sight

Our eyes are our windows to the world. According to Julian Jaynes in *The Origin of Consciousness in the Breakdown of the Bicameral Mind*, a book on the origin of human consciousness, we appear to occupy a space just behind the eyes (which, of course, does not exist). From this vantage point, we seem to peer out at the world. But quite the opposite is true: light enters the eyes and falls on photoreceptor cells, initiating the train of nerve impulses from which the brain creates the world that we "see."

Embryonic induction The eyes, in their development, display a series of **embryonic inductions** that finally integrate separate components into a marvelously functioning pair of organs. In fact, the earliest discovery of embryonic induction was in eye development.

Retina and optic nerve Early in development, even before the anterior neuropore (the opening at the cephalic end of the neural tube) has closed, a pair of lateral outpocketings arise from the walls of the forebrain (the portion destined to become the **diencephalon**). These **optic vesicles** grow out toward the overlying ectoderm on the sides of the developing head. As they do this, the portion of each closest to the wall of the brain constricts to form a relatively narrow **optic stalk**. Each optic vesicle develops a ventro-lateral inpocketing, which converts it into a two-layered **optic cup** with a break in its wall (called the **choroid fissure**). This fissure is carried back along the optic stalk to provide a pathway for the growth of the optic nerve toward the brain (Figure 17.7). The optic cups give rise to the **retinas**. The inner layer of the cup—the **nervous layer**—develops **photoreceptor cells** (called **rods and cones**), which receive light energy and change it into nerve impulses. The outer layer of the cup—the **pigmented layer**—becomes increasingly pigmented as development proceeds, so as to absorb light not intercepted by the nervous layer.

The cells of the nervous layer of each retina send processes called axons back along the choroid fissure of the optic stalk, which become the **optic nerve** (cranial nerve II). The retina may be considered an extension of the walls of the brain, and the optic nerves nerve tracts in the central nervous system.

Lens and cornea As the optic vesicles form the optic cups, they exert an inductive influence on the overlying ectoderm. That is, the **lens** of the eye forms from this surface ectoderm, which otherwise would have produced epidermis. Rather than differentiating along the epidermis pathway and producing the protein keratin (a fibrous protein produced

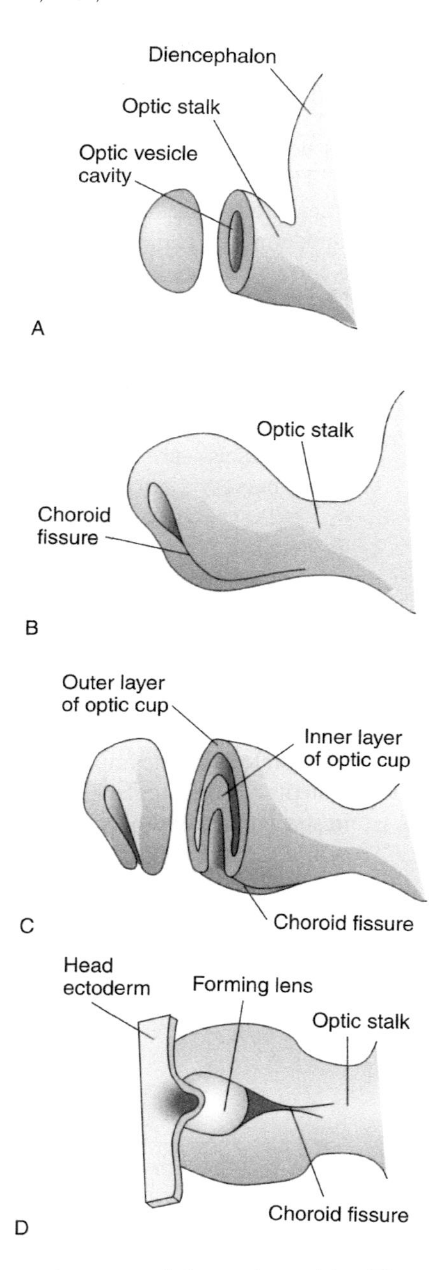

FIGURE 17.7 Development of the optic vesicle. Diagrams of transverse views of the development of the optic vesicle. Although the optic vesicles are paired and bilaterally symmetrical, the development of a single optic vesicle is shown here. (A) By the fourth week, the optic vesicles arise as outgrowths of the lateral walls of the brain (prosencephalon). (B) Week 5. The portion of the optic vesicle immediately adjacent to the wall of the brain becomes constricted to form the optic stalk. (B and C) Week 5. Invagination (inpocketing) transforms the optic vesicle into a double-walled optic cup and forms a groove along the optic stalk. (D) Week 5. Ventral view. The optic vesicles induce the formation of the lenses from surface ectoderm.

by cells of the epidermis) in large amounts, the same cells differentiate along the lens pathway and accumulate large amounts of proteins called **crystallins**. These proteins must be arranged so that the lenses become tissues that are transparent to light.

The lens of the eye, in turn, induces the formation of the cornea (the transparent anterior portion of the eyeball) of the eye from the surface ectoderm. The cornea must also be transparent, but, in addition, it must have a correct curvature, for it does most of the refracting of light. The cornea brings light into sharp focus on the retina, which results in sharp vision.

The rims of the retina give rise to the **iris** and the muscles that control them. Unlike most muscles, which arise from mesoderm, the muscles of the retina arise from ectoderm. Other components of the eye—other muscles, the **sclera** (tough white outer connective tissue of the eyeball), blood vessels—arise from mesoderm. Cranial nerves III, IV, and VI (all motor nerves) control (innervate) these eye muscles.

Hearing

To perceive vibrations propagated through the air, natural selection has provided us with a high-fidelity, stereophonic receiver, comprising our ears and brain. Sound waves, gathered into our external ear canals by the pinnae, cause first our eardrums and then our middle ear ossicles to vibrate, setting in motion fluid movement in the cochlea of the inner ear. This movement causes stereocilia in the cochlea to bend, resulting in a train of nerve impulses being sent (via cranial nerve VIII) to the brain, which interprets the impulses as sound. In addition, the inner ears possess systems of semicircular canals. These canals allow us to sense our position in space, and consequently maintain balance as we go about our various activities, by detecting fluid motion in three axes (x, y, and z) (Figure 17.8).

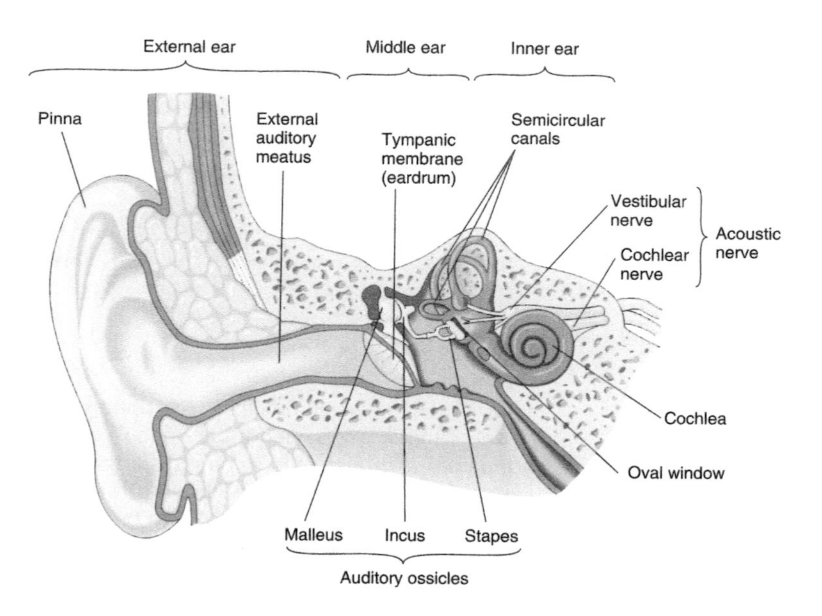

FIGURE 17.8 The anatomy of the adult ear. The adult ear is divided into an external ear, a middle ear, and an inner ear.

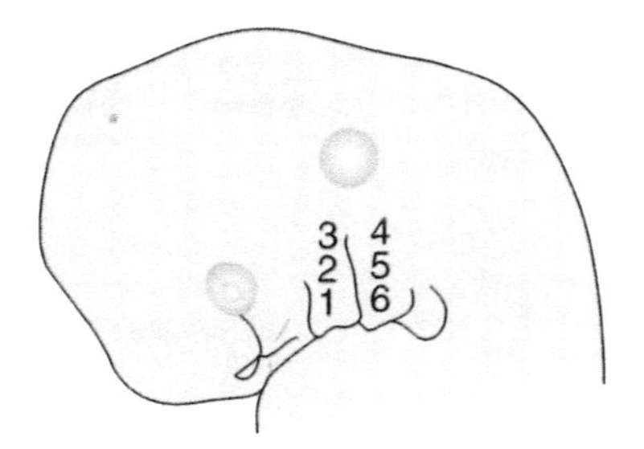

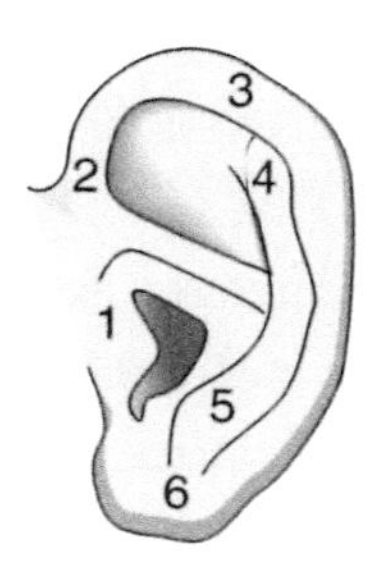

FIGURE 17.9 The development of the pinnae of the ear. The numbers 1 to 6 show the correlation between the embryonic tubercles and the parts of the pinnae of the adult ear.

Outer ear The pinnae arise from surface tissue (**tubercles**) flanking the hyomandibular furrows, each of which gives rise to an outer ear canal (**external auditory meatus**) (Figure 17.9).

Middle ear The three bony ossicles of the middle ear are derived from ossified cartilage of the upper ends of the first two branchial arches. The chambers of the middle ears (the **tympanic cavities**, which are frequent sites of ear infections, or **otitis media**) arise from the expanded distal portions of the first pharyngeal pouches; the narrow proximal portions give rise to the **eustachian tubes** (Figure 17.10).

Inner ear The inner ear is derived from the **otic vesicles** (**auditory vesicles**), which flank the hindbrain. In the same way that the optic vesicles change the pathway of ectodermal cell differentiation from epidermis to lens, the hindbrain alters the pathway of ectodermal cell differentiation from epidermis to inner ear. The otic vesicles give rise to the **membranous labyrinths** (tortuous anatomic structures, so called because of their epithelial lining and labyrinthine chambers), which will become incorporated into similarly shaped **bony labyrinths** in the temporal bones. Each membranous labyrinth gives rise to three structures—**endolymphatic duct, vestibule**, and **cochlea**—in a dorsal to ventral direction (Figure 17.11).

The endolymphatic duct fills the space between the membranous and bony labyrinths with endolymphatic fluid. The vestibule gives rise to three structures: the **semicircular canals** (concerned with awareness of positional changes), a dorsal **utriculus** (to which the semicircular canals attach), and a ventral **sacculus**. Both the utriculus and the sacculus develop a **macula** region. These maculae are concerned with awareness of static position (see Figure 17.11).

Smell

Few sensations evoke memories of a previous experience as acutely as does the sense of smell. The scent of grape jam can rapidly transport me back to a cherished childhood experience. This is a common experience that may be triggered by any of a variety of odors. The sense of smell is a primitive one, and the pathways (**olfactory nerves**) are very short

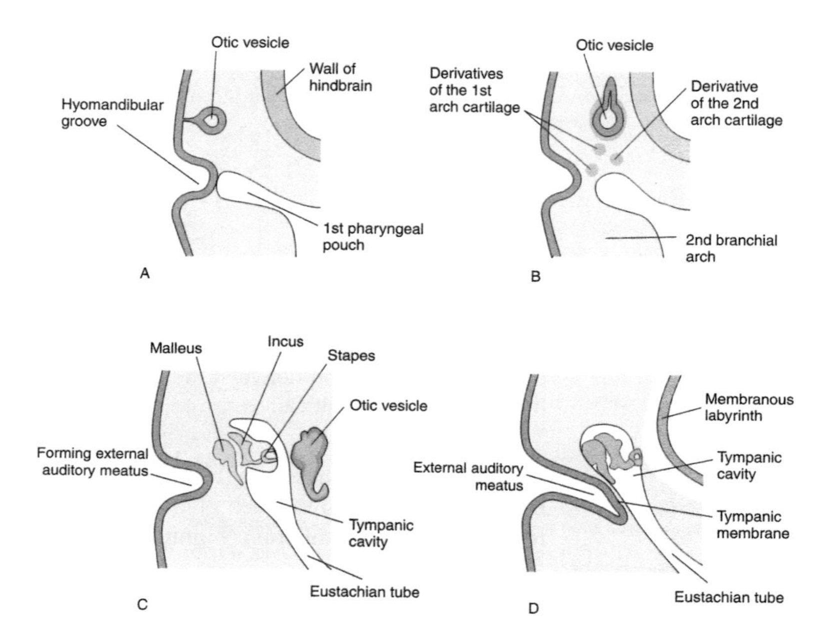

FIGURE 17.10 Stages in the development of the middle ear. Diagrams of transverse sections through the head at the level of the hindbrain. Although occurring bilaterally symmetrically, structures on only one side of the head are shown. (A) Week 4. (B) Week 5. (C) A later stage. (D) The final stage in development of the middle ear. (B through D) Note the development of the bony ossicles—malleus, incus, and stapes—from cartilage of the first and second branchial arches. Also, note the development of the tympanic cavity from the first pharyngeal pouch. The external auditory meatus develops from the hyomandibular groove.

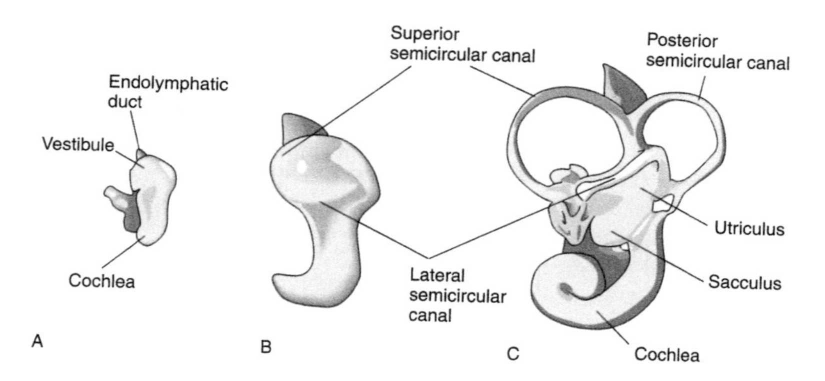

FIGURE 17.11 Development of the internal ear. (A and B) The three regions of the membranous labyrinth: endolymphatic duct, vestibule, and cochlea. (C) The vestibule gives rise to three structures: semicircular canals, a dorsal utriculus, and a ventral sacculus.

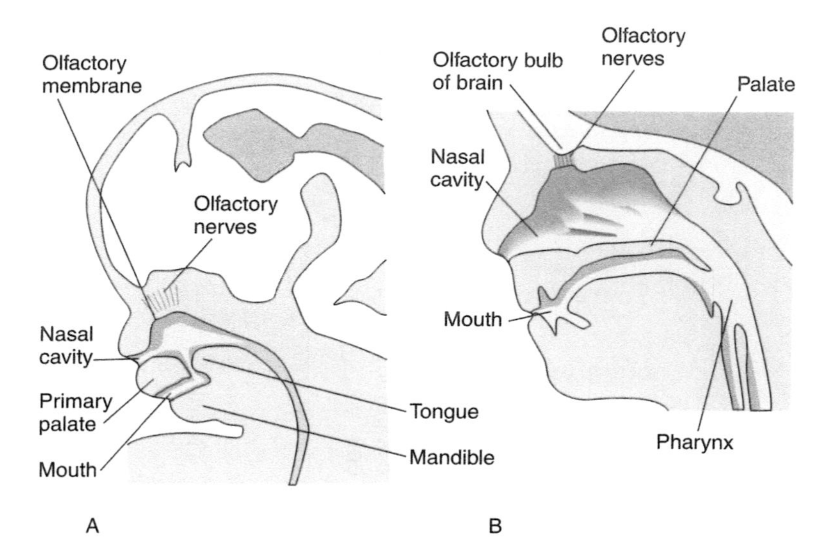

FIGURE 17.12 The nasal chambers. (A) Diagram of a sagittal section of the head of an embryo at 7 weeks. (B) Diagram of a sagittal section through the head of a fetus. Note in both diagrams the short path taken by the olfactory nerves from the nasal cavity to the brain.

between the **olfactory epithelium** lining the nasal passages and the **olfactory lobes** of the brain (Figure 17.12).

The external **nares** (nose openings) originate as a pair of thickenings, the **nasal placodes**, on the surface of the head. As these sink into the head and the surrounding tissue grows forward, **olfactory** pits are established. Cells in the epithelial lining of these pits send processes the short distance back to the wall of the brain. The processes from each olfactory pit collectively make up an olfactory nerve.

Taste

The sense of taste probably evolved to enable animals to distinguish between nutritive and harmful substances. Probably, the variety of tastes to which most animals are exposed is very limited. Not so with humans! Not only does each human culture have a fair variety of tastes, but, because of the development of commerce between cultures, modern humans are presented with an almost infinite variety of tastes from which to pick and choose. In the affluent countries of the world, this has frequently led to overindulgence and obesity.

Development of the tongue The tongue arises from a number of elevations in the floor of the early pharynx. Of these elevations, the **lateral lingual swellings**, anterior to the **foramen cecum** (the site in the floor of the pharynx from which the thyroid diverticulum arose; see Figure 17.1), give rise to the anterior two-thirds (oral part) of the tongue. The **hypobranchial eminence**, posterior to the foramen cecum, gives rise to the posterior third (pharyngeal part) of the tongue (Figure 17.13A and B).

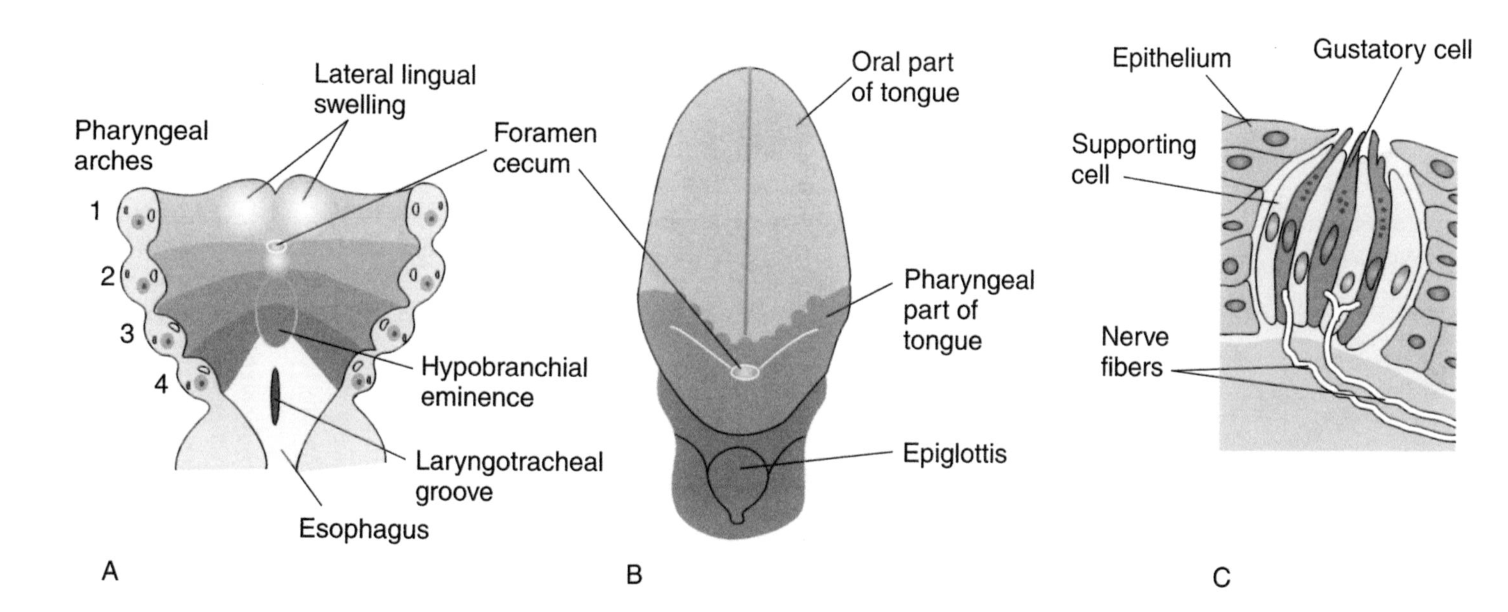

FIGURE 17.13 The tongue. (A and B) Diagrams of the embryonic (4 weeks) and adult tongue showing the developmental origin of its parts. Note the oral and pharyngeal parts of the tongue and their relationship to the foramen cecum. (C) Diagram of a section through a taste bud. Note the taste (gustatory) cells and the supporting cells.

Taste buds Taste is the responsibility of taste buds found on the surface of the tongue. They are specialized for sweet, salty, bitter, and sour tastes, with the four types of taste buds restricted to different parts of the tongue. Local thickenings of the tongue epithelium are the first indications of the taste buds. Although considered to be derived from endoderm, some taste buds are located in presumably ectodermal areas. Some of the cells in the epithelial thickenings differentiate into slender **taste cells**, whereas others become nonsensory, columnar **supporting cells** (see Figure 17.13C).

Our sense of touch

Unlike the special sensory organs responsible for vision, hearing, taste, and smell, the sense of touch belongs to the category of *general sensory organs*. This category includes free nerve terminations and encapsulated nerve endings.

Free nerve terminations simply move in among the cells of epithelial or connective tissues and develop branches. By the end of the third month, these nerve endings are invading the epidermis. This category (free nerve terminations) also includes Merkel's tactile discs, each composed of a Merkel cell and the leaflike expansion of a nerve terminal. Because of their association with the ends of nerves, Merkel cells are thought to play a role in sensation—as skin **mechanoreceptors**, sensitizing the skin to touch. By the end of the fourth month, Merkel cells appear in the epidermis.

Encapsulated nerve endings have the nerve terminations wrapped in connective tissue. These nerve endings differ in (1) the branching patterns of the nerve endings, (2) the organization (amount and arrangement) of the connective tissue, and (3) the shape and size of the organs. These nerve endings include the tactile **corpuscles of Meissner**. These corpuscles begin their development at 4 months and complete it about 1 year after birth.

Embryonic induction: An important phenomenon in development is that of cascades of embryonic inductions. This is beautifully exemplified by the development of the eye. The retinas of the eyes are originally parts of the wall of the brain early in development. They eventually become pairs of outgrowths (evaginations) of the brain, which become, by ingrowths (invaginations), double-walled structures; that is, the retinas. These developing retinas induce the formations of lenses from the surface ectoderm of the head of the embryo. The paired lenses, in turn, induce the formations of the corneas of the eyes

In Chapter 12 the concept of epithelial–mesenchymal interactions was introduced in the context of skin development. Here, we consider an additional example of such interactions in the development of the tooth. The tooth arises from an interaction between oral epithelium and mesenchyme, with the epithelial component becoming the enamel organ and the mesenchyme becoming the dental papilla. The enamel organ contains ameloblasts, cells that give rise to the enamel of the developing tooth, while the dental papilla contains odontoblasts, cells derived from migratory neural crest cells, which give rise to the dentin of the developing tooth.

Study questions

Mouth and throat

1. What is the stomodeum? What separates it from the end of the foregut?
2. Where and when does the oropharyngeal membrane break down, opening the mouth for the first time?
3. Where is the pharynx located?
4. Which branchial arches play an important role in the development of the face and to what do they give rise?
5. What two kinds of glands arise from the third and fourth pairs of pharyngeal pouches?

The face

6. The first indication of the nose is a pair of thickenings called what? These thickenings sink below the surface of the developing head and are surrounded by outgrowths to form what?
7. What do the nasal cavities open into in the interior of the head?
8. Where do the olfactory epithelia arise? What sensory cells do these epithelia contain and what do their nerve fibers back to the brain collectively constitute?
9. What migration do the developing ears undergo?
10. What gives rise to the external philtrum of the upper lip, the incisor teeth, and the primary palatine process?
11. What does the hard palate subdivide? From what do the soft palate and uvula arise?
12. If fusion of the palatal shelves does not happen, what birth defect results?

Five senses

13. What are the two layers of the optic cup? To what does each one give rise? In each layer, cell differentiation gives rise to what cells?
14. The lens of the eye induces the formation of what from head surface ectoderm?
15. The three bony ossicles of the middle ear are derived from what? From what do the tympanic cavities arise?
16. The elevations of the floor of the pharynx that give rise to the tongue surround what opening? What arises from this opening?
17. What two kinds of cells differentiate in the taste buds?
18. What two kinds of nerve endings are exhibited by general sensory organs?

Critical thinking

1. Describe the role of embryonic induction in early eye development.
2. Distinguish between special sensory organs and general sensory organs.
3. From a phylogenetic point of view, why are branchial arches also called gill arches?

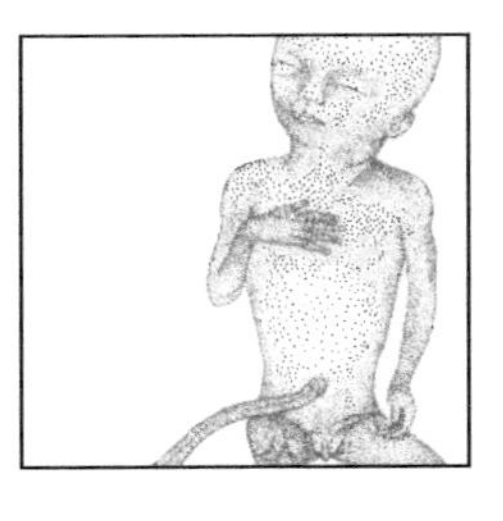

CHAPTER EIGHTEEN

The urinary and reproductive systems and the external genitalia

CHAPTER OBJECTIVES

After studying this chapter, you should be able to:

1. Explain why the origin and development of the urinary and reproductive systems are considered together.
2. Beginning with intermediate mesoderm, discuss the origin of the kidneys in human development, including the pronephros, mesonephros, and metanephros.
3. Beginning with the splanchnic mesoderm, discuss the origin of the gonads—both testes and ovaries.
4. Discuss the role of the sex chromosomes in human gender development.
5. Beginning with intermediate mesoderm, discuss the origin of the male and female reproductive ducts.
6. Explain the origin of the male and female sex glands.
7. Beginning with the indifferent condition, describe the development of the external genitalia of men and women.

The urinary and reproductive systems are considered together because of their close structural relationship in both time and space. Both systems arise simultaneously from the **intermediate mesoderm**. In fact, the developing reproductive system scavenges parts from the early urinary system as progressively more advanced versions of the urinary system develop. The external genitalia arise on the surface of the body. Because the urethra (canal from the bladder), running through the penis (part of the male external genitalia), is a urogenital duct and the female urethra and vagina open into the vestibule (part of the female external genitalia), it is reasonable to include development of the external genitalia here.

In this chapter, we begin with the development of the urinary system, which provides a foundation for development of the reproductive system. Last, we consider the development of the external genitalia, to which both the urinary and reproductive systems lead.

Urinary system

Our urinary system allows us to eliminate **nitrogenous wastes**, as well as other kinds of wastes. Our bodies require and metabolize a variety of nitrogen-containing chemical compounds—proteins, nucleic acids, coenzymes, the "heme" (iron-containing) part of the hemoglobin molecule, and others. This metabolism produces nitrogen-containing waste products that must be removed from the body. The urinary system becomes functional in the fetus before birth, and the fetus urinates into the same amniotic fluid that bathes it.

Kidneys

During human development, three kinds of **nephroi** (kidneys) appear, the pronephros, the mesonephros, and the metanephros, in that order. They have some common features:

- They develop in pairs from the intermediate mesoderm.
- They grow progressively from cephalic to caudal locations, bilaterally disposed across the midline.
- All three are composed of tubes.

The kidneys are located in the lower back—the location of an illegal "kidney punch" in boxing. They develop from the intermediate mesoderm in a region that becomes the back, where the somatic and splanchnic mesoderm diverge. The cavity between these mesoderms becomes the region of the coelom that becomes the peritoneal (abdominal) cavity. Consequently, the adult kidneys occupy a dorsal (back), retroperitoneal (behind the peritoneum) position.

The first of the three nephroi to arise is the **pronephros**, located farthest toward the cephalic end of the embryo. A series of tubes arises at the levels of the cephalic-most somites, from intermediate mesoderm, on each side of the body. One pair of **pronephric tubules** forms per pair of somites (Figure 18.1). The ends of these tubules initially grow dorsad (upward) and then bend caudad (backward). The tubules on each side then join to form the paired **pronephric ducts**. The pronephric ducts grow toward the caudal end of the embryo (beyond the level of the pronephric tubules) (see Figure 18.1), but the pronephric tubules degenerate shortly after they appear.

The second type of kidney to arise is the **mesonephros**. Another series of tubules arises from intermediate mesoderm on each side of the body, this time caudal to the level of the pronephros. These **mesonephric tubules** grow dorsad and join the pronephric ducts, which will now be referred to as **mesonephric ducts**. These ducts extend back to the

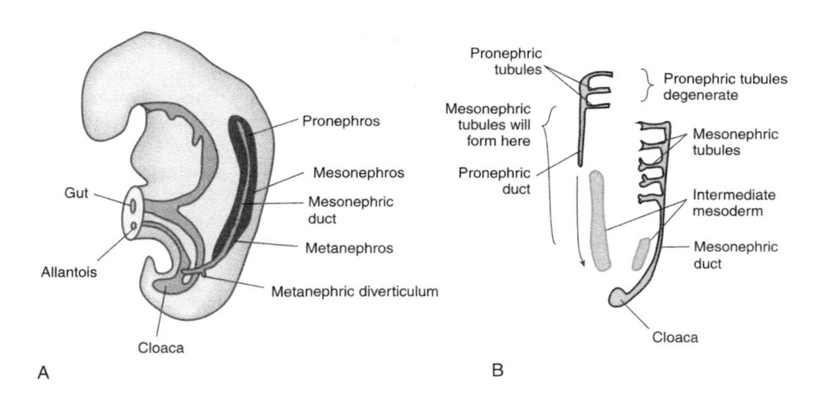

FIGURE 18.1 Kidney development: pronephros and mesonephros. (A) Diagram of an embryo shows the intermediate mesoderm (shaded), along the length of which are indicated the relative positions where the pronephros, mesonephros, and metanephros will develop. Note that the pronephros is the most cephalic and the metanephros is the most caudal of the three. Note that although intermediate mesoderm is bilaterally distributed across the midline; this diagram shows only one side of the embryo. (B) Composite diagram shows development of the pronephros (pronephric tubules and pronephric duct) on one side and development of the mesonephros (mesonephric tubules and mesonephric duct) on the other side. The bilaterally positioned mesonephros replaces the bilaterally positioned pronephros. Mesonephric tubules replace degenerating pronephric tubules, and the pronephric ducts become the mesonephric ducts.

level of the cloaca, where they bend ventrad and open into the portion of the cloaca that becomes the **urogenital sinus** (see Figure 18.1). The mesonephros functions during the late embryonic and early fetal periods.

Arising from the caudal-most portion of the intermediate mesoderm, the **metanephros** is the last of the three types of kidneys. Two structures are initially involved in inductive interactions and ultimately become structurally and functionally integrated into the definitive kidney: the **metanephric diverticulum** and the **metanephrogenous mesoderm**. The metanephric diverticula grow cephalically from the mesonephric ducts, cephalic to the level at which the ducts enter the cloaca. The paired metanephric diverticula give rise to the **ureters, calyces**, and **collecting tubules** of the kidneys. As the metanephric diverticula grow, they gather about their blind ends intermediate mesoderm—the metanephrogenous mesoderm, which gives rise to the **uriniferous (urine-carrying) tubules** of the kidneys (Figure 18.2). The portions of the mesonephric ducts caudal to the level of the metanephric diverticula are incorporated into the wall of the cloaca, part of which becomes the **urinary bladder**.

There is an interesting parallel between the succession of the three types of kidneys in **ontogeny** (the development of the individual human) and the evolution of kidneys in **phylogeny** (the evolution of vertebrates). In fish, the pronephric kidneys—and in amphibians, the mesonephric kidneys—are the functional kidneys. In humans, the metanephros is the functional kidney in the adult.

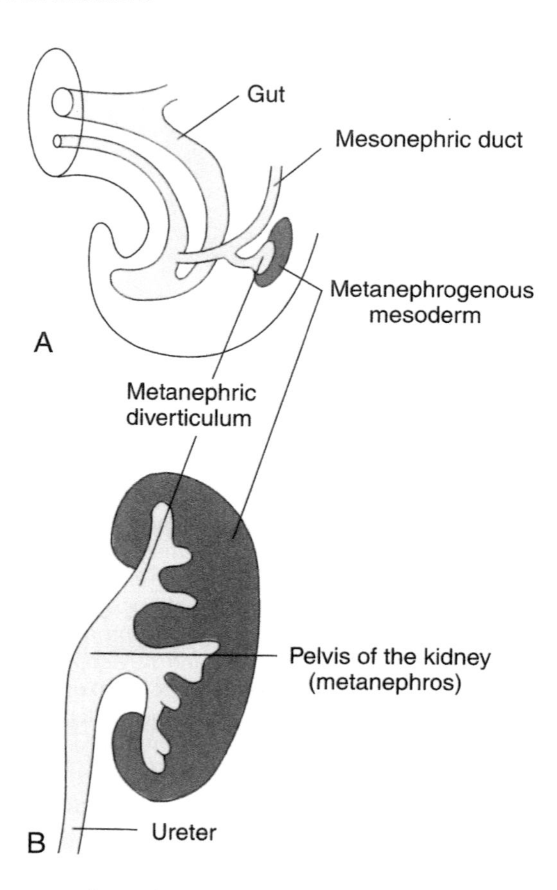

FIGURE 18.2 Kidney development: metanephros. (A and B) Diagrams illustrating development of one of the two definitive metanephric kidneys. Notice that two sources of tissue contribute to the formation of each of the metanephric kidneys: the metanephric diverticulum, which arises as an outgrowth of the caudal portion of the mesonephric duct, and metanephrogenous mesoderm, derived from the intermediate mesoderm.

Blood and urine

The kidneys undergo their development so as to service the circulatory system, removing nitrogenous wastes from the blood. In fact, all three types of kidneys have a relationship with small blood vessels of the circulatory system. The most intimate and efficient relationship of the three is the relationship between the metanephros and circulatory system. Renal arteries and veins, respectively, supply blood to and remove blood from the kidneys. The arterial vessels become smaller and smaller as they approach the uriniferous tubules of the kidneys. Finally, the **glomerulus** (a tuft of capillaries) fits into **Bowman's capsule** (an inpocketing of each uriniferous tubule). Waste products and useful substances pass from the blood into uriniferous tubules. Beyond Bowman's capsule, blood vessels form a network of capillaries around the uriniferous tubule to absorb some water and useful materials from the **glomerular filtrate** formed at Bowman's capsule (Figure 18.3). Urine, formed in this way, makes its way downstream to the urinary bladder.

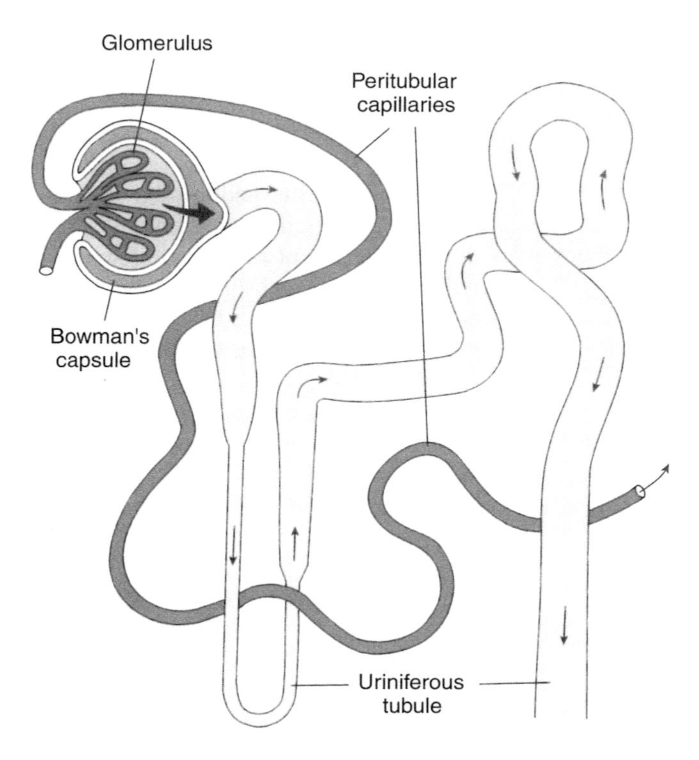

FIGURE 18.3 Relationship between kidneys and the circulatory system. Diagram shows the relationship between blood vessels, especially the glomerulus and peritubular capillaries, and the uriniferous tubule of the metanephric kidney.

Reproductive system

The reproductive system is important both to the individual and to the species. The **gonads** (ovaries and testes) provide a microenvironment for the formation of **gametes** (eggs and sperm). A system of ducts associated with the gonads provides a conduit through which the gametes can reach the site of fertilization. Recall that the germ cells (which give rise to the gametes) arise outside of the embryonic body on the yolk sac; they then migrate into the developing gonads. During fertilization, the gametes pass on the genetic information necessary for the development of the next generation.

The gonads are also the sites of production of the **sex hormones** (or **sex steroids**). **Testosterone**, produced by the Leydig cells of the testes, is necessary for the development of male secondary sex characteristics, such as muscular build and facial hair. **Estrogen**, produced by follicle cells of the ovaries, is necessary for development of female secondary sex characteristics, such as curvaceous body and higher-pitched voice. **Progesterone**, originally coming from the pregnant woman's ovaries and subsequently from the placenta, is necessary for the establishment and maintenance of pregnancy (progesterone means "for pregnancy").

Earlier, we discussed how the adults' reproductive systems lead to the formation of the embryo. Here, we consider components of the *embryo's* developing reproductive system: gonads, ducts, and glands. The development of the external genitalia is considered later in the chapter.

Gonads

The gonads—testes and ovaries—are the only sites in the human body where meiotic (reductional) cell division occurs. All other cell divisions in the body are of the mitotic (conservative) variety. It is here where reduction of chromosome number and generation of genetic diversity occur. It is also here where mutations can have an effect on the next generation. Therefore, when a person receives radiation therapy, the gonads are shielded from the harmful effects of x-rays.

The gonads (both male and female) arise from **splanchnic mesoderm** (the inner layer of the lateral plate mesoderm). During the embryonic period, aggregations of mesenchyme form out of splanchnic mesoderm behind the **peritoneum**, the lining of the peritoneal cavity. As the gonads develop, they are suspended in the peritoneal cavity by a double layer of peritoneum (Figure 18.4). The membranous support for a testis is the **mesorchium**; for an ovary, it is the **mesovarium**.

On about the 38th day after fertilization, germ cells arising extraembryonically on the yolk sac begin to arrive in the rudimentary gonads. If the developing individual has a normal female sex chromosome makeup (XX), the rudimentary gonad normally develops into an ovary, which is a solid type of organ. The stroma of the organ eventually provides follicle cells, and the germ cells eventually provide eggs.

If the developing individual's chromosomes are XY, the rudimentary gonad normally becomes a testis. Stromal cells provide seminiferous tubules (where sperm is produced), with Sertoli cells (somatic "nurse" cells) within them and Leydig cells (somatic testosterone-producing cells) between them. Germ cells again provide gametes—spermatozoa.

The basic mammalian developmental pathway is female. If a Y chromosome is present, then the developmental pathway is shifted to

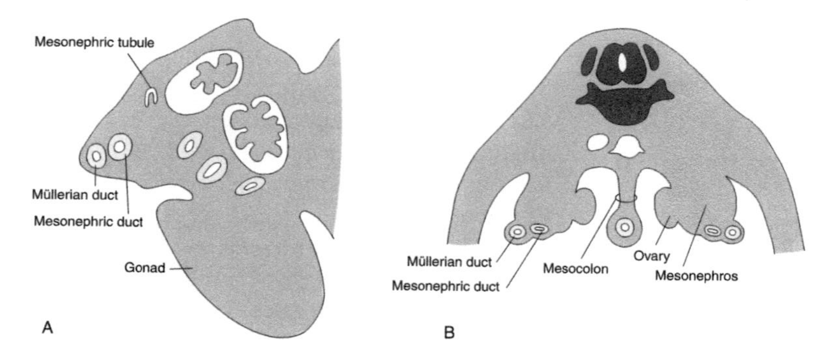

FIGURE 18.4 Origin of the gonads. (A) Diagrammatic transverse section through the cephalic mesonephric level of a 7-week embryo. The gonads (one of which is shown here) arise from splanchnic mesoderm in close proximity to the developing mesonephros. (B) Diagrammatic transverse section through the embryo showing the bilateral origin of the ovaries.

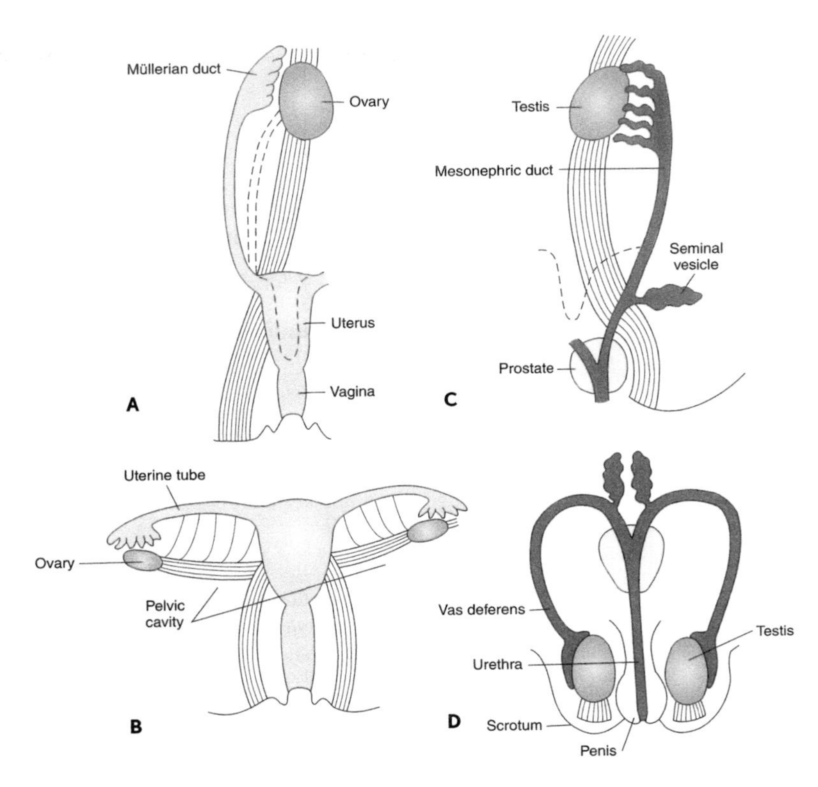

FIGURE 18.5 Descent of the gonads. Schematic diagrams show the descent of the gonads from the sites of their embryologic origins to their final positions. (A, B) The ovaries descend into the pelvic cavity. (C, D) The testes descend into the scrotum. (Redrawn after Rana, M.W. *Human Embryology Made Easy*. Harwood Academic Publishers, 1998: Figure 15.2.)

male development because of the presence of male-determining genes. If these genes on the Y chromosome are somehow lost, an XY individual will develop into a female. Conversely, if one of the X chromosomes (derived from the father) bears the male-determining genes (owing to an anomalous crossover event between X and Y chromosomes in the father's testes), an XX individual will develop into a male. This illustrates the difference between **phenotypic sex** (the sex the individual appears to be) and **genetic sex** (the chromosome makeup).

During fetal life, the testes descend into an outpocketing of the body wall called the **scrotum**, which maintains a temperature slightly lower than the normal body temperature. Normal human sperm development requires this lower temperature. Although less obvious, the ovaries also descend during development, not into an outpocketing of the body wall, but from a relatively higher position in the peritoneal cavity into the **pelvic cavity** (Figure 18.5).

Ducts

The following are the ducts of the male reproductive system:

- Seminiferous tubules
- Rete tubules

- Tubules of the epididymides (plural of epididymis)
- Vasa deferentia
- Ejaculatory ducts
- Urethra

The primary function of the ducts is to convey sperm from the seminiferous tubules, where they are formed, to the back (vault) of the vagina (see Figure 5.2). In addition, during their stay in the duct system, sperm undergo a transformation ("physiologic ripening"). This is required for sperm to become capable of fertilizing an egg (however, also see Chapter 21).

The following are the ducts of the female reproductive system (see Figure 5.9).

- Fallopian tubes
- Uterus
- Vagina

The **fallopian tubes** are the normal site of fertilization (one or the other following a given ovulation). If fertilization occurs, the earliest developmental processes take place in the fallopian tube. Then, implantation, embryonic development, and fetal development take place in the **uterus**. The **vagina** serves as the birth canal for the fetus under the propulsive force of uterine muscle contractions. The vagina is also the female organ of copulation.

The ducts of both males and females arise from intermediate mesoderm, close to the splanchnic mesoderm from which the gonads arise. Two pairs of ducts arise side by side from the intermediate mesoderm: the **mesonephric ducts** and the **müllerian ducts** (see Figure 18.4A). The mesonephric ducts, as previously discussed, contribute to the formation of the mesonephros.

As the mesonephros degenerates in the male, some of it is scavenged for use by the reproductive system. The cephalic portion of the mesonephric duct and some of the mesonephric tubules become parts of the epididymides, whereas much of the rest of the mesonephric ducts becomes the **vasa deferentia** and **ejaculatory ducts**. The origin of the urethra, which is both a urinary and a reproductive duct in males, is discussed later when we consider the origin of the external genitalia.

In the female, the mesonephros degenerates without giving rise to any important structures, although some vestigial (imperfectly formed) structures may persist. Müllerian ducts, on the other hand, persist in the female. Their cephalic ends remain separate, giving rise to the paired fallopian tubes, and their caudal ends fuse, giving rise to the uterus and vagina.

Note that among mammals, the degree of müllerian duct fusion varies. For example, in humans, such fusion produces a single uterus with a single uterine cavity. In mice, less fusion results in a uterus partially divided into two uterine horns; in kangaroos, a lack of müllerian duct fusion results in two vaginas!

Auxiliary sex glands

Males Four kinds of male auxiliary sex glands are found:

- Two seminal vesicles
- The prostate
- Two bulbourethral glands (**Cowper's glands**)
- Multiple glands of Littre

The functions of these glands generally are to provide the liquid portion of the semen (**seminal plasma**), which protects and nourishes the sperm, and to provide secretions for lubrication at the time of copulation (see Chapter 5). The testes may also be considered a type of gland, because they produce and secrete hormones, as do the various endocrine glands (e.g., pituitary and thyroid).

The epithelial (secretory) portion of the **prostate gland** arises from numerous outgrowths of the endodermal lining of the prostatic portion of the urethra, whereas other elements of this gland arise from surrounding mesenchyme. The paired **bulbourethral glands** also arise as endodermal outgrowths of the urethra, at a level between that of the prostate gland and that of the glands of Littre. The paired **seminal vesicles** arise as caudal outgrowths from the caudal ends of the mesonephric ducts (Figure 18.6).

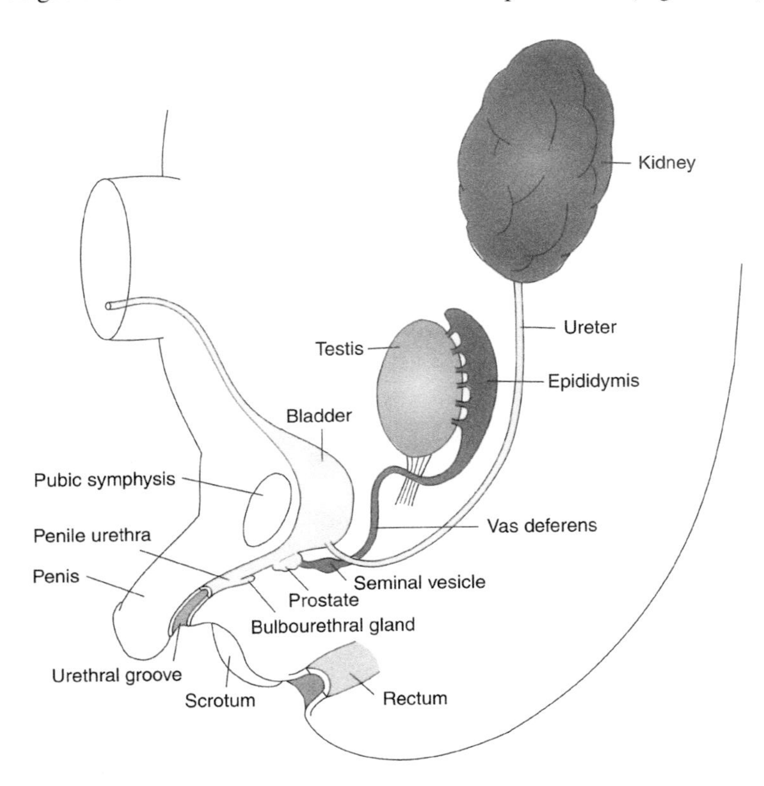

FIGURE 18.6 Origins of the male auxiliary sex glands: Diagram of the reproductive and urinary structures of a 3-month fetus. Note that the testis has not yet descended. Also, note the origins of the prostate gland, a seminal vesicle, and a bulbourethral gland.

The **glands of Littre** arise as numerous outgrowths of the penile urethra along the lumen of the penis.

Females The female auxiliary sex glands are Skene's glands and Bartholin's glands, the functions of which are to provide lubrication at copulation (see Chapter 5). To the extent that they synthesize and produce hormones, the ovaries are also a type of gland.

Outgrowths of the urethra give rise to **Skene's glands**, which are homologous to the male prostate glands. Outgrowths from the urogenital sinus give rise to the **glands of Bartholin**, which are homologous to the male bulbourethral glands.

External genitalia

In the male, the external genitalia consist of the penis and scrotum (see Figure 5.1). The **penis** is the male sex organ and also delivers urine to the outside of the body. The penile urethra (running through the penis) is therefore a true urogenital duct. The **scrotum** contains the male gonads and maintains them (by smooth muscle contraction or relaxation) at a temperature (94°F) lower than that of body temperature (98.6°F/37°C), a necessary condition for normal spermatogenesis.

The female external genitalia, the **vulva**, includes the following (Figure 18.7):

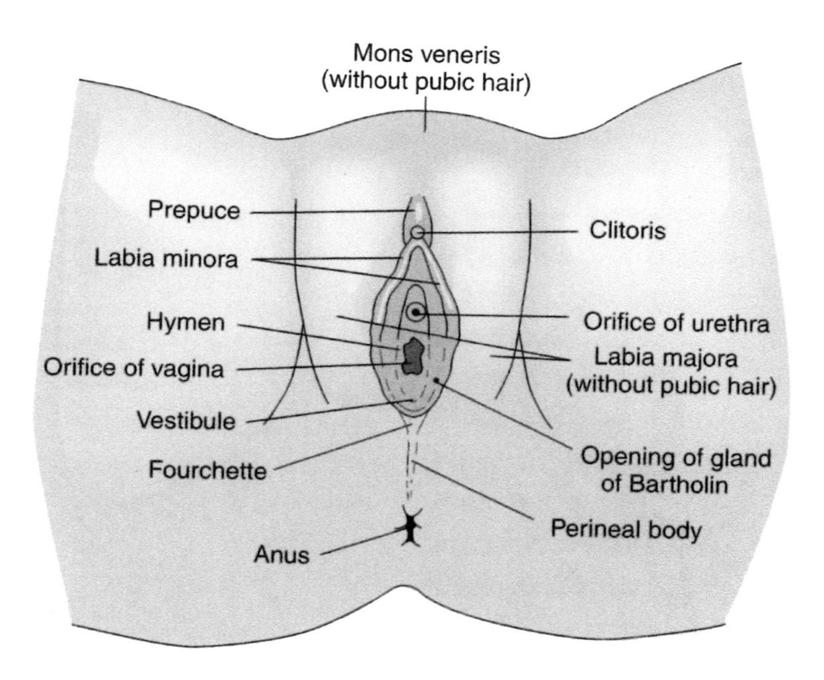

FIGURE 18.7 The vulva.

- Mons veneris
- Clitoris
- Labia majora
- Labia minora
- Fourchette
- Vestibule

The vestibule is an almond-shaped space—bounded cephalically by the mons, laterally by the labia, and caudally by the fourchette—into which the vagina, urethra, Skene's glands, and glands of Bartholin open. Unlike the male urethra, the female urethra is not a urogenital duct, since it functions only to carry urine from the urinary system. The vagina carries the products of the female reproductive system, which include the monthly menstrual fluids and an infant during childbirth. The vagina is also the female organ of intercourse.

As with the gonads, the external genitalia start out in an indifferent condition. That is, the earliest rudiments do not give any indication of the sex of the embryo or early fetus. Initially, in both sexes, the external genitalia consist of a **genital tubercle** and a pair of urogenital folds, the **labioscrotal folds**, surrounding the **urogenital sinus** (Figure 18.8).

By the 12th week of development (early fetal stage), it is possible to determine the sex of the fetus by observing the external genitalia. Remember that the basic pathway of human development is female and when there is no Y chromosome, this destiny is fulfilled. However, when a Y chromosome is present, development is diverted into a male direction, as was the case with the gonads. In either case, the point of departure is the genital tubercle and the labioscrotal folds.

Development of the male

In the development of a male, the genital tubercle gives rise to the penis. As the penis forms (which involves growth of the genital tubercle and fusion of the urogenital folds), a portion of the urogenital sinus is incorporated into the penis as the penile urethra. Fusion of the labioscrotal swellings gives rise to the scrotum, into which the testes descend during the late fetal period (see Figure 18.8).

Development of the female

Growth of the genital tubercle is less marked in the female, giving rise to the diminutive **clitoris**. Thus, the female clitoris is homologous to the penis. The labioscrotal folds do not fuse in the developing female, but rather grow into the labia majora. Thus, the **vestibule** is the persistent urogenital sinus (into which both vagina and urethra empty) (see Figure 18.8). As indicated by the name of the folds ("labioscrotal"), the labia majora are homologous to the male scrotum. The penile urethra is homologous to the vestibule, because both are derived from the urogenital sinus.

It may be useful to recall that the urogenital sinus arose from the **cloaca** when it was divided by the **urogenital septum** into the urogenital sinus and the anal canal (see Figure 16.6).

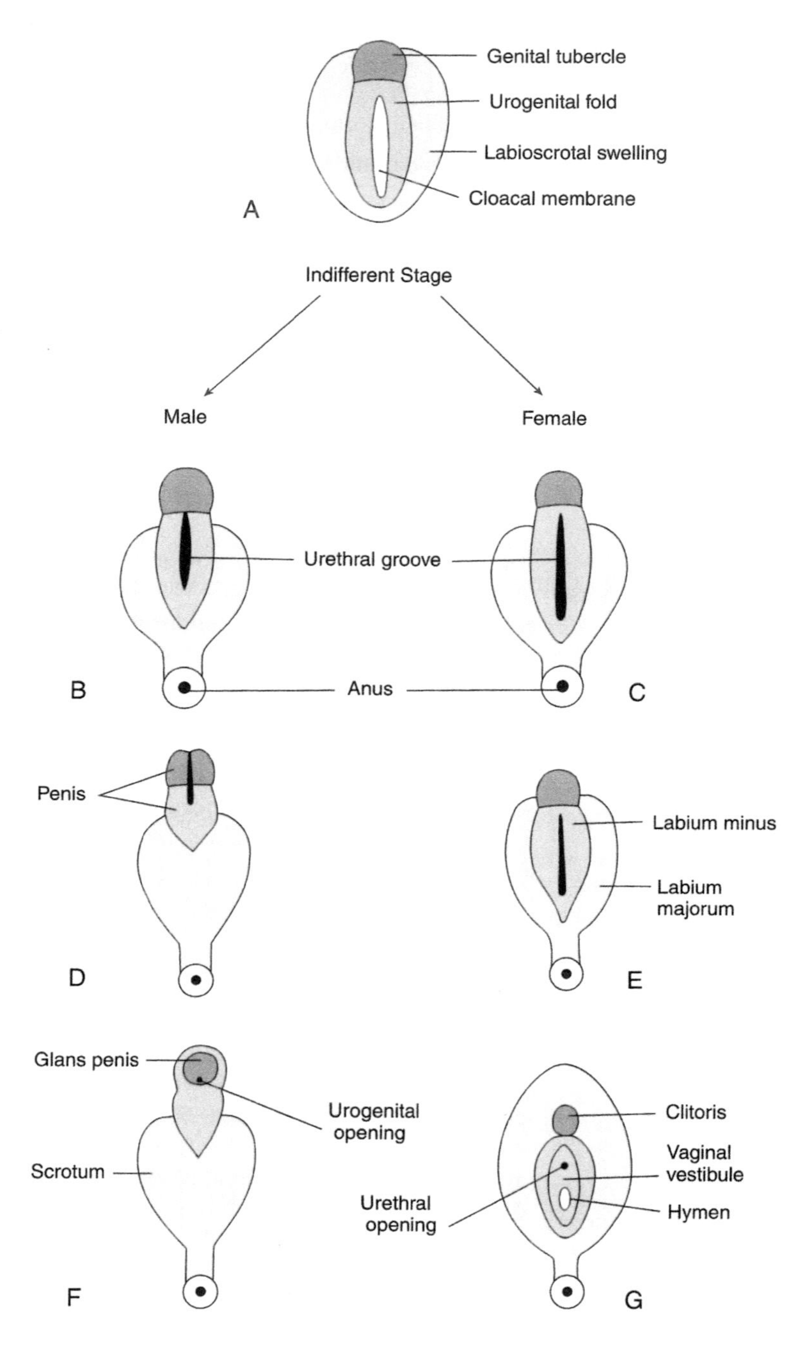

FIGURE 18.8 Development of the external genitalia: Diagrams showing the development of the male and female external genitalia from an initially indifferent state. (A) Four to 7 weeks—the indifferent state. (B, D, and F) Nine to 12 weeks—the development of the penis and scrotum from the indifferent state. (C, E, and G) Nine to 12 weeks—the development of the vulva from the indifferent state. (Redrawn after Moore, K.L. and T.V.N, Persaud. *Before We Are Born*, 4th edition. WB Saunders, Philadelphia, 1993, 219.)

Study questions

Urinary and reproductive systems

1. Why are the urinary and reproductive systems considered together? Both systems arise from what?
2. What are the three kinds of kidneys that appear during development? What features do they have in common?
3. From what do the metanephric diverticula arise and to what do they give rise?
4. From what does the urinary bladder arise?
5. Where do the gonads arise and what will come to suspend each kind of gonad in the peritoneal cavity?
6. What is the basic mammalian developmental pathway regarding gender? What will a Y chromosome do to this pathway?
7. From what do parts of the epididymis, vasa deferentia, and ejaculatory ducts arise? In the female, to what do the müllerian ducts give rise?
8. Skene's glands and the glands of Bartholin are homologous to what in the male?

External genitalia

9. What are the external genitalia of the male and what are their functions? What are the external genitalia of the female and what are their functions?
10. When is it possible to determine the sex of the fetus by observation of its external genitalia?
11. During male development, the genital tubercle and the labioscrotal folds give rise to what? From what does the penile urethra arise? During female development, to what do the genital tubercle and the labioscrotal folds give rise? From what does the vestibule arise?
12. What in the female is homologous to the following in the male: penis, scrotum, penile urethra?

Critical thinking

1. Explain the statement that the developing reproductive system scavenges parts from the urinary system.
2. What is an interesting parallel between kidney development in ontogeny and phylogeny?
3. Describe the relationship between the circulatory system and the metanephric kidney.

PART III

Society and human development

Our individual development, both before and after birth, is primarily the concern of our parents and perhaps our extended family. But from the beginning of our education to the collection of our last social security check, we are a concern of society. Indeed, the interaction between developing humans and society is even more fundamental than this. Because the treatment of birth defects (Chapter 19) can be so financially debilitating for an individual family, society has a role to play in their prevention and treatment.

We Americans no longer live on a frontier in which having many children is necessary both to provide labor and to offset infant and childhood mortality. In industrialized nations, this has led to the increased use of birth control (Chapter 20). On the other hand, reproductive technologies hold the promise of biological parenthood to couples who are otherwise infertile (Chapter 21).

Societies are always concerned with communicable diseases, and a society's vitality and survival depend on this concern and subsequent action. An important category of communicable diseases is sexually transmitted diseases (Chapter 22).

Another area in which human development affects society is in our very language. Many terms dealing with development, such as "surrogate mother" and "test tube baby," have become part of the lexicon of the average citizen (Chapter 23).

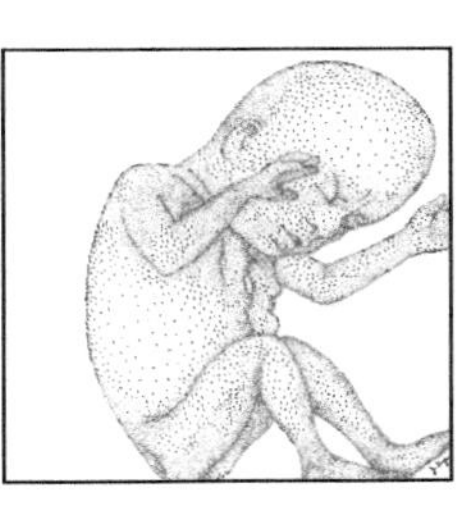

CHAPTER NINETEEN

Birth defects

CHAPTER OBJECTIVES

After studying this chapter, you should be able to:

1. Give examples of birth defects due to chromosomal abnormalities of both sex chromosomes and autosomes.
2. Give examples of birth defects due to abnormalities of both chromosome number and chromosome structure.
3. Give examples of nonchromosomal birth defects.
4. Discuss teratology and the three general categories of teratogens.
5. Explain the concept of "critical periods in development" and the practical consequences of their existence.
6. Discuss the various methods of detection of birth defects and what might cause a woman to choose one method over the others.
7. Give the earliest time in pregnancy when each of the various methods of detection of birth defects can be used.

The degree of severity of birth defects ranges from relatively minor to absolutely devastating. Some persons with birth defects require institutionalization. The expression of a birth defect may be as subtle as the lack of a single enzyme, as in Tay–Sachs disease. An infant might be born with **syndactyly** (two fingers fused together), which is easily corrected by minor surgery (Figure 19.1). At the other extreme, an infant born with **anencephaly** lacks the upper portion of the brain, which results in death (Figure 19.2). See Appendix A and Appendix B for lists of common birth defects.

The causes of birth defects

Perhaps because having children is such a momentous event, a good deal of superstition has traditionally surrounded the appearance of a birth

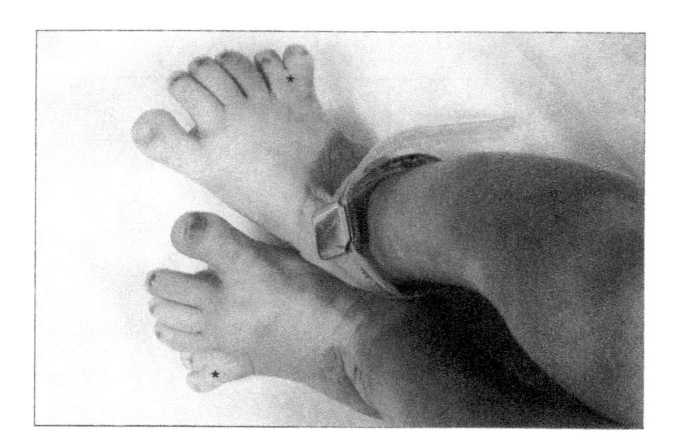

FIGURE 19.1 Photograph of an infant with syndactyly in which the toes are fused together. This is often corrected by surgery. (Courtesy of Jack Wolk, MD, The Danbury Hospital.)

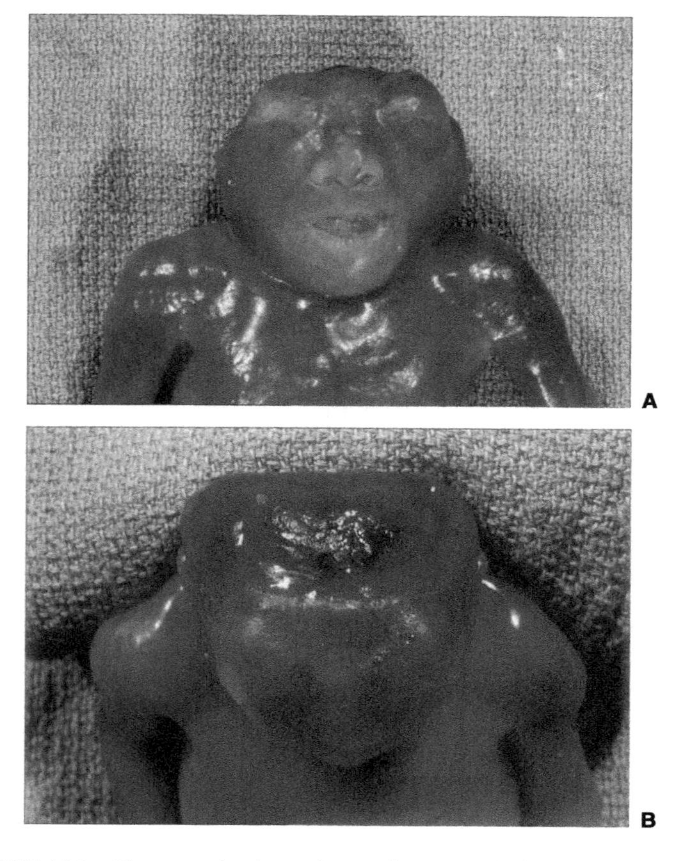

FIGURE 19.2 Photograph of an infant with anencephaly (most of the brain is missing). This birth defect is not consistent with life after birth (A) Face-on view. (B) View of the top of the head. (Courtesy of Jack Wolk, MD, The Danbury Hospital.)

defect. Today, biology can link almost all birth defects to a causative factor or factors. A birth defect may be caused by a tiny (point) mutation in DNA inherited from the parents or by a chromosomal defect of some sort. Birth defects may also be caused by chemicals, biological agents, and physical forces such as radiation.

Fortunately, modern medicine has developed methods of detecting birth defects *in utero*. Although parents often elect abortion in cases of severe deformities, modern medicine is able to treat some birth defects, either after birth or (more recently) by the application of surgery or medicine to the fetus while it is still in the womb.

Chromosomal abnormalities

Most *potential* birth defects never become birth defects as such because nature screens out most defective conceptuses before birth. This results in what were once more commonly referred to as **miscarriages**, but now are generally referred to as spontaneous abortions. Examination of spontaneous abortuses has revealed that many have **chromosomal abnormalities**. The earlier in the pregnancy the miscarriage occurs, the more likely it is accompanied by a chromosomal abnormality (Table 19.1).

It should not be surprising that chromosomal abnormalities result in miscarriages or birth defects, because the chromosomes carry the genetic material passed from one generation to the next. Defective genes result in defective development. In response to this knowledge, modern hospitals have **cytogenetics** departments responsible for analysis of tissue or fluid samples for chromosomal defects. Cytogenetics came into being as a distinct biological discipline at the dawn of the twentieth century by the

Table 19.1 Percentage of spontaneous abortions with chromosomal abnormalities at different times during pregnancy

Stage of pregnancy	Percent with chromosomal abnormalities
1st trimester	50–60
2nd trimester	34
Stillborn	5
Live-born	0.5

Source: Data tabulated from Simpson, J.L. et al. *Genetics in Obstetrics and Gynecology*. Grune & Stratton, New York, 1982.

Note: Spontaneous abortions eliminate most of the conceptuses with chromosomal abnormalities, so that only 0.5% of live-born infants have chromosomal abnormalities. On the other hand, considering the number of humans born every day, this is still a considerable number of people born with chromosome abnormalities.

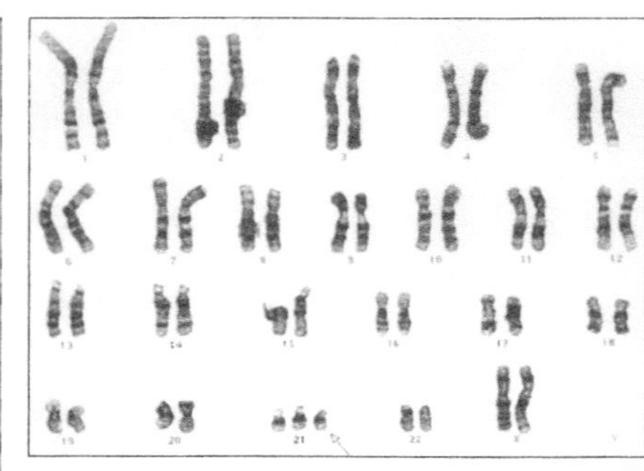

FIGURE 19.3 (A) Photograph of a young girl with Down's syndrome. (A, M. Coleman/Visuals Unlimited.) (B) Karyotype of a patient with Down's syndrome; note the extra number 21 chromosome. (B, courtesy of Jacqueline Burns, PhD, The Danbury Hospital.)

marriage of **genetics** (the study of heredity) and **cytology** (the study of cells). It was not until 1959 that a birth defect (**Down's syndrome**) was shown to have an underlying chromosomal abnormality (**trisomy 21**) as its cause (Figure 19.3).

Chromosomal abnormalities may affect either the sex chromosomes or the autosomes (all chromosomes other than sex chromosomes). In either case, the involved chromosomes may be defective either in number or in structure.

Sex chromosomes

Defects involving sex chromosomes are interesting because the individual carrying the abnormality may be phenotypically (outwardly) either a male or a female and the sex chromosome involved may be either an X or a Y chromosome. Some males are known to have the sex chromosome makeups XYY or XXY, the latter of which gives rise to **Klinefelter's syndrome** (Figure 19.4). People with Klinefelter's syndrome have a male phenotype and small testes lacking sperm, are sterile, and are often mildly mentally retarded. Such men are often tall.

Some females are known to have XXX or XXXX chromosome makeups. Such females are called **metafemales**. More interesting perhaps are those females conceived with an XO sex chromosome makeup. This shorthand indicates only one sex chromosome—the gene-rich X chromosome. The consequences of the XO is abnormality ranging from early spontaneous abortion to essentially normal females, who exhibit **Turner's syndrome** to a greater or lesser extent (Figure 19.5). People with Turner's syndrome are female with short stature, often exhibiting a webbed neck and the absence of sexual maturation. Such women are sterile. The range of consequences might be due to the number and type

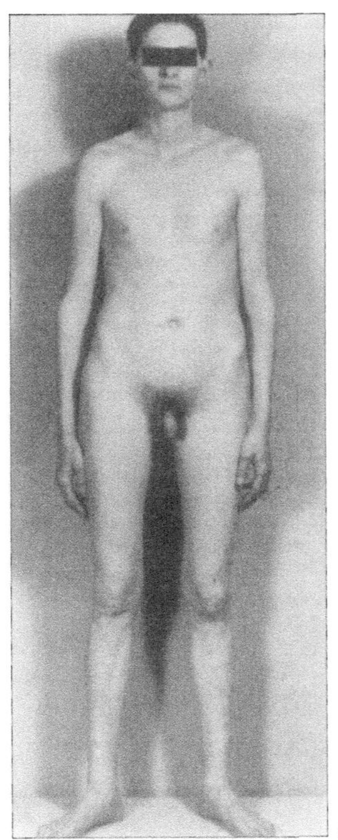

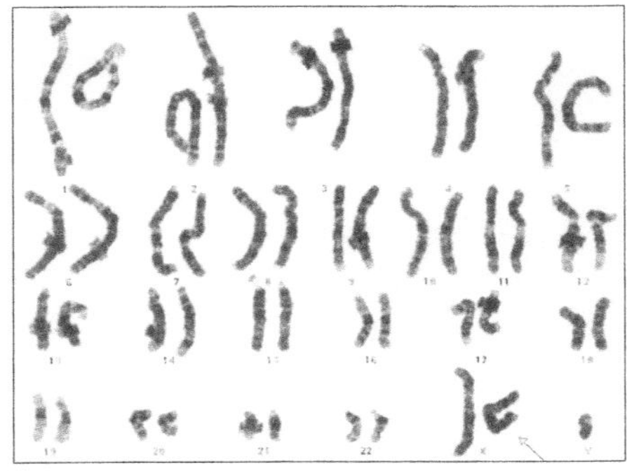

FIGURE 19.4 (A) Photograph of an adult with Klinefelter's syndrome. (B) Karyotype of a patient with Klinefelter's syndrome; note the three sex chromosomes. (A, Moore, K.L. and T.V.N. Persaud. *Before We Are Born*, 4th edition. 1993: Figure 9.7A, p. 125, with permission from WB Saunders; B, courtesy of Jacqueline Burns, PhD, The Danbury Hospital.)

of harmful genes carried on the single X chromosome, which no longer can be masked by good copies of such genes on a second X chromosome. YO is not found because it is not viable, even into the earliest stages of human development.

Autosomes

Normally, the cells of the body (excluding gametes) have two copies of each autosomal chromosome, numbered from 1 to 22. **Monosomy** is the presence of only a single copy of an autosome—and **trisomy** is three copies of a particular autosome. The best-known autosomal chromosome defect is trisomy of autosome 21 or **trisomy 21**. This is the causative factor of **Down's syndrome**, a condition affecting a significant number of hospitalized mental patients. People with Down's syndrome may be either male or female and are mentally retarded. This is a semilethal condition, and many of those affected die as children or young adults. Abnormalities of the head and face, skin, abdomen, chest, and limbs often are consequences of Down's syndrome. Other possible consequences are heart defects, leukemia, and, generally, a poor immune system. Down's

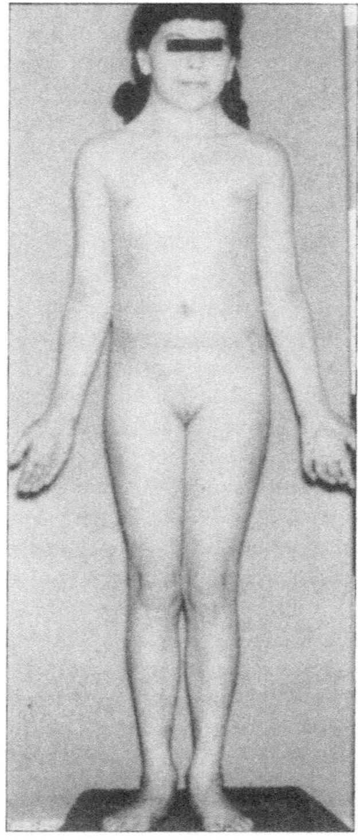

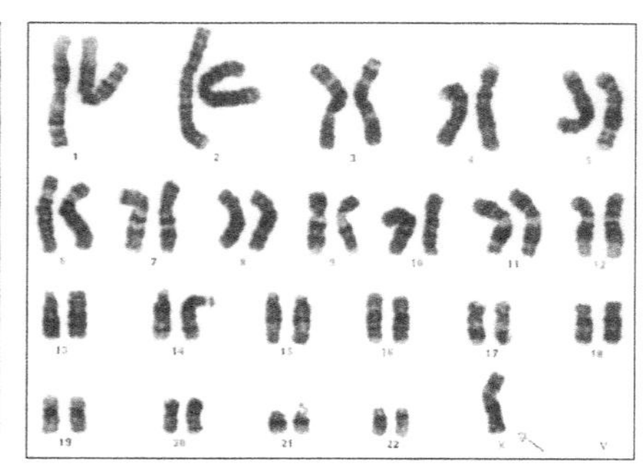

FIGURE 19.5 (A) Photograph of a girl with Turner's syndrome. (B) Karyotype of a patient with Turner's syndrome; note the presence of only one sex chromosome. (A, Moore, K.L. and T.V.N. Persaud. *Before We Are Born*, 4th edition, 1993: Figure 9.3, p. 121, with permission from WB Saunders; B, courtesy of Jacqueline Burns, PhD, The Danbury Hospital.)

syndrome has an enormous impact on the families involved and on society as a whole.

The only other two trisomies that occur with any frequency in live births are trisomy 13 (Patau's syndrome) and trisomy 18 (Edward's syndrome), but their consequences are dramatic and lethal, causing death early in postnatal life. Among spontaneous abortuses, trisomies of various autosomes are common and the leading cause of first-trimester miscarriage (Tables 19.2 and 19.3). Although found among spontaneous abortuses, autosomal monosomies are essentially nonexistent among live-born infants. Having only one copy of the genes found on an autosome is not sufficient to support postnatal life, presumably because of unmasked, deleterious copies of genes.

Chromosome number

All the previous examples involve abnormalities of chromosome numbers, but they have involved an abnormal number of a *single* chromosome, a condition called **aneuploidy**. How do aneuploid chromosomal abnormalities come about? One possibility is abnormal cell division, in which the two daughter cells do not receive equal numbers

Table 19.2 Types of chromosomal abnormalities in spontaneous abortions

Anomaly	Percent of abortions (N = 1863)
Trisomy	52 (see Table 19.3)
45,X	18
Triploidy	17
Tetraploidy	6
Other (mainly translocations)	7

Source: Modified from Thurman, E. *Human Chromosomes: Structure, Behavior, Effects*, 2nd edition. Springer-Verlag, New York, 1986.

Table 19.3 Percentage of spontaneous abortions in trisomy

Chromosome involved in trisomy	Percent of spontaneous abortions (N = 669)
1	0.0
2	4.9
3	0.6
4	2.5
5	0.2
6	0.5
7	4.0
8	3.9
9	2.7
10	2.0
11	0.3
12	1.0
13	4.6
14	4.6
15	7.7
16	32.3
17	0.6
18	5.1
19	0.2
20	2.7
21	9.4
22	10.2

Source: Modified from Thurman, E. *Human Chromosomes: Structure, Behavior, Effects*, 2nd edition. Springer-Verlag, New York, 1986.

Note: Except for chromosome 1, trisomies of all of the autosomes are found in spontaneous abortuses. Perhaps trisomy 1 causes spontaneous abortion so early in development that it is not detected.

of chromosomes. However, no one is sure what causes abnormal cell division. If the two chromatids during anaphase of mitosis or during anaphase II of meiosis or the two chromosomes of anaphase I of meiosis do not go to opposite poles of the spindle, we have the possibility of both monosomic and trisomic conditions (Figure 19.6). If we knew more about normal chromosome movement during cell division, we would have a better grasp on what occasionally goes wrong.

Human chromosomal abnormalities can also involve entire extra copies of the *normal sets* of chromosomes, a condition called **polyploidy**—more than the normal number of sets of chromosomes. Recall that a gamete has a single set of 23 chromosomes, the **haploid** number, and all the other cells in your body have two sets of 23 (the **diploid** number). A cell with three sets is said to be **triploid** and one with four sets is said to be **tetraploid**. Some spontaneous abortuses are triploid or tetraploid, but these conditions do not seem to be compatible with live births.

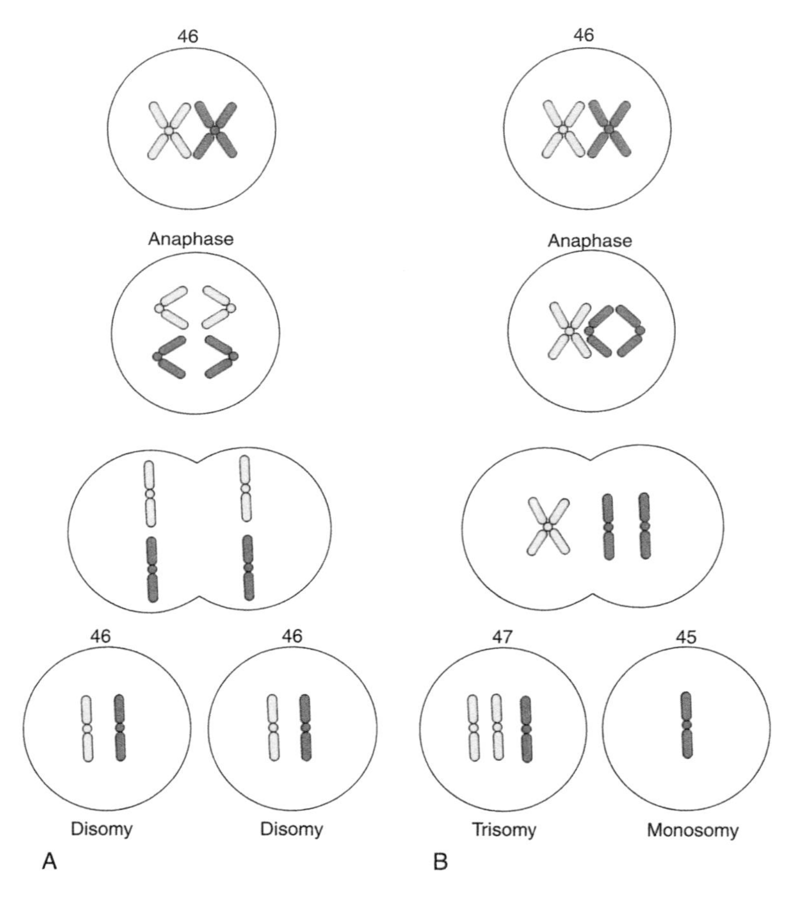

FIGURE 19.6 Origin of aneuploidy. (A) Normal cell division. Each daughter cell receives an equal number of chromosomes. (B) Abnormal cell division. One daughter cell is missing a chromosome (monosomy), and the other daughter cell has an extra chromosome (trisomy). (Redrawn after Rana MW: *Human Embryology Made Easy*. Harwood Academic Publishers, 1998: Figure 1.3.)

Reported cases of live-born human **polyploids** seem to have been not pure polyploids, but **mosaics**. A mosaic is an individual who has more than one type of cell, as far as chromosome number is concerned, making up his or her body. Some of these cells are polyploid and some are diploid, with the diploid cells somehow being able to "rescue" the polyploid cells.

What is the origin of polyploidy? A couple of possibilities have been proposed. One is **polyspermy**, a condition in which the egg is fertilized by more than a single sperm. If one extra sperm is involved, the zygote is triploid. If two extra sperm are involved, a tetraploid zygote results (Figure 19.7). Another possibility is that if, during oogenesis (formation of the egg), the oocyte does not undergo normal division and give off the polar bodies with their contained chromosomes, then the retained polar body will provide an extra set of chromosomes (Figure 19.8).

Chromosome structure

An individual may have the correct number of chromosomes, but they may not be the correct chromosomes. The chromosomes may instead have **structural abnormalities**. For example, a birth defect known as ***cri du chat*** (cry of the cat) syndrome is caused by a defect on chromosome 5 in which a small piece is missing from its short arm (Figure 19.9). Babies with *cri du chat* do not sound like a human infant when crying, but make a sound more akin to a mewing cat. More significantly, these babies are mentally retarded and short-lived.

What causes structurally abnormal chromosomes? One possibility is **chromosome breakage**, which is a common action of a teratogen (see following section). Both x-rays and viruses can cause chromosome breakage. At this point, it should be interjected that a piece of chromosome

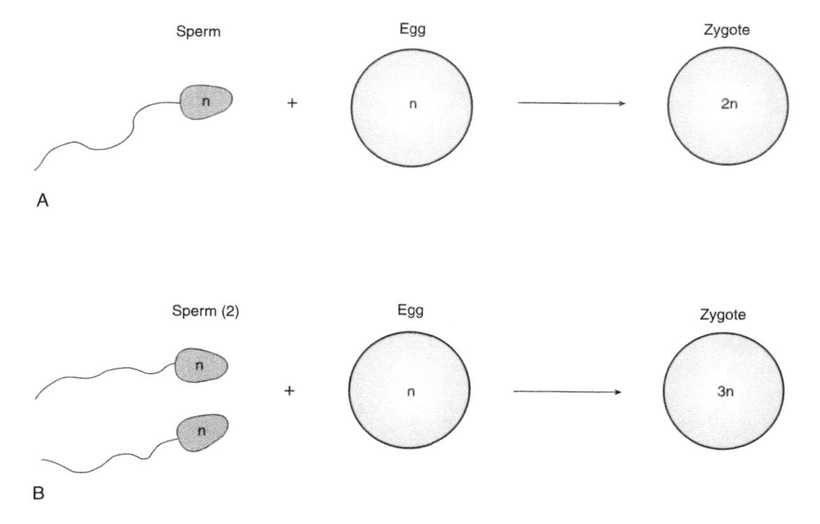

FIGURE 19.7 Origin of polyploidy by polyspermy. If an egg is fertilized by two spermatozoa, the resulting zygote will be triploid. (A) Normal fertilization, in which one spermatozoon fertilizes one egg, forming a diploid zygote. (B) Polyspermy, in which two spermatozoa fertilize one egg, forming a triploid zygote. n = haploid, 2n = diploid, 3n = triploid.

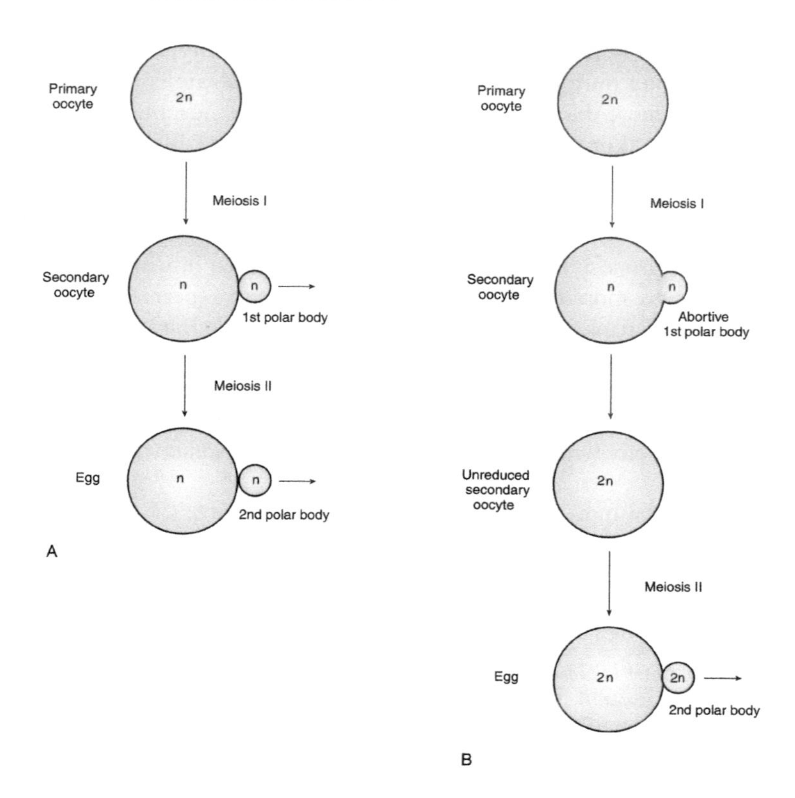

FIGURE 19.8 Origin of polyploidy by polar body restitution. (A) Normal oogenesis, resulting in a haploid egg. (B) Oogenesis with restitution nucleus formation in the abortive formation of the first polar body, resulting in a diploid egg. Fertilization of such an egg would result in a triploid zygote. n = haploid, 2n = diploid, 3n = triploid.

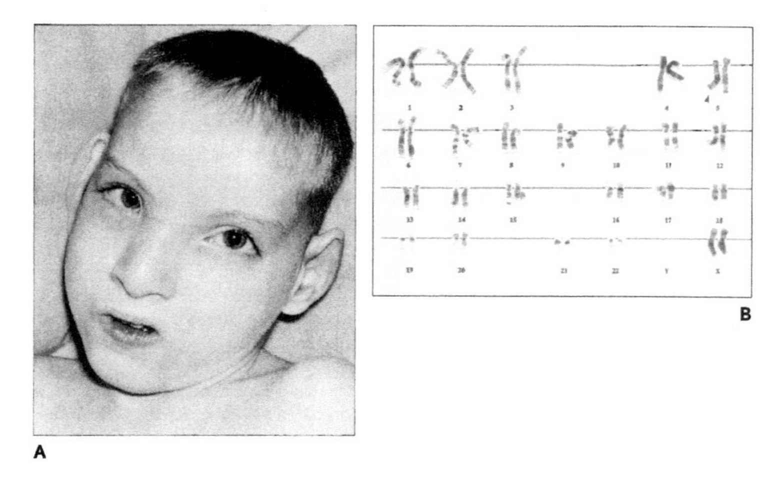

FIGURE 19.9 (A) Photograph of a child with ***cri du chat*** syndrome. (B) Karyotype of a patient with ***cri du chat*** syndrome; note the normal number of chromosomes, but a deletion of the short arm of chromosome 5. (A, Gardner, E.J. et al. *Principles of Genetics,* 8th ed. John Wiley & Sons, New York, 1991:Figure 18.18, p. 502; B, courtesy of Teresa L. Yang-Feng, PhD, Yale University School of Medicine.)

without an attached centromere will not migrate to a spindle pole and is lost from the nuclei of the daughter cells.

Nonchromosomal birth defects

Not all birth defects involve chromosome defects that are visible through the microscope. More sensitive tests are required to detect gene mutations. Birth defects due to gene mutations (as opposed to chromosome mutations such as aneuploidy and polyploidy) do not involve chromosomal abnormalities in the usual sense. Birth defects due to gene mutations include enzyme deficiencies (such as Tay–Sachs disease), hemophilia, albinism, and red-green color-blindness.

Teratology is the study of abnormal development. The agents that cause abnormal development (**teratogens**) may be chemical, physical, or biological. Perhaps the best-known chemical teratogen is thalidomide, which, when administered to pregnant women at a particular time, resulted in **phocomelia** (severe limb abnormalities) in their offspring (Figure 19.10). The use of the antibiotic tetracycline can cause stained

FIGURE 19.10 Photograph of a child with phocomelia. This birth defect is characteristic of children born to mothers who took thalidomide during the critical period for limb development in the first trimester (see Figure 19.12). (From Lewis, R. *Beginnings of Life*. William C. Brown, Dubuque, IA, 1992: Figure 10.6, p. 203, with permission from The McGraw-Hill Companies.)

teeth. Although the effects are much less dramatic than thalidomide, tetracycline should be avoided by expectant mothers.

Radiation is a physical agent that may be teratogenic. X-rays are a common example. Biological teratogens include viruses. For example, serious harm can result from the exposure of a pregnant woman to German measles (rubella) virus. Her infant's eyes may be affected by chorioretinitis and cataracts.

Other birth defects are of unknown origin. Researchers have not discovered genetic or environmental causes for **spina bifida** (a defect in the closure of the vertebral column) (Figure 19.11), **hydronephrosis** (dilatation of the pelvis and calyces of the kidney secondary to urinary tract obstruction), and **hydrocephaly** (an abnormal increase in the

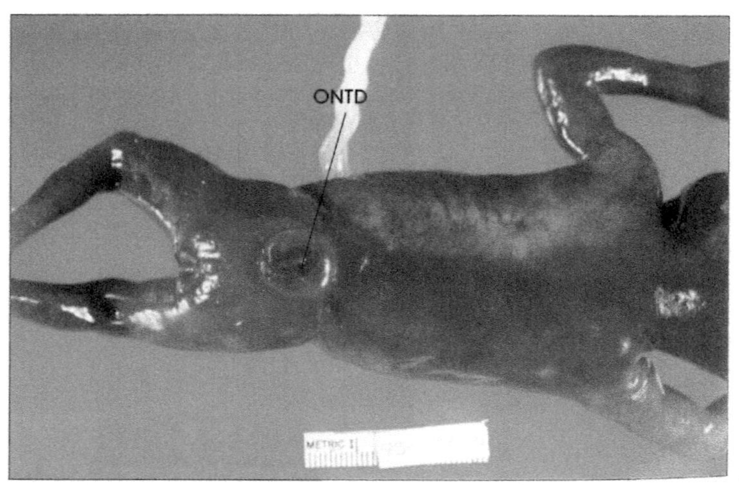

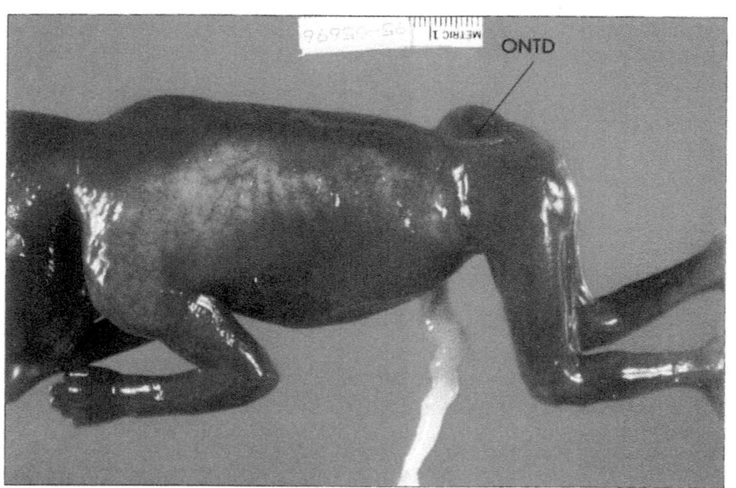

FIGURE 19.11 Photograph of a child with spina bifida, an open neural tube defect. (A) Surface view. (B) Side view. (Courtesy of Jack Wolk, MD, The Danbury Hospital.)

volume of cerebrospinal fluid within the skull secondary to obstruction in the ventricles of the brain, that is, in the aqueduct of Sylvius). The likelihood of spina bifida occurring may be significantly decreased by the adequate intake of folic acid by the expectant mother.

Timing of birth defects

An important concept in the study of birth defects is that of "critical periods in development" (Figure 19.12). In general, the most critical period during development is from the beginning of the third week through the end of the eighth week. Before the third week, a "teratologic insult" will have an all-or-none effect: either the embryo will be killed outright or the embryo will be able to completely compensate for the damage done. It is during the rest of the embryonic period, when all the organ systems are forming, that teratogens have their most dramatic effects. The fetal period is not exempt from the effects of teratogens. However, the abnormalities induced during the fetal period generally are less dramatic.

Unfortunately, the time when the pregnant woman is uncertain that she is pregnant—between her first and second missed periods, the second through sixth weeks of embryonic development—is the time when her conceptus is most at risk for the severe effects of teratogens. Because alcohol, nicotine, and some drugs are known to cause birth defects, behavioral and occupational exposure to these agents by women of reproductive age is of great concern to both the individuals directly involved and to society at large.

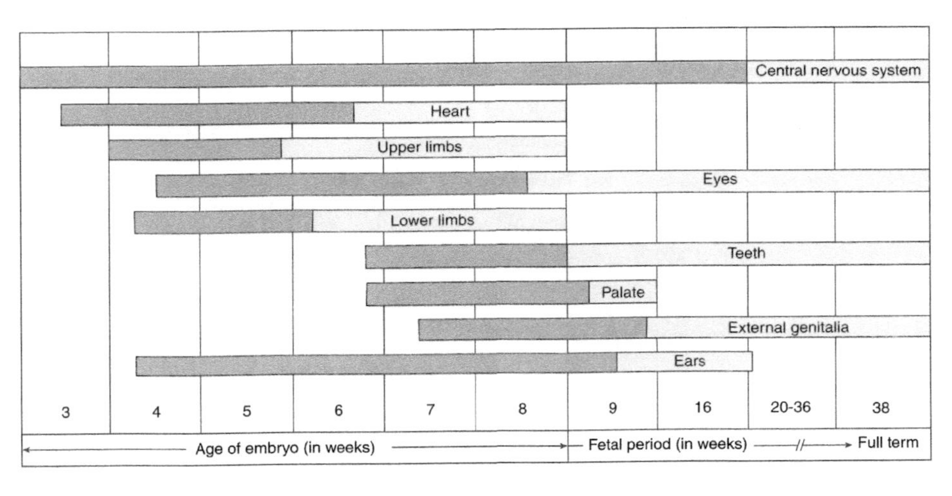

FIGURE 19.12 An important concept in trying to understand birth defects is that of "critical periods in development." Different organs and structures are most vulnerable to teratogens during specific periods in development. Most of the embryonic period, 3 weeks to 8 weeks, is very critical for most organs and structures because this is the time when most organs and structures are forming. The darker color indicates periods more sensitive to birth defect-causing agents (teratogens). (Adapted from Moore, K.L. and T.V.N. Persaud. *Before We Are Born*, 4th edition. WB Saunders, Philadelphia, 1993.)

Detection of birth defects

For most of human history, the developing baby has been unseen and unattainable, as if developing behind a shroud. One of the milestones of the twentieth century has been the ability to see and get at the developing baby without interrupting the pregnancy or causing birth defects.

Ultrasonography

Down through the ages, one of the biggest questions at childbirth, sometimes carrying great political importance, has been, "Is it a boy or a girl?" Many parents now know the answer to this question before birth, owing to the increasing use of **ultrasound**. In fact, the earliest photo in a baby book may very well be a prenatal photograph (**sonogram**) (Figure 19.13).

Of course, ultrasound was not created simply to answer the "boy or girl" question. Examination of the body with high-frequency sound waves has been a boon to medicine, especially to obstetrics. A wealth of useful obstetric information is available from watching a "movie" of the developing baby. Major anatomic birth defects such as anencephaly and spina bifida may be detected. The size, and therefore the growth rate, of the fetus can be charted to detect whether intrauterine growth retardation is occurring. The condition of the placenta also may be evaluated.

Fetoscopy

Sometimes greater detail than can be provided by ultrasonography is needed for detection of birth defects. Ultrasound images may not show minor anatomic defects. Moreover, if the obstetrician wishes to obtain a sample of the baby's blood from the umbilical cord, ultrasound may not be precise enough to guide the needle. **Fetoscopy** (direct viewing of the fetus) can accomplish both of these feats. However, it is a much more invasive and risky procedure, because an observation scope and fiber optics must be inserted into the amniotic cavity (Figure 19.14).

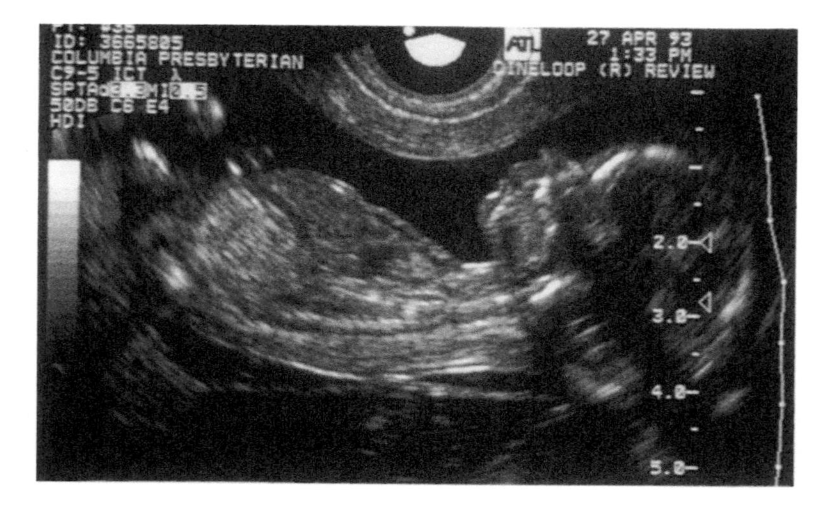

FIGURE 19.13 Photograph of a sonogram. (Courtesy of Gerard Foye, MD, The Danbury Hospital.)

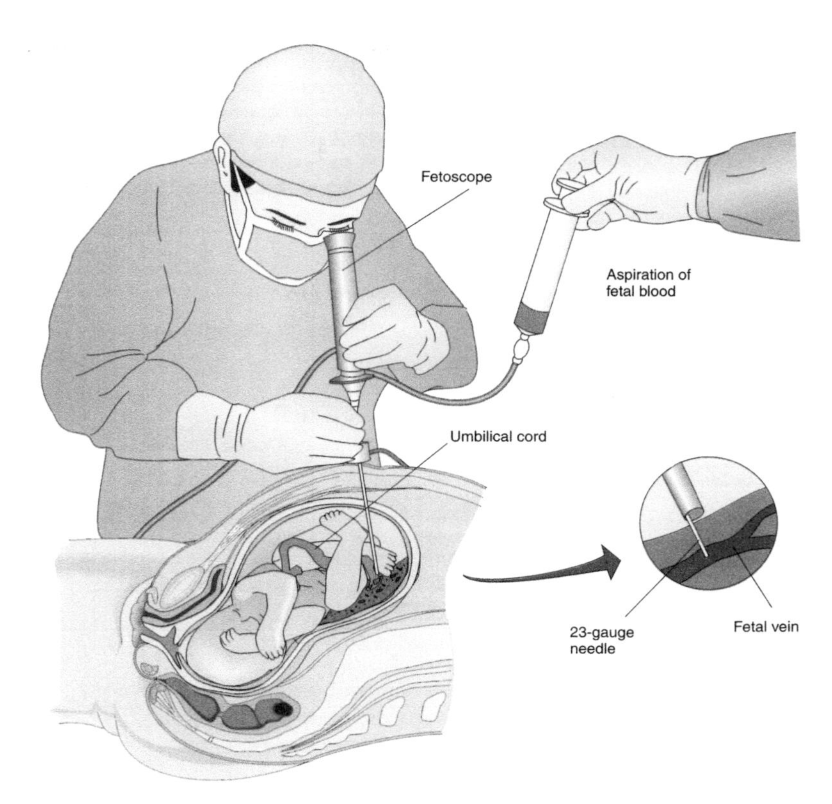

FIGURE 19.14 This drawing shows what fetoscopy involves. Using fiber optics and a fetoscope, the obstetrician is able to directly observe the fetus in its amniotic cavity.

Amniocentesis

Although some kinds of birth defects are morphologic (involving defects in structure), many are much more subtle, ranging from enzyme deficiencies (inborn errors of metabolism) to mental aberrations. These more subtle defects are not revealed by either ultrasonography or fetoscopy, and their detection requires obtaining physical specimens of the fetus. These specimens—cells and fluids—are obtained by the invasive procedure called **amniocentesis**. Amniocentesis involves placing a **cannula** (tube) through the abdominal wall and the uterine wall of the mother, thereby gaining access to the amniotic cavity. A hypodermic needle may then be passed through the cannula and a sample of **amniotic fluid** withdrawn (Figure 19.15).

Amniotic fluid consists mostly of fluid and fetal cells. The fluid portion may be analyzed by biochemical methods and will reveal, for example, the presence of **alpha-fetoprotein** (AFP). The presence of this protein in amniotic fluid, normally restricted to the central nervous system, indicates the probability of an **open neural tube defect** (**ONTD**), such as anencephaly or spina bifida.

Cells in amniotic fluid may be maintained in cell culture under conditions that will allow them to grow and divide. These cells

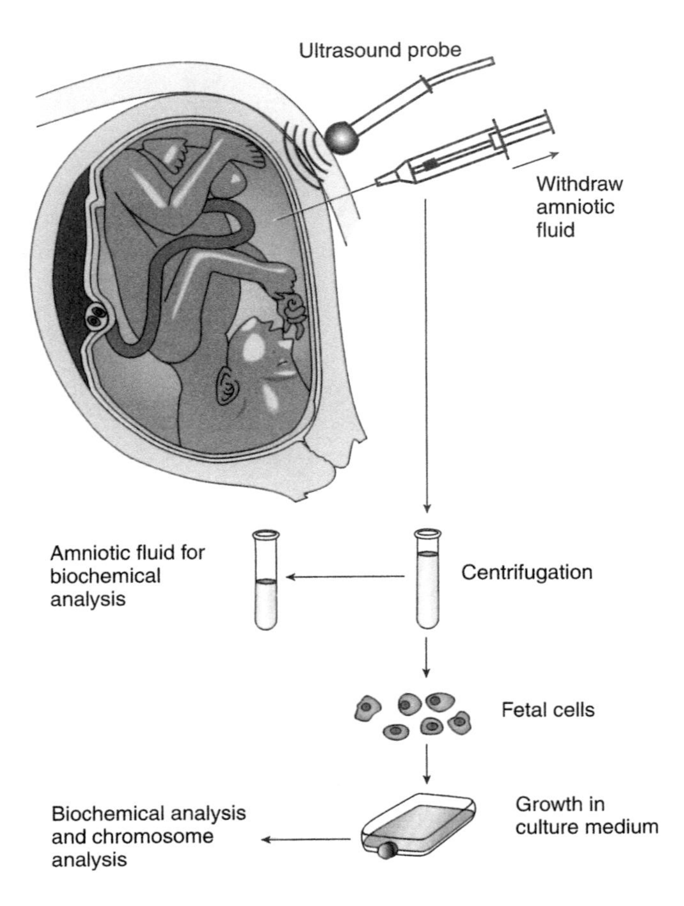

FIGURE 19.15 Diagram illustrating the technique of amniocentesis. The amniotic fluid obtained by this technique contains both cells and liquid for prenatal diagnosis.

(**amniocytes**) are fetal cells primarily shed from fetal skin. Once dividing fetal cells are obtained, the cytogeneticist can obtain fetal chromosomes and analyze them for a variety of fetal chromosomal abnormalities. Down's syndrome, Turner's syndrome, Klinefelter's syndrome, *cri du chat* syndrome, and the genetic sex of the fetus are only a few examples of what may be determined by chromosomal analysis of fetal cells.

Numerous biochemical abnormalities may be determined by analysis of amniotic fluid. Tay–Sachs disease, Hurler's disease, phenylketonuria (PKU), and Lesch–Nyhan disease are but a few examples (Table 19.4). The amniotic fluid may also be tested for its content of lecithin and sphingomyelin, the ratio of which is an indicator of fetal lung maturity. A premature delivery is postponed as long as possible to allow the fetal lungs to develop to a stage compatible with extrauterine survival.

Table 19.4 Some metabolic diseases that can be diagnosed prenatally from chorionic villi cells or cultured amniotic fluid cells

Anencephaly	Muscular dystrophy, X-linked
Cystic fibrosis	Niemann–Pick disease
Diabetes mellitus	Phenylketonuria
Erythroblastosis fetalis	Progeria
Gaucher's disease	Spina bifida
Lesch–Nyhan syndrome	Tay–Sachs disease
Marian's syndrome	Xeroderma pigmentosum
Hurler's syndrome	

Source: Adapted from Stine, G.J. *The New Human Genetics*. WC Brown Publishers, Dubuque, IA, 1989.

Chorionic villus sampling

Early during the embryonic period, the surface of the **chorionic vesicle** is covered by fingerlike extensions called **chorionic villi**. These contain dividing fetal cells, which may be obtained by a nonsurgical, transvaginal procedure. With the guidance of ultrasound, a cannula is passed through the vagina, through the cervical canal, into the uterine cavity, and close to the villi of the chorionic vesicle. An instrument passed through the cannula is used to suction off (aspirate) some of the villi (Figure 19.16), from which dividing fetal cells, and therefore fetal chromosomes, may be obtained.

Of the prenatal diagnostic techniques considered here, chorionic villus sampling (CVS) is the most recent, most widely used, technique. Its advantage is that it can be done much earlier in pregnancy than amniocentesis, which cannot be done until the beginning of the second trimester with the accumulation of sufficient amniotic fluid. On the other

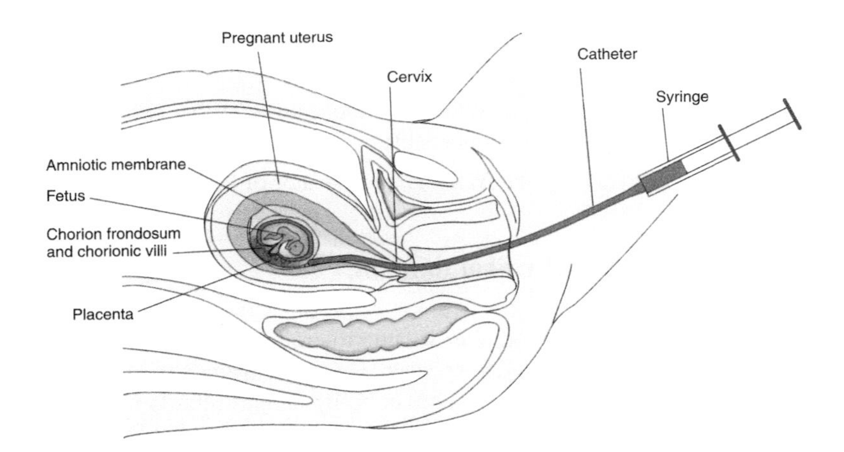

FIGURE 19.16 Diagram shows the setup for the procedure of chorionic villus sampling. Not shown are other instruments used, such as the ultrasound probe (for guiding the catheter) and the speculum (for making the uterine cervix accessible).

hand, although providing fetal chromosomes for analysis, CVS does not provide amniotic fluid and is therefore more limited in its diagnostic value. However, if parents choose an induced abortion as a result of CVS, it will come much earlier in the pregnancy than one based on amniocentesis.

Fetal cells from maternal blood

For many years, it has been known that fetal cells circulate in the mother's bloodstream during pregnancy. For just as long, the possibility of using these cells has been a goal for those seeking a noninvasive technique for prenatal diagnosis. That is, the goal is to cull fetal cells from the pregnant woman's circulatory system and analyze them. The possibility of using these cells for this purpose remains a goal for the future.

Study questions

1. What is teratology? What is cytogenetics?
2. Generally, what types of agents may act as teratogens? Give examples of each type.
3. What is perhaps the best-known chemical teratogen?
4. What was the first birth defect shown to be caused by a chromosome abnormality?
5. What are the two categories of chromosomes that may show abnormalities? Give examples.
6. What are the two general ways in which chromosomes may be abnormal? Give examples.
7. What is a mosaic?
8. What is meant by the concept of critical periods in development? What is the most critical period during development?
9. What is fetoscopy? What are its advantage and disadvantage over ultrasound?
10. What is amniocentesis? What two general parts of amniotic fluid, obtainable by amniocentesis, may be analyzed to detect birth defects?
11. What is an indicator of fetal lung maturity?

Critical thinking

1. We do not tolerate chromosomal abnormalities very well as a species. Human chromosomal abnormalities are very often accompanied by mental retardation. Why do you think this may be so? (Hint: Consider the complexity of the human brain and the role of chromosomes.)
2. Why should it not be surprising that chromosome abnormalities would result in miscarriages or birth defects?
3. The fetal cells that circulate in the bloodstream of the expectant mother may be of what use?

CHAPTER TWENTY

Birth control

CHAPTER OBJECTIVES

After studying this chapter, you should be able to:

1. Explain the difference between contraception and contragestion as general methods of birth control.
2. Discuss the methods of birth control for which the man is primarily responsible, including potential side effects.
3. Discuss the methods of birth control for which the woman is primarily responsible, including potential side effects.
4. List in decreasing order of efficacy the methods of birth control in current use.
5. Discuss the various methods of induced abortion and when during pregnancy they are usually used.

Birth control is a topic with both personal and societal implications. For many people, birth control is a matter of practical necessity—a means of temporarily or indefinitely postponing having children. Birth control has also become an issue that sparks public debate. Many organized religions take stands against all or certain forms of birth control. On the other hand, popular science fiction writer Isaac Asimov (At stake: 500,000,000 years of life. *National Wildlife*. April–May, 1972) warned that humankind's increasing numbers and burgeoning technology have cast a shadow across the future. Perhaps with the human population exploding, birth control is becoming less of a personal matter and more of a public concern.

Having studied the physiologic basis of procreation from gametogenesis to implantation in the uterus and ultimately to birth, you can appreciate what a complex process human development is and that there are a number of alternatives for interfering with this process. The basic means of limiting procreation are:

- Abstinence—not engaging in sexual intercourse
- Interference—if engaging in sexual intercourse, interfering with the union of the gametes
- Prevention of implantation—after fertilization by the gametes, preventing the conceptus from implanting into the uterus and establishing a pregnancy
- Termination—if a pregnancy is established, terminating it

Abstinence

The only 100% effective means of preventing pregnancy is to abstain completely from sexual intercourse or any other sexual activity in which semen may enter the female reproductive system. This seems fairly obvious, but there are also birth control methods that rely on periodic abstinence to prevent pregnancies. These methods are practiced most often by couples whose religious beliefs prohibit use of the artificial birth control methods, described later in this chapter.

Fertility awareness methods

If a couple avoids sexual intercourse on the days of the month when the woman is most likely to get pregnant, the couple can avoid an unplanned pregnancy. **Fertility awareness methods** identify the fertile days of the month. These methods are the rhythm (calendar) method, the basal body temperature method (BBT), and the cervical mucus method.

Rhythm method

The traditional and well-known "natural" birth control method is the rhythm method. This method uses a timekeeping system in which the woman avoids intercourse at around the time of ovulation. However, the menstrual cycles of a given woman are often variable; therefore, it is difficult to accurately predict the time of ovulation. Because of the large margin of error, the rhythm method is often unsuccessful.

The rhythm method involves (1) determining the fertile part of the woman's menstrual cycle by subtracting 18 days from her shortest cycle and 11 days from her longest cycle (the shortest and longest cycles are determined from measurement of the length of the woman's carefully recorded last six cycles) and (2) abstaining from sexual intercourse during the resulting days. For example, if her longest cycle is 33 days ($33 - 11 = 22$) and her shortest cycle is 27 days ($27 - 18 = 9$), the couple should abstain from intercourse on days 9 through 22, counting from the first day of her menstrual periods.

Basal body temperature

The BBT method is based on an increase in the BBT at the time of ovulation, which may be determined by the careful use of a special expanded-scale BBT thermometer. However, because sperm introduced into the female reproductive tract may remain capable of fertilization for 2 days, it is necessary to avoid sexual intercourse *before* ovulation occurs. It is recommended that a woman avoid intercourse or use a barrier method during at least the first half of her cycle until 3 days after her BBT has risen.

Cervical mucus method

The cervical mucus method is based on a change in the consistency of cervical mucus around the time of ovulation. Mucus may not be seen until a few days after menstrual bleeding has stopped. When mucus first appears, it is sticky and ranges in color from yellow to white. As the time of ovulation approaches, the mucus increases in amount, becomes clearer in color and wetter (resembling raw egg white), and can be stretched between two fingers. After a woman's fertile time (around midcycle), the mucus usually becomes sticky again and decreases in amount. In some women, no mucus is seen for the remainder of their menstrual cycle. Because vaginal infections, some drugs, and sexual intercourse all can affect the normal pattern of a woman's mucus, the cervical mucus method, by itself, is not here recommended.

It is strongly advised that a woman discuss all three of these fertility awareness methods with a health care professional.

Preventing fertilization (contraception)

Apart from refraining from sexual activity, the most direct method of preventing birth is to interfere with the union of the sperm and egg. Various methods are available for preventing fertilization, collectively termed **contraception** ("against conception"). Although many people today agree that birth control is a concern of both men and women and that both should share responsibility, various forms of contraception require either the man or the woman to take the initiative. These methods can interfere either physically or chemically with the meeting of the gametes.

Methods requiring the initiative of the man

From the man's point of view, contraception requires preventing the sperm from getting from the cauda epididymis to the ampulla of the fallopian tube, where fertilization occurs. This is achieved by preventing the sperm from entering the female reproductive system.

Barrier methods The most dramatic method of blocking the pathway of sperm is the surgical procedure called **vasectomy**, which prevents sperm from entering the ejaculate and, therefore, from leaving the male reproductive system. Vasectomy involves cutting and tying the cut ends of the vas deferens, the sperm duct (Figure 20.1). Vasovasostomy, the reversal of vasectomy by the surgical rejoining of the cut ends of the vas deferens, is often not successful. Thus, vasectomy should be entered into as a permanent change.

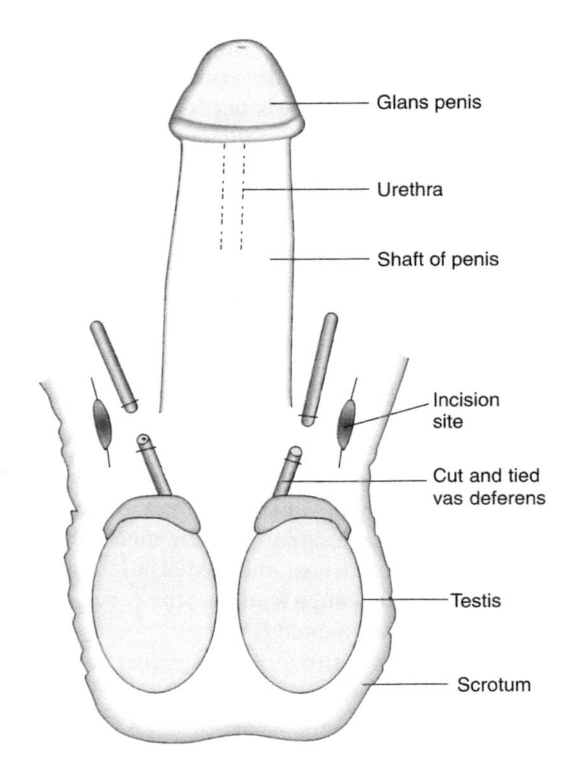

FIGURE 20.1 Illustration of the procedure of vasectomy. By cutting the vas deferens, the sperm no longer have a conduit to reach the urethra and the outside of the body (see Figure 5.5).

A more common (and less drastic) method is to use a **condom**, which interferes with the trajectory of a sperm-laden ejaculate so that it is not deposited in the vagina (Figure 20.2). As a method of both birth control and disease prevention, condoms are effective. However, the effectiveness of condoms varies according to several factors:

1. The condom must be used properly. This means that the condom should be unrolled over the entire length of the penis, that the penis should be erect when it is withdrawn from the vagina, and that the condom should be held in place to avoid leakage of semen into the vagina.
2. A fresh condom should be used rather than an outdated one. Body heat and other factors can cause the condom to fail over time.
3. A latex condom should be used. Natural membrane condoms do not protect against sexually transmitted diseases.
4. Use of a spermicide (nonoxynol-9) kills both sperm and the HIV virus.

Keep in mind that the use of a condom is not 100% effective (see Table 20.1), either for birth control or for prevention of sexually transmitted diseases, including AIDS.

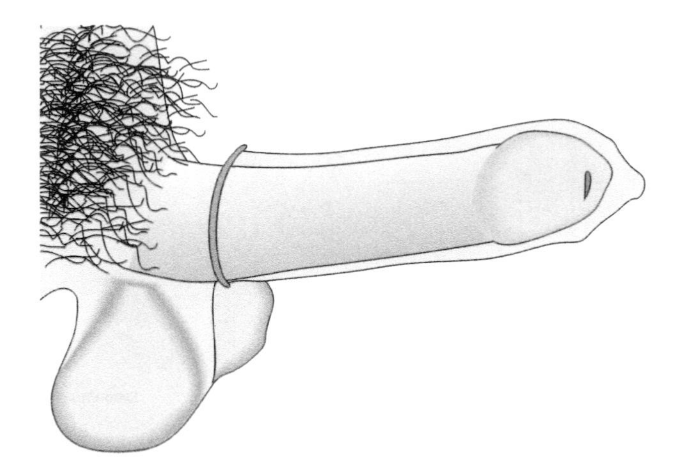

FIGURE 20.2 Illustrates proper placement of a condom over the penis.

Withdrawal method Although **coitus interruptus** (withdrawal) attempts to achieve the same birth control objective, it is less reliable. There are two possible explanations for this method's *failure*:

- Men frequently have some dribbling of mucus mixed with semen before ejaculation.
- A great deal of self-control is required to withdraw the penis from the vagina when ejaculation is about to occur.

Coitus interruptus has another major disadvantage: it provides no protection against the transmission of sexually transmitted diseases.

Chemical methods Every so often, there are news reports about the work of scientists who are trying to develop birth control pills for men. Currently, the search for male contraceptives includes research involving a hormonal gel, which is rubbed onto the upper arms and shoulders of men once a day; a male pill; and a nonhormonal method, wherein a polymer gel is injected into the vas deferens.

Methods requiring the initiative of the woman

Once the sperm enter a woman's body, a man no longer controls the fate of the sperm as they swim toward the egg. However, a woman may use different strategies to prevent the union of sperm and egg. These strategies include the use of physical barriers and chemicals.

Barrier methods **Diaphragms** and **cervical caps** physically prevent sperm from entering the cervical canal. They are soft rubber barriers that cover the cervix. These barriers are always used with a spermicidal cream or jelly, which immobilizes sperm. A diaphragm has the shape of a shallow dome (Figure 20.3). By squeezing its flexible rim, it is inserted into the vagina and over the cervix. The smaller, but deeper, cervical cap fits snugly onto the cervix. A pelvic examination by a gynecologist is necessary to determine the correct size of the diaphragm or cervical cap for an individual woman. A gynecologist or other health care professional

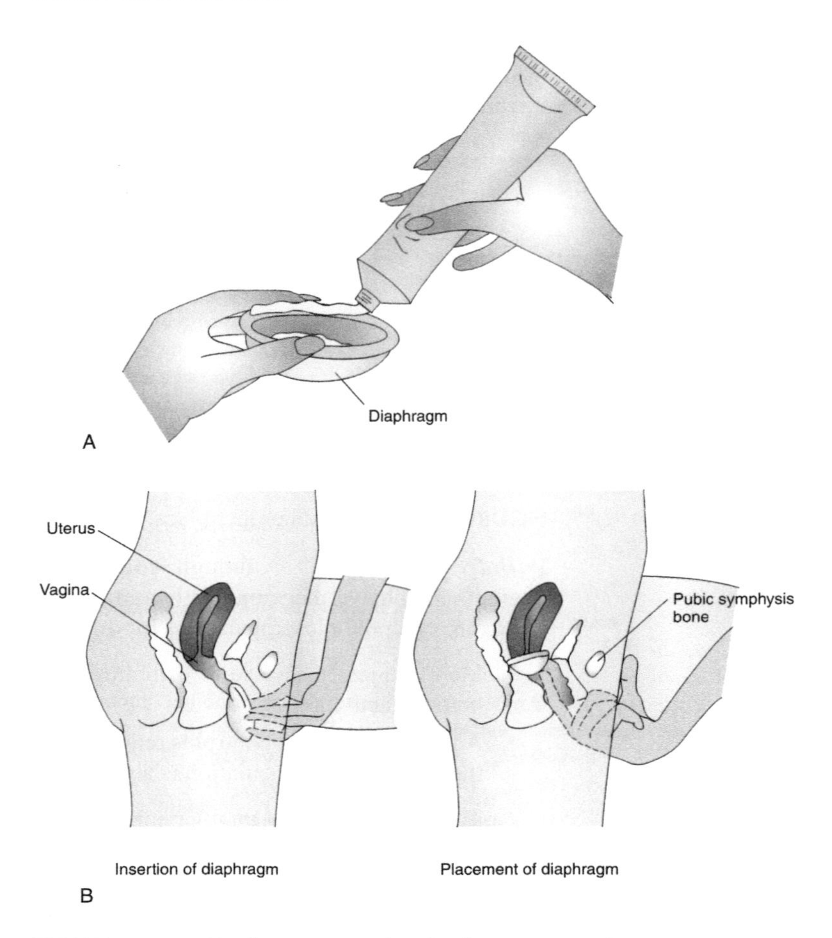

FIGURE 20.3 Diaphragm. (A) The diaphragm is used with spermicidal cream or jelly applied to its concave side. (B) Proper insertion means that the diaphragm completely covers the cervix of the uterus where it projects into the back of the vagina.

should provide instructions for the proper use, insertion, and removal of these barriers.

A physical barrier in the female analogous to vasectomy in the male is the surgical procedure of **tubal ligation**, in which the fallopian tubes are cut and ligated (tied) (Figure 20.4). This removes the egg's pathway for union of the gametes and also prevents sperm from reaching the normal site of fertilization in the upper end of the fallopian tube. Because tubal ligation involves manipulation inside the pelvic cavity, it is not, like the vasectomy, a simple office procedure. Tubal ligation is 99.6% effective.

Chemical methods

Spermicide Spermicidal preparations are always used in conjunction with diaphragms and cervical caps to increase their effectiveness. Other spermicidal preparations are used as foams without a barrier device. The spermicidal foam makes the vagina even less hospitable to sperm

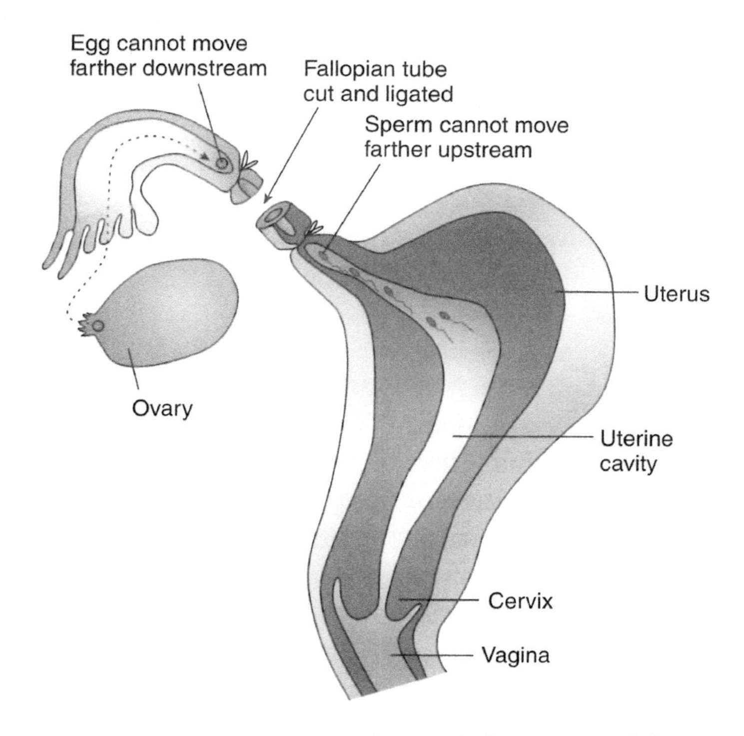

FIGURE 20.4 Tubal ligation. Interferes with the passage of the gamete (egg) down the fallopian tube to the uterine cavity and prevents sperm from reaching the upper end of the fallopian tube.

than it normally is. This is because the foam is a physical barrier to the entry of sperm into the cervical canal and the spermicide immobilizes and kills the sperm. The foam, which has the consistency of shaving cream, is placed into the vagina with an applicator. Although foam is theoretically 94% effective for 30 minutes when used correctly, its *actual* use effectiveness is 78%, owing to incorrect use, inconsistent use, and mechanical failure. Vaginal spermicides are also available in the form of creams, jellies, suppositories, and film, but spermicidal foam is the easiest to use (Figure 20.5).

Birth control pill Another chemical approach to birth control is to use "the pill," one of various hormonal preparations that interfere with the production of eggs. Birth control pills use synthetic estrogens and progesterones (progestins) that work to inhibit release of follicle-stimulating hormone and luteinizing hormone from the pituitary gland. In turn, inhibition of release of these hormones prevents ovulation. Because hormones are powerful chemicals, it is not surprising that birth control pills have side effects—headaches, depression, and possible increased risk of some cancers. Also, women taking birth control pills who are older than 35 years and who smoke and have other risk factors, such as untreated hypertension, have an increased risk of death from cardiovascular disease, that is, heart attacks and strokes.

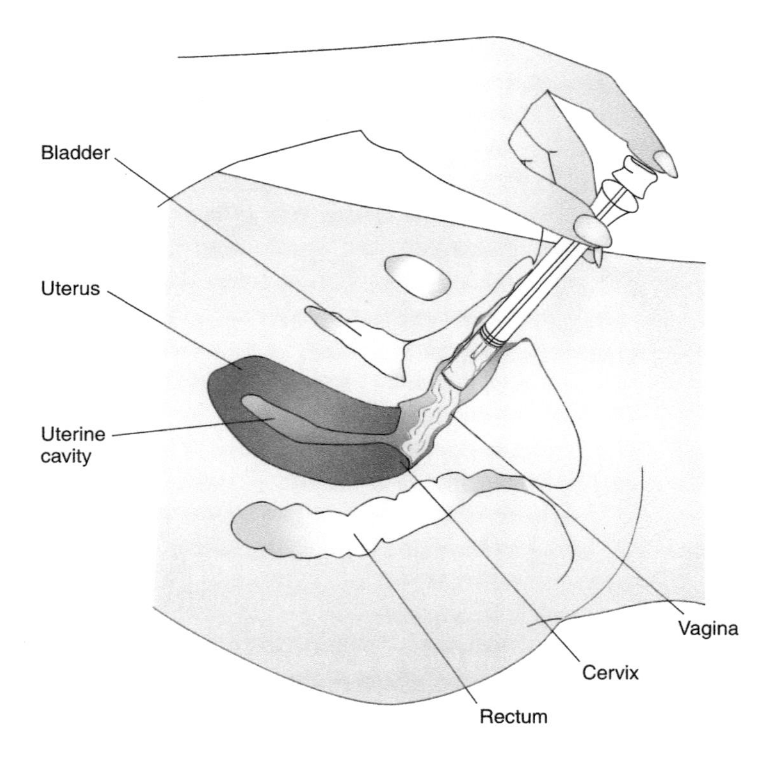

FIGURE 20.5 Proper application of vaginal foam or jelly.

Some people attribute the beginning of the "sexual revolution" to the availability of the pill. Use of the pill is not obtrusive to the act of sexual intercourse and also gives women a heightened sense of security regarding the avoidance of an unwanted pregnancy. These two aspects of use of the pill made it easier for women to engage in premarital sexual activity. Of course, the pill does not prevent or even decrease the likelihood of contracting a sexually transmitted disease. In this regard, use of the pill is not practicing safe sex. The realization of this drawback, especially in the era of AIDS since the early 1980s, has dampened the sexual revolution.

Other methods Newer methods of chemical control of fertility use progestins introduced into the body by other routes. Two methods of applying hormonal control of reproduction are (1) intramuscular injections of medroxyprogesterone (Depo-Provera) every 3 months and (2) implants of levonorgestrel (Norplant), in the form of flexible, hormone-containing capsules, surgically placed under the skin of the upper arm. These capsules are effective for 5 years. Norplant was removed from the U.S. market because of serious side effects.

Preventing implantation (contragestion)

After the gametes have united in conception, contraception in the strict sense is no longer an option. The next step in limitation of procreation

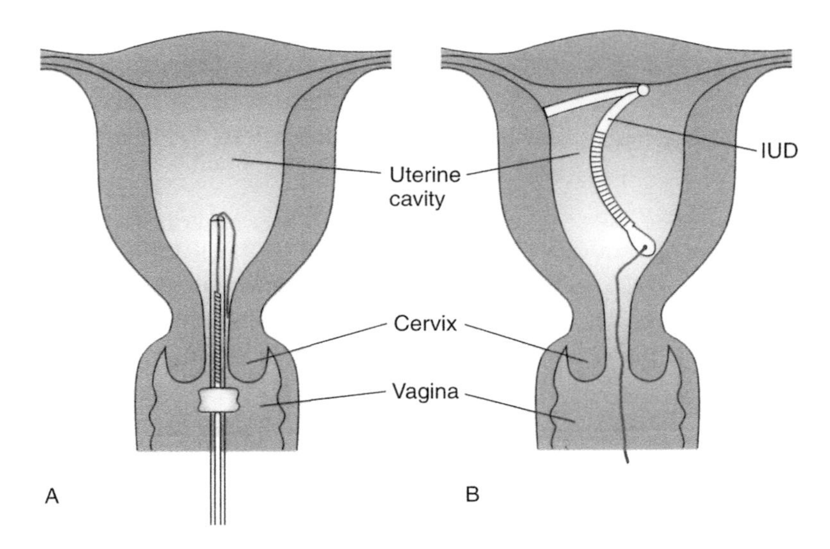

FIGURE 20.6 Insertion of the intrauterine device is normally done by an obstetrician and inhibits implantation of the conceptus into the lining of the uterus. (A) Insertion of the IUD. (B) IUD in position.

is **contragestion** ("against gestation"). In contrast to contraception ("against fertilization"), contragestion prevents gestation (pregnancy) by preventing the new embryo from implanting in the uterus. Again, both chemical and physical methods are available.

Chemical contragestion

Hormones are as essential to pregnancy as they are to fertilization. Therefore, interfering with hormonal balance can prevent pregnancy. **RU 486 (mifepristone)** interferes with progesterone's function of preparing the endometrium for receiving the conceptus so that implantation and pregnancy initiation do not occur.

Emergency contragestion drugs ("morning after" pills) have been available in the United States since the late 1960s. Patients who face an unwanted pregnancy resulting from rape, failure of contraception, or unprotected sex may take a high dose of certain common birth control pills. These can reduce the risk of pregnancy by more than 75% when taken within 72 hours after sex. The most commonly used morning after pills are norgestrel and levonorgestrel. A deliberate overdose of these pills provides emergency contragestion. They prevent pregnancy by rapidly elevating a woman's hormone levels, which prevents the fertilized egg from implanting into the endometrium.

Physical contragestion The **intrauterine device (IUD)** is a plastic or metal device which, when inserted into the uterine cavity by a physician, makes the endometrium unreceptive to implantation by the conceptus (Figure 20.6). The two types of IUDs that are approved for use in the United States are the Copper T and the Progesterone T. Although exactly how IUDs prevent pregnancy is not known, there are two ways in which IUDs probably affect pregnancy: (1) by changing the lining of the uterus, thereby making it unable to support a fertilized egg and (2) by somehow changing the sperm or egg or both, thereby preventing fertilization.

Pregnancy may be prevented by the insertion of the Copper T intrauterine device up to 5 days after unprotected intercourse and is more effective than the use of minipills, reducing the risk of pregnancy by more than 99%.

IUD use poses a number of serious risks, though small, for the user. Possible serious side effects include puncture of the uterus—a medical emergency—and pelvic inflammatory disease. Other, more common and less serious, side effects include cramps and irregular bleeding. Such side effects caused withdrawal of IUDs from the market in the United States in 1987, but safer versions have recently become available.

Pregnancy termination (abortion)

For years, the word "abortion" has caused hotly contested political debates, and the issue of abortion will continue to rage for some time. In this context and in the context of this chapter, "abortion" refers to induced abortion, which is the intentional termination of a pregnancy for any of an endless variety of reasons. Spontaneous abortion (**miscarriage**) is almost always unwanted by the couple to whom it happens.

In 2014, 652,639 legal induced abortions were reported to the CDC from 49 reporting areas.

Abortions are performed as a procreation control method, as well as to protect women whose pregnancy threatens their well-being. The availability of legal abortions was secured by the Supreme Court's *Roe v. Wade* decision, which limited states' rights to outlaw abortion based on the trimester into which the pregnancy has progressed. As we will discuss later, the trimester distinction is based on the viability of the fetus (ability of the fetus to survive outside the womb) as well as the methods used in each trimester.

Abortion methods

The method of abortion depends on the stage of the pregnancy. Most abortions performed early in the first trimester (between 3 and 5 weeks after fertilization) use a **suction method** in which the cervix of the uterus is dilated and the conceptus is removed by suction. Later in the first trimester (after 5 weeks), the most common method is **dilation and evacuation** (D & E). Here, the cervix is dilated, instruments are used to remove the fetus, and the endometrium is removed by suction. The most common method of inducing abortion during the second trimester is to induce uterine contractions by injecting a **hypertonic salt solution** or a **prostaglandin solution** into the amniotic fluid.

Third-trimester abortions are rare. All abortions have some degree of risk of death. This risk is low during early pregnancy, but increases as the procedure is carried out later in a pregnancy. Most states limit third-trimester abortions (consistent with *Roe v. Wade*) to cases in which the risk of continuing the pregnancy is great (and consequently outweighs the risk of the abortion).

Partial-birth abortion **Partial-birth abortion (PBA)** sounds gruesome and is. More properly termed **intact dilation and evacuation**, this procedure is usually performed between 4 and 7 months of pregnancy. The cervix is dilated, and the fetus is pulled out feet first. A cannula is inserted into the base of the fetus's skull, and the brain is sucked out, causing the head to collapse and pass out of the cervix and vagina more easily. One does not have to be pro-life to find this procedure repugnant. So why is it done? It is done to protect the life or health of the mother.

Timing and number of abortions Most abortions are performed sometime during the first 2 months of pregnancy (when the embryo is 1 inch long or less). More than 90% of abortions are performed during the first trimester. Second-trimester abortions performed each year represent about 10% of the total. It has been estimated that less than 1% of abortions are performed during the third trimester. (Because the total number should add up to 100%, it is clear that these are estimates.) Although the exact number of PBAs performed is impossible to estimate with accuracy, the executive director of the National Coalition of Abortion Providers estimated a total of 3,000 to 4,000 annually in the United States.

Comparing methods

As shown by Table 20.1, birth control methods (and variations on the methods) vary widely in effectiveness. In addition, different methods of birth control involve different degrees of risk, especially surgical and chemical methods. The required involvement of a physician in each of these potentially hazardous methods serves to reduce the hazards and educate the user. Like all surgical procedures, vasectomy and tubal ligation carry inherent risks. Vasectomy may be done as an office procedure, but tubal ligation is an operation that requires the use of general anesthesia. In a modern hospital, the risk associated with anesthesia is not zero, but it is very low. Also, whenever one goes "under the knife," as in vasectomy or tubal ligation, there is some risk of infection.

Various side effects accompany use of "the pill," as is true of many medications. The hormones found in the pill are potent biological molecules that exert dramatic effects on the body. (Any woman can testify to the effects of the hormones that regulate and produce the "symptoms" of the normal menstrual cycle.)

IUD use has been blamed for various serious medical problems in highly publicized cases. For instance, the Dalkon Shield, available in the United States during the early 1970s, produced pelvic inflammations and spontaneous abortions—in some cases resulting in death.

In the controversies surrounding birth control, we can generalize that the earlier the interference with the biology of reproduction occurs, the more people will find it acceptable. Such attitudes reflect the various

Table 20.1 Comparison of birth control methods (BCMs)

Type	Method	Mechanism	Advantages	Disadvantages	Pregnancies per year per 100 women[a]
	No BCM				80
Barrier and Spermicidal	Condom	Worn over penis, keeps sperm out of vagina	Protection against sexually transmitted diseases	Disrupts spontaneity, can break, reduces sensation in male	3–10
	Condom and spermicide	Worn over penis, keeps sperm out of vagina, and kills sperm that escape	Protection against sexually transmitted diseases	Disrupts spontaneity, reduces sensation in male	2–5
	Diaphragm and spermicide	Kills sperm and blocks uterus	Inexpensive	Disrupts spontaneity, messy, needs to be fitted by doctor	3–17
	Cervical cap and spermicide	Kills sperm and blocks uterus	Inexpensive, can be left in 24 hours	May slip out of place, messy, needs to be fitted by doctor	5–20
	Spermicidal foam or jelly	Kills sperm and blocks vagina	Inexpensive	Messy	5–22
	Spermicidal suppository	Kills sperm and blocks vagina	Easy to use and carry	Irritates 25% of users, male and female	3–15
	Contraceptive sponge	Kills, blocks, and absorbs sperm in vagina	Can be left in for 24 hours	Expensive	5–16

(*Continued*)

Table 20.1 (*Continued*) Comparison of birth control methods (BCMs)

Type	Method	Mechanism	Advantages	Disadvantages	Pregnancies per year per 100 women[a]
Hormonal	Combination birth control pill	Prevents ovulation and implantation, thickens cervical mucus	Does not interrupt spontaneity, lowers risk of some cancers; decreases menstrual flow	Raises risk of cardiovascular disease in some women, causes weight gain and breast tenderness	0–10
	Minipill	Blocks implantation, deactivates sperm, thickens cervical mucus	Fewer side effects	Weight gain	1–10
	Norplant	Inhibits ovulation			~1
	Depo-Provera	Inhibits ovulation			~1
Behavioral	Rhythm method	No intercourse during fertile times	No cost	Difficult to do, hard to predict timing	13–21
	Withdrawal	Removal of penis from vagina before ejaculation	No cost	Difficult to do	9–25
Surgical	Vasectomy	Sperm cells never reach penis	Permanent, does not interrupt spontaneity	Requires minor surgery	0
	Tubal ligation	Egg cells never reach uterus	Permanent, does not interrupt spontaneity	Requires surgery, entails some risk of infection	0
Other	Intrauterine device	Prevents implantation	Does not interrupt spontaneity	Severe menstrual cramps; increases risk of infection	15

Source: Adapted from Lewis, R. *Beginnings of Life*. Wm. C. Brown, Dubuque, IA, 1992, p. 209.

[a] The lower figures apply when the contraceptive device or technique is used correctly. The higher figures take into account human error.

opinions about when an individual human life begins. The public debate about abortion is being fought between people who believe that life begins at conception and people who think that life does not begin until viability of the fetus or even birth. Biology does not provide an answer to this philosophical question.

Study questions

Birth control

1. What is the only 100% effective means of preventing pregnancy?
2. What do fertility awareness methods identify? Name them.
3. What is the objective of contraception?
4. What does contraception require for the male? List three ways of achieving this and explain how it is achieved for each method.
5. Females use different strategies to prevent the union of sperm and egg. List four and explain how gamete union is prevented in each method.
6. Contraception is to the prevention of conception as contragestion is to the prevention of ___________.
7. Name a chemical and a physical method of contragestion.
8. What are "morning after" pills, what do they contain, and how do they prevent pregnancy?
9. What is another name for miscarriage?
10. What are the most common abortion methods used during (a) early first trimester, (b) late first trimester, and (c) second trimester? What is true of third-trimester abortions?
11. What are three methods currently being explored for male contraception?

Critical thinking

1. Describe the difference between miscarriages, induced abortions, and partial-birth abortions.
2. Birth control in the United States is legal, a personal choice, and almost always assisted. Should the same be true of suicide?

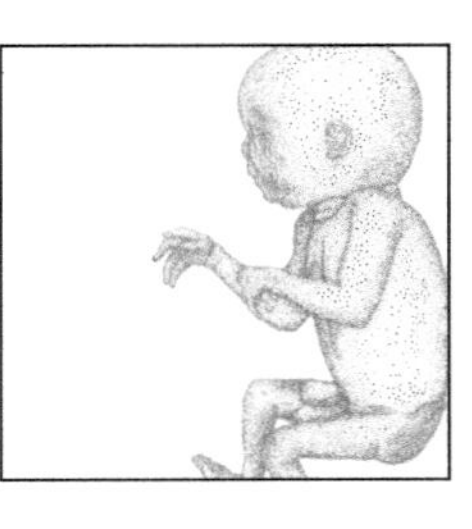

CHAPTER TWENTY-ONE

Reproductive technology

CHAPTER OBJECTIVES

After studying this chapter, you should be able to:

1. Explain what in vitro fertilization (IVF) is, why it is used, and why it usually involves embryo transfer.
2. Discuss the possible fates of embryos produced by IVF and the possible undesirable consequences of each.
3. Compare gamete intrafallopian transfer (GIFT) and zygote intrafallopian transfer (ZIFT) with IVF.
4. Distinguish among genetic mothers, gestational mothers, and surrogate mothers.
5. Explain intracytoplasmic sperm injection (ICSI) and nonsurgical sperm aspiration (NSA), including the purpose of these reproductive technologies.
6. Discuss the advantages of preimplantation embryo diagnosis for expecting parents who are carriers of known genetic defects.
7. Explain cloning as it pertains to the development of mammals and the current status of human cloning.

In July of 1978, a healthy baby girl was born in England. Although this was not in itself a newsworthy event, Louise Brown's birth will be remembered because she was the first child conceived outside her mother's body and brought to full term as a healthy infant. Since then, thousands of children have been born worldwide as a result of in vitro fertilization and other reproductive technologies.

In this chapter, we consider reproductive technologies, one of which (in vitro fertilization) has been around for some time. The others—gamete intrafallopian transfer, zygote intrafallopian transfer, pregnancy after menopause, balloon catheterization, intracytoplasmic sperm injection, nonsurgical sperm aspiration, and preimplantation embryo diagnosis—are recent. Finally, we will consider cloning.

Infertility's role in reproductive technology

It is worth pointing out that much of reproductive technology is concerned with treatment of infertility. A couple's infertility may be caused by infertility of either the man or the woman (or both). Furthermore, infertility may be due to genetic, developmental, or environmental causes. There is an old saying: If your parents did not have children, neither will you. Well, even if your parents were fertile, you may experience temporary or permanent infertility.

A man may be genetically infertile, as in Klinefelter's syndrome, and simply not make any sperm (**azoospermia**). A developmental cause of infertility in the man occurs when the testes do not descend into the scrotum (**cryptorchidism**). If not surgically corrected, no sperm can be produced. Moreover, a hot bath may make a man temporarily infertile by raising the temperature of the testes; this constitutes an environmental cause.

A woman may experience anovulation, wherein eggs are not released from her ovaries. Her fallopian tubes may be blocked as a result of endometriosis (the presence of endometrial tissue in abnormal locations, including the fallopian tubes), in which case some of the lining of the uterus moves into the fallopian tubes. Moreover, sexually transmitted disease often leads to pelvic inflammatory disease, a major cause of occlusion of the fallopian tubes. Finally, immunologic problems may be the culprit, in which, for example, a woman produces antibodies against her husband's sperm.

Some cases of infertility are idiopathic (no cause can be found) for both men and women. This is the case for about 10% of infertile couples.

Specific reproductive technologies

In vitro fertilization

Several methods of artificial birth control seek to prevent the rendezvous of gametes in the ampulla of the fallopian tube. However, sometimes nature prevents this meeting of sperm and egg, much to the consternation of a couple seeking to conceive a child. A common cause of this unintentional contraception is a previous infection of the woman's fallopian tubes, resulting in scarring and physical blockage of these pathways to conception.

However, this blockage is no longer an absolute bar to the union of the couple's gametes. If both partners are able to produce healthy gametes, it is now possible to bring them together outside the woman's body, using **in vitro fertilization (IVF)**. *In vitro* is a Latin phrase meaning literally "in glass." The procedure generally takes place in plastic, but the Latin vocabulary has not kept pace with scientific advances!

A couple's decision to undergo IVF can be an emotional time for both partners. However, the man's physical role in IVF is trivial, beginning

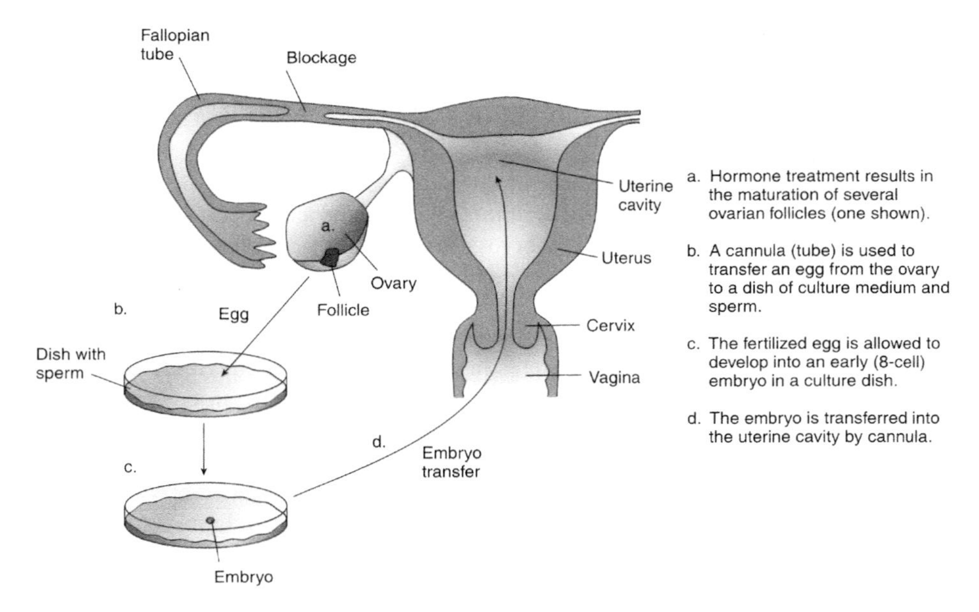

FIGURE 21.1 In vitro fertilization. This simplified diagram shows the bringing together of the egg and sperm in a laboratory dish so that in vitro fertilization may occur. Also shown is transfer of the resulting embryo back into the woman's uterus.

and ending with self-induced ejaculation (**masturbation**) into a specimen container. The role of the woman is more complicated and potentially dangerous, since it includes surgery—**laparotomy** and **laparoscopy**—to obtain the eggs. To increase the chances of obtaining eggs, the woman takes fertility drugs that cause her to **superovulate** (ovulate more than one egg during a single menstrual cycle). Ultrasound is used to monitor the size of her ovarian follicles, so that the eggs may be obtained from the follicles just before they would normally be ovulated from the surface of the ovary.

Once both parents' gametes have been collected, the sperm and eggs are brought together in a plastic culture dish containing a medium with all the factors—food, amino acids, vitamins, and so on—necessary to support fertilization. If the sperm successfully fertilize one or more eggs, the zygotes are maintained *in vitro* through the earliest stages of cleavage, until eight-cell embryos are obtained. These early embryos are transferred back into the woman's uterine cavity by a nonsurgical, transvaginal route. If everything goes according to plan, a normal pregnancy ensues (Figure 21.1).

Variations of In Vitro Fertilization

There are some variations on the theme of IVF. Once the embryos are obtained, what happens next varies according to the wishes of the biological parents.

1. The embryos may be transferred back into the biological mother whose ovaries produced the eggs, in a procedure called **embryo transfer** (also called **ET**).

2. The embryos may be transferred into a different woman who agrees to act as a **surrogate mother** and carry the pregnancy to term.
3. The embryos may be frozen for an indefinite period of time for embryo transfer at some future date.

All these possibilities have presented undesirable consequences. For example, unwanted multiple births may occur because, as is the general practice, several embryos are transferred into the uterus to increase the chances of a single pregnancy. Surrogate motherhood has resulted in the legal dilemma whereby the surrogate mother refuses to surrender the infant to its biological parents, as in the famous "Baby M" case. Moreover, some interesting cases are associated with **frozen embryos**. In one instance, the biological parents of frozen embryos were both killed in a plane crash, resulting in the legal question of whether the frozen embryos should be considered heirs of the parents' inheritance. In another case, the biological parents of frozen embryos divorced and entered into a legal battle over custody of the embryos.

The media, ever prone to catch-phrases, has referred to babies resulting from IVF as "test tube babies." One journalist has even referred to frozen embryos as "popsicle babies."

Science has not yet learned to maintain human development from conception to term *in vitro* and so a "brave new world" in which natural parents are no longer needed is far from reality. See the section on the artificial placenta in Chapter 10.

Gamete intrafallopian transfer and zygote intrafallopian transfer

Embryo transfer after IVF does not always result in the successful implantation of the embryo into the woman's uterus. One technique to increase the chances of implantation is to bring together the parents' gametes in the fallopian tubes. **Gamete intrafallopian transfer (GIFT)** begins with gathering of sperm and eggs, as if in preparation for IVF. Next, first the eggs and then the sperm are loaded into a catheter (a hollow tube), which is used to transfer the gametes into the fallopian tubes. In GIFT, fertilization occurs more naturally than with IVF, and the embryo formed subsequently enters the uterus by the normal route (Figure 21.2).

Another procedure not involving embryo transfer, but that does rely on IVF, is **zygote intrafallopian transfer (ZIFT)**. In ZIFT, zygotes are transferred into the fallopian tubes on the day after fertilization occurs in the culture dish, instead of waiting for the zygotes to develop into embryos.

Pregnancy after menopause

Postmenopausal women do not usually get pregnant because their ovaries no longer produce eggs. However, recent successes demonstrate that the other reproductive organs of older women are capable of carrying a normal pregnancy to term. In these cases, the women were recipients of embryo transfers produced from their husband's sperm and eggs from donor women. Although the recipients were not the genetic mothers of the offspring, they have been the "gestational" mothers, and their

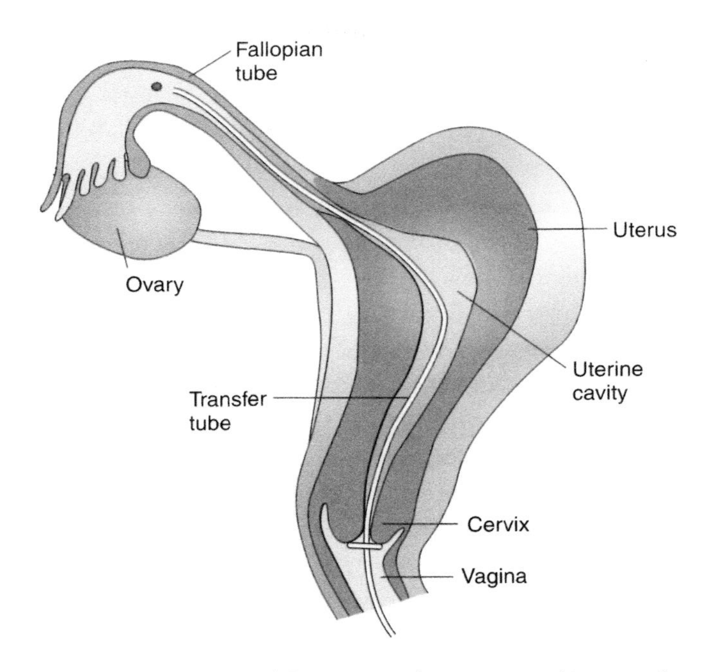

FIGURE 21.2 Gamete intrafallopian transfer (GIFT). In this procedure, the sperm and eggs are not brought together in a laboratory dish as with IVF, but are brought together in the woman's fallopian tubes. If fertilization and early development occur normally, the embryo enters the uterus by the normal route.

husbands have been the genetic fathers. Because the "other women" in these cases are "only" providing the eggs for these pregnancies, they are not making the physical and psychological contributions of surrogate mothers. Presumably, custody battles will be less prevalent with this procedure than with surrogate motherhood.

Balloon catheterization

Balloon catheterization involves inserting an inflatable catheter into a blocked passageway and inflating it to make passage possible. For some years, this technique has been used to open clogged arteries. More recently, balloon catheterization is being used to open blocked fallopian tubes. This procedure can be done in a doctor's office.

Intracytoplasmic sperm injection

In the same way that a woman may produce eggs but be unable to conceive through natural methods, a man may produce too few sperm, morphologically aberrant sperm, or immobile sperm. **Intracytoplasmic sperm injection** (**ICSI**) resolves the problem of male infertility by injecting a single spermatozoon into a single egg's cytoplasm (Figure 21.3). This technique bypasses the complex changes that occur when gametes meet under normal conditions. Recall that millions of sperm are ejaculated, but few make it to the egg, and only one (generally) can fertilize it.

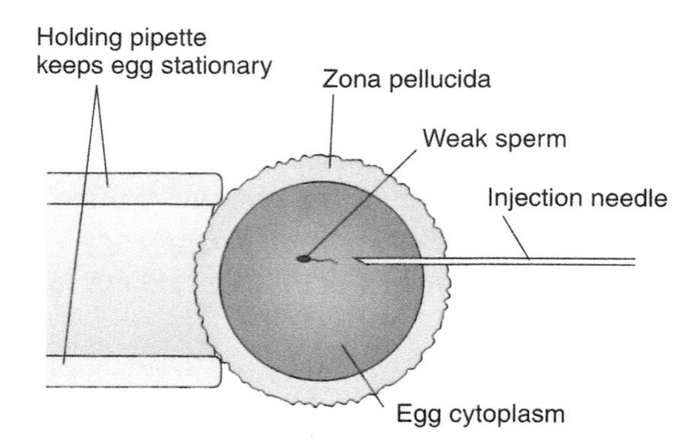

FIGURE 21.3 Intracytoplasmic sperm injection (ICSI). By this procedure, a spermatozoon that is incapable of reaching the egg or penetrating the egg coats or the plasma membrane of the egg may still deliver its genetic "payload" to the egg.

In a sense, ICSI is to male infertility as IVF is to female infertility. IVF allows eggs to bypass blocked fallopian tubes, whereas ICSI allows an otherwise incompetent spermatozoon's genetic payload to reach the egg's cytoplasm.

Nonsurgical sperm aspiration

Another technique to address the problem of male infertility is **nonsurgical sperm aspiration** (**NSA**). Used for men with no spermatozoa or only dead spermatozoa in their ejaculates, this technique aspirates the spermatozoa directly from the sperm ducts through a thin needle. The success of NSA is surprising because normally sperm undergo a "physiologic ripening" process as they pass through the epididymis, which is bypassed in this technique. Work with NSA shows that sperm from the vasa efferentia (ducts that conduct sperm away from the testis) fluid are capable of IVF and are even better than sperm aspirated from the corpus epididymis. A related technique is **microsurgical epididymal sperm aspiration** (**MESA**) (Figure 21.4). Perhaps the human epididymis is not so critical for the development of functional sperm.

NSA is combined with ICSI to help otherwise infertile men achieve biological fatherhood. This method is also an alternative to the frequently unsuccessful attempt to reverse vasectomies (vasovasostomy). In addition, men who cannot ejaculate because of medical conditions, such as multiple sclerosis or removal of the prostate gland, may be helped by this technique.

Preimplantation embryo diagnosis

So far, our examples of reproductive technologies involve the parents of the new individual. **Preimplantation embryo diagnosis** (**PiED**) involves the embryo created by in vitro fertilization. In typical IVF, the embryo is maintained *in vitro* after fertilization until it is transferred into the recipient womb. While the IVF embryo develops in the culture dish, it undergoes cleavage divisions resulting in several blastomeres (cells); usually, eight-celled embryos are transferred into the womb during embryo transfer.

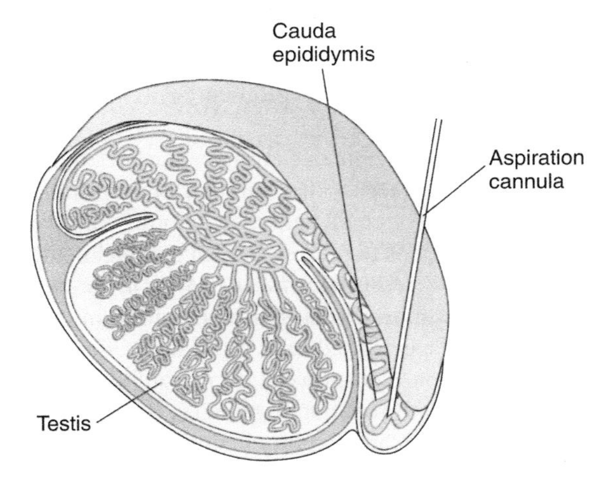

FIGURE 21.4 Microsurgical epididymal sperm aspiration (MESA). If a man's ejaculate is missing sperm, various parts of the epididymis may serve as a source of sperm for intracytoplasmic sperm injection.

Experiments with animals, such as sea urchins and amphibians, as well as the natural occurrence of armadillo identical quadruplets and human identical twins, suggest that early blastomeres are totipotent. That is, each is able to give rise to a complete organism and each is expendable. Therefore, it is not surprising that a blastomere may be removed from an early embryo without harm to the embryo or to the organism that will arise from it. See the section on regulation in Chapter 8.

This removed blastomere may be subjected to molecular genetic analysis, thanks to a powerful technique, **polymerase chain reaction (PCR)**, which greatly increases the blastomere's DNA. Molecular genetic analysis allows the detection of a number of genetic diseases, such as cystic fibrosis, Tay–Sachs disease, and Lesch–Nyhan syndrome. If a biopsied embryo is found to be carrying the gene mutation for such a disease, it is not selected for embryo transfer, thus avoiding a possible elective abortion after the initiation of a pregnancy.

Occurring before implantation in the uterus, PiED is the earliest form of prenatal diagnosis for genetic disorders.

Cloning

The world's attention has long been captured by the possibility of cloning humans. **Cloning** is the act of creating organisms from preexisting organisms by asexual means. The resulting organisms are genetically identical with the original organisms. A familiar example of cloning is making cuttings of plants such as geraniums to produce new geraniums. Cloning may be accomplished with animals also, either naturally (some animals reproduce asexually, such as planarian worms) or artificially (e.g., by shaking apart a four-cell sea urchin embryo to produce four sea urchins). See the section on regulation in Chapter 8.

Some have made *unsubstantiated* claims that human cloning has occurred.[*] But science certainly moved a step closer when, in early 1997, Ian Wilmut's group in Scotland succeeded in artificially cloning a sheep ("Dolly").

Is human cloning on the horizon? Is the "brave new world" just around the corner? It seems that for now, human cloning is conceivable, technically very challenging, and morally of great concern. Governments are taking steps to prevent human cloning, but the future of human cloning is open.

Toward Making Sperm in the Lab[†,‡]: By manipulating their environment, stem cells can be made to differentiate into various kinds of specialized cells, for example, nerve cells, muscle cells, heart cells, and so on. Not surprisingly, researchers would try to have stem cells differentiate into functional gametes, that is, sperm and eggs. Recently, researchers in China reported getting mouse embryonic cells to undergo, in culture, spermatogenesis up to the spermatid level. Recall that spermatids are the product of meiosis, so these researchers were able to have pluripotent stem cells undergo meiosis in culture, quite a feat in itself. They first induced the mouse embryonic stem cells to differentiate into primordial germ cells (PGCs), then exposed these PGCs in culture to testosterone in an attempt to duplicate the environment in the testes. As you know, although spermatids are products of the meiosis portion of spermatogenesis, they have not yet undergone the spermiogenesis portion of spermatogenesis to produce functional, flagellated sperm. Therefore, the Chinese researchers injected the spermatids into mouse ova by a technique similar to ICSI (see ICSI), and the resulting embryos were transferred into female mice. Six of the 200 or so embryos that were so transferred into female mice developed and were born; these pups then developed into fertile adult mice.

As impressive as this technical tour de force is, there are a number of caveats. Is this achievable with humans as part of, for example, IVF? Are the mice produced by this technique genetically normal? Did the mouse spermatids produced undergo normal meiosis regarding meiotic recombination, thus avoiding increased rates of aneuploidy? Could these results be achieved starting with induced pluripotent stem cells rather than starting with embryonic stem cells, thus making the technique more acceptable for human application? (See Chapter 6, "Gametogenesis.")

Functional Eggs Created from Adult Female Cells[§,¶]: Japanese researchers have repeatedly made functional eggs, resulting in healthy live births, from adult mouse cells. Both embryonic stem cells and induced pluripotent stem cells transformed into oocytes. Once the cells reached the stage of PGCs (primordial germ cells), they were placed in the ovaries of female mice to finish the process; that is, they had to rely

---

[*] See Rorvik, D., *In His Image: The Cloning of a Man*. Philadelphia: JB Lippincott, 1978.

[†] Zhou, Q. et al. Complete meiosis from embryonic stem cell-derived germ cells *in vitro*. *Cell Stem Cell*, doi:10.1016/j.stem.2016.01.017, 2016.

[‡] Shaikh-Lesko, R. Researchers devise a technique for creating gametes from murine embryonic stem cells, *The Scientist*, February 25, 2016.

[§] Hayashi, K. et al. Offspring from oocytes derived from *in vitro* primordial germ cell-like cells in mice, *Science* 338:6109 (2012).

[¶] Fox, C. Functional eggs created from adult female cells, *Bioscience Technology* (2013).

on the reproductive systems of the animals to finish the job. About 40 PGCs naturally exist in mouse embryos, and dogma from the twentieth century dictates that oogonia, derived from PGCs, do not exist at all in adult mammalian females. (See Chapter 6, "Gametogenesis.")

Study questions

1. What is the objective of in vitro fertilization? How is this achieved?
2. What technique is used to monitor the size of a woman's ovarian follicles so that mature eggs may be harvested?
3. What are the roles of embryo culture and embryo transfer in IVF?
4. Why may unwanted multiple births result from IVF and embryo transfer?
5. To what extent is gamete intrafallopian transfer (GIFT) similar to IVF and to what extent is it different from IVF? What two developmental processes occur more naturally with GIFT?
6. To what extent is zygote intrafallopian transfer (ZIFT) similar to GIFT and to what extent is it different from GIFT?
7. Distinguish among genetic mothers, gestational mothers, and surrogate mothers.
8. Male infertility often results from a deficit in the number or quality of sperm produced. How has intracytoplasmic sperm injection (ICSI) helped such men become biological fathers?
9. What are the objectives of nonsurgical sperm aspiration (NSA) and microsurgical epididymal sperm aspiration (MESA)?
10. What is removed from the embryo in preimplantation embryo diagnosis (PiED)? What technique is used to increase its tiny amount of DNA?
11. What is cloning? What are natural and artificial methods of animal cloning?

Critical thinking

1. Reproductive technology has presented some initially unexpected problems. Give examples involving surrogate motherhood and frozen embryos.
2. The media, with its penchant for catch-phrases, has come up with "test tube babies" and "popsicle babies." Explain how both of these phrases are misleading.
3. Is there irony in a society that decides to pursue both induced abortion and cloning? Explain.
4. What, if any, are imperative reasons for cloning humans? What, if any, are imperative reasons for not cloning humans?

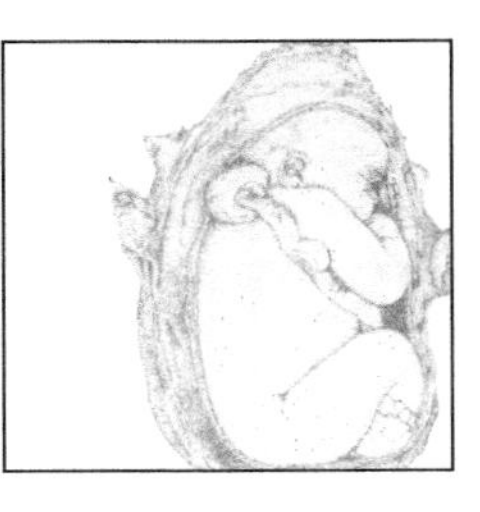

CHAPTER TWENTY-TWO

Sexually transmitted diseases

CHAPTER OBJECTIVES

After studying this chapter, you should be able to:

1. Discuss the costs of sexually transmitted diseases (STDs) in both financial and human terms.
2. Explain what AIDS is, what causes it, how it is transmitted, and why it is a difficult disease to treat.
3. Explain the following STDs, their symptoms, mode of transmission, and treatment: gonorrhea, genital herpes, chlamydia.
4. In a general way, explain why AIDS and genital herpes are more difficult to treat than syphilis, gonorrhea, and chlamydia.

The United States has a high rate of sexually transmitted diseases (STDs). These diseases include herpes, which, like AIDS, is caused by a virus and untreatable with antibiotics, and the bacterial infections syphilis, gonorrhea, and chlamydia, which can be treated with antibiotics. The Centers for Disease Control and Prevention (CDC) 2016 report states that in the United States there were 1.59 million cases of chlamydia (up 4.7% from 2015), 468,514 cases of gonorrhea (up 18.5% since 2015), and 27,814 cases of syphilis (up 17.6% since 2015) (Figure 22.1). The total HIV diagnosis for 2016 was 39,782. STDS can affect the developing fetus (Table 22.1).

https://www.cdc.gov/std/stats16/infographic.htm?s_cid=tw_SR_17003#STDreport

https://www.cdc.gov/hiv/pdf/library/reports/surveillance/cdc-hiv-info-sheet-diagnoses-of-HIV-infection-2016.pdf

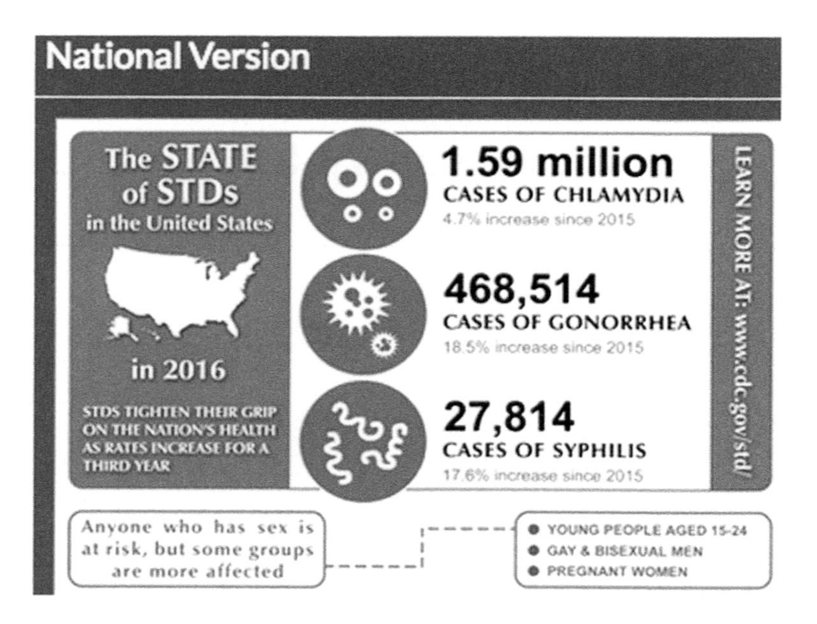

FIGURE 22.1 Sexually transmitted disease surveillance. (Available at https://www.cdc.gov/std/stats16/infographic.htm?s_cid=tw_SR_17003#STDreport)

Table 22.1 Effects of STDs on the fetus

Disease	Effects on fetus
AIDS	Exposure to AIDS virus and other infections
Chlamydia infection	Prematurity, blindness, pneumonia
Genital herpes	Brain damage, stillbirth
Gonorrhea	Blindness, stillbirth
Syphilis	Miscarriage, prematurity, birth defects, stillbirth

Source: Modified from Lewis, R. *Beginnings of Life*. Wm. C. Brown, Dubuque, IA, 1992.

Importance of treatment and prevention

Before the advent of modern medicine, syphilis was a much-feared lethal disease. Today, AIDS is immediately recognized as a feared lethal disease. Moreover, without treatment, any STD can result in undesired complications, so the importance of treating syphilis, gonorrhea, and chlamydia with antibiotics cannot be overstressed.

Herpes, though not curable, can be controlled by antiviral agents and is not generally fatal. The human immunodeficiency virus (HIV) is much more difficult to control by available drugs; although its progress can be slowed by azidothymidine; zidovudine (AZT) and new drugs such as protease inhibitors, it is almost always fatal.

Women and infants bear a disproportionate burden of complications from STDs, including infertility and various types of cancers. As appropriate to a book about life before birth, we will briefly review these diseases, then discuss their effects on women, especially pregnant women. When you see how devastating these diseases can be, you will appreciate the discussion of prevention. Keep in mind that many resources are available to educate you about and protect you from STDs. Although seeking advice and treatment can be difficult, it is important to do so if you are at risk.

Acquired immunodeficiency syndrome

Acquired immunodeficiency syndrome (AIDS) was probably the most feared disease of the late twentieth century. The HIV epidemic currently affects nearly 37 million people worldwide. Although hope for a cure springs eternal (and should), as of now, there is effective treatment of patients with AIDS, but a cure seems to be over a distant horizon. The current role of medicine in AIDS treatment is to extend life until a cure becomes available.

Cause

The two chief routes of infection by the HIV virus, which causes AIDS, are direct introduction into the circulatory system by intravenous drug use or blood transfusions and indirect introduction into the circulatory system by sexual intercourse, either vaginal or anal. The HIV virus belongs to a family of viruses called **retroviruses**. Most organisms on earth have DNA as their genetic material. However, some viruses, including the retroviruses, have ribonucleic acid (RNA) as their genetic material.

Biology of acquired immunodeficiency syndrome

The use of RNA as genetic material disrupts the functioning of the cells that the viruses infect. We have seen in this book repeated examples of cell differentiation (which results in specific proteins made by specific genes). When genes are expressed, their DNA creates RNA as an intermediary or messenger in the production of proteins (see Figure 4.4). The so-called "central dogma of molecular biology" is that the direction of flow of information is DNA-to-RNA-to-protein. Making RNA from DNA is called **transcription**, and making protein from RNA is called **translation**.

HIV and other retroviruses violate this flow of information because they get it backwards. Having genes of RNA, they make an enzyme called reverse **transcriptase**. When HIV infects a cell, its RNA makes DNA that becomes incorporated into the DNA of the infected cell. In the DNA (or genome) of the infected cell, the viral DNA may just mark time and not do much else. However, for reasons not clearly understood, the viral DNA at some point in time becomes activated and begins to make many copies of the HIV viruses. These multiple copies of the virus cause the infected cell to burst and die, releasing many viruses to find other cells to infect.

The cell death caused by viruses results in diseases in adults and birth defects during development. Biomedical researchers constantly strive to find drugs that will interfere with the life cycle of viruses. In the meantime, the HIV virus remains particularly onerous because of its targeting of certain immune cells (**T cells**) for infection.

Acquired immunodeficiency syndrome and the immune system Our immune system, in general, is composed of two types of cells: **B lymphocytes** (B cells) and **T lymphocytes** (T cells). B cells make antibodies to protect us by means of what is called **humoral immunity**. T cells aggressively destroy foreign germs and home-grown cancer cells and, in addition, orchestrate the response of the immune system to an onslaught of infectious agents. There are various kinds of T cells according to their role in the immune defense system. One kind of T cell important in the regulation of the immune system is the CD4 T cell. This is precisely the type of cell infected by the HIV virus. By destroying these CD4 cells, the HIV virus weakens our immune system and, although this does not kill people directly, it makes people vulnerable to potentially fatal infections and cancers.

Symptoms of acquired immunodeficiency syndrome

The symptoms of patients with AIDS include unexplained tiredness; unexplained fever; shaking chills; soaking night sweats; long-lasting swollen glands (over several weeks); weight loss not due to dieting; white spots on the tongue or in the mouth; persistent diarrhea; unexplained dry cough; and pink or purplish flat or raised blotches that don't go away, found on or under the skin or inside the mouth, nose, eyelids, or anus. However, do not panic if you have some of these symptoms! Many of these symptoms may be due to other less severe diseases, from the common cold to Lyme disease. Nevertheless, see your physician if you have these symptoms.

Syphilis

Until the beginning of the twentieth century, syphilis was a much-feared disease with good justification: it was often fatal. Syphilis starts as a localized infection, but the bacteria that cause syphilis spread throughout the body. The disease progresses through stages ultimately resulting in blindness, dementia, and death. In the past, medical treatment for syphilis consisted of harsh chemicals, but the cure was almost as bad as the disease: arsenic, once used in the treatment of syphilis, is a toxic substance with major side effects. Today, early-stage syphilis may be easily treated with antibiotics, but an untreated syphilitic infection is still likely to prove fatal.

Stages

Syphilis, a chronic infectious disease caused by a spirochete (spiral bacterium), *Treponema pallidum*, can be either acquired (usually during sexual intercourse) or congenital (passed from mother to child *in utero*). The disease passes through three distinct stages, which are progressively more serious and progressively more difficult to treat.

The **first stage** of syphilis is characterized by local (at the site of infection) formation of **chancres** (lesions, usually ulcers). These persist for 2 to 6 weeks and then heal spontaneously. During this time, *T. pallidum* is found in the chancres.

The **second stage** of syphilis affects the entire body, resulting in ulcerous skin eruptions. Usually appearing about 6 weeks after the chancre heals, bacteria (*T. pallidum*) are found in many tissues, including the aqueous humor (fluid) of the eye and the cerebrospinal fluid. Symptoms include sores, followed by a rash, mild fever, fatigue, sore throat, hair loss, and swollen glands.

After lapsing into a **latent stage** (the absence of clinical symptoms), up to one-third of patients develop the third stage of syphilis—**tertiary syphilis**—ending with blindness, dementia, and death.

Gonorrhea

Gonorrhea is a *reportable* disease (i.e., a disease that, when diagnosed by a physician, must be reported to the health authorities) in the United States. Gonorrhea (or "the clap") is caused by the gonococcal bacterium *Neisseria gonorrhoeae*. Consequently, the disease is treatable with antibiotics (penicillin), but treatment is most effective when begun early. Unfortunately, this is possible only if the infected person knows about the disease (is not asymptomatic) and is willing to seek treatment.

Gonorrhea affects the mucous membranes, primarily in the genital and urinary tracts. Symptoms may include a discharge from the penis, vagina, or rectum and burning or itching during urination. However, women often have no symptoms and, when gonorrhea is left untreated, it can lead to pelvic inflammatory disease (PID), which is discussed later in this chapter. In men, the infection can cause sterility by scarring in the vas deferens, which inhibits the travel of sperm. In either men or women, if gonorrhea spreads to the bloodstream, it can infect joints, heart valves, and the brain.

Genital herpes

Unlike the bacterial infections, syphilis, gonorrhea, and chlamydia, herpes is caused by a virus and is therefore not curable with antibiotics. Consequently, once infected with herpes, you are infected for life and exhibit recurring symptoms or "outbreaks" of blisters in the affected area. Fortunately, medications (the antiviral drug acyclovir) have been developed to treat these symptoms.

The **herpes simplex viruses** (HSVs) are DNA viruses (unlike the HIV retroviruses), and they include two distinct but closely related viruses, HSV-1 and HSV-2. Both viruses can cause genital herpes when spread through sexual contact: oral-genital contact with a partner with type 1 herpes labialis (cold sores) can result in genital herpes.

Genital herpes causes the eruption of small blister-like vesicles on the skin or mucous membranes. An outbreak of genital herpes usually begins with tingling, burning, itching, and soreness. This is followed by red patches on the involved areas, followed in turn by the appearance of vesicles or blisters, which are quite painful. In women, the blisters may extend into the vagina to the cervix and into the urethra. These form shallow ulcers, which join together and generally heal in 10 days, with scarring. In patients with HIV infection or other causes of reduced immunity, the lesions may persist for weeks or longer. Other symptoms include difficult, painful urination; vaginal discharge; pain in the legs, buttocks, or groin; enlarged lymph glands; fever; and a general ill feeling.

Chlamydia

Chlamydia is the most widespread STD in America. The incidence of chlamydia is highest among sexually active women under the age of 20. This bacterial illness, caused by the microorganism *Chlamydia trachomatis* and transmitted by sexual intercourse, is often asymptomatic.

Among women who do exhibit symptoms, these symptoms may include increased vaginal discharge, painful urination, unusual vaginal bleeding, bleeding after sex, and lower abdominal pain. Infected men may notice a burning sensation while urinating or a urethral discharge or both. Also, inflammation of the testicles may lead to discomfort during intercourse.

Chlamydia is the easiest STD to treat. A single dose of an antibiotic may eliminate the disease within 1 week. Of course, as with any disease, you must know you have the disease before you seek a cure. Given the high percentage of people (especially women) who do not exhibit any symptoms, it is understandable that many cases go untreated. Anyone diagnosed with chlamydia should inform all sexual partners so that they may seek treatment.

Untreated women may develop complications from chlamydia, including PID, ectopic pregnancy, infertility, and dangerous complications during pregnancy and birth.

Sexually transmitted diseases and women

If one point should be obvious from our discussion of STDs, it is that STDs can have a disproportionate effect on women. Symptoms are often concealed, making diagnosis more difficult. In addition, STDs can lead to PID, which can lead to infertility and ectopic pregnancy. Also, a woman who carries an STD may pass the disease to her child before or after birth.

Pelvic inflammatory disease

PID is an infection that may affect the fallopian tubes, the uterus, ovaries, lining of the pelvis, and the broad ligament of the uterus. PID results most frequently from sexual relations with an infected partner. The bacterial agents most commonly responsible are *C. trachomatis, N. gonorrhoeae, Mycoplasma hominis*, and other such bacteria.

PID can cause infertility from scarring of the fallopian tubes, which may or may not be overcome by reproductive technology. Twenty percent of women with PID become infertile. In addition, almost 10% of women who have had PID experience a potentially fatal ectopic pregnancy as their first post-PID pregnancy.

Ectopic pregnancy

An ectopic pregnancy occurs outside the uterus. The most common site for an ectopic pregnancy is the fallopian tube (although it can occur at other pelvic sites, such as the bowel). An ectopic pregnancy occurs in 1 in every 100 pregnancies. Risk factors for a slightly higher occurrence are an intrauterine device, the "morning after" pill, the minipill, tubal damage caused by infection (PID) or salpingitis (inflammation of the fallopian tube), tubal ligation or its reversal, *in vitro* fertilization, gamete intrafallopian transfer (GIFT), and a previous ectopic pregnancy. Whatever the cause, the conceptus does not make it through the fallopian tube and into the uterine cavity.

Symptoms of an ectopic pregnancy are abnormal vaginal bleeding 1 to 2 weeks after a missed menstrual period, lower abdominal pain on the side of the ectopic pregnancy (which may be associated with a feeling of light-headedness or a desire to move the bowels), and, if the fallopian tube ruptures, severe abdominal pain and fainting.

Because a urine pregnancy test may be negative, diagnosis includes confirmation of pregnancy by the use of a blood (serum) pregnancy test. This test measures the rate of rise of serum alpha-hCG (human chorionic gonadotropin), which is detectable 24–48 hours after implantation. In a normal pregnancy, the level doubles every 2 days, with a peak within 10 weeks after the last menstrual period. An ectopic pregnancy is associated with an impaired hCG production—an increase in serum alpha-hCG of less than 66% over 2 days. If the blood serum test is positive, an ultrasound scan can establish whether the conceptus is in the uterus. However, if no pregnancy can be seen in the uterus with ultrasound, laparoscopy (an optical technique that allows the gynecologist to see the pelvic organs and observe a pregnancy in the fallopian tube) in conjunction with a positive blood pregnancy test is the only way of confirming a diagnosis of ectopic pregnancy.

Historically, tubal pregnancy, a medical and life-threatening emergency, has been treated by removing the involved fallopian tube. Recently, because of the availability of sensitive tests, gynecologists have been able to make earlier diagnoses and save the involved tube. If the conceptus is removed from the fallopian tube using laparoscopy early enough, the woman will avoid major surgery. If one tube is removed and the woman is fertile, she may still choose biological motherhood.

Sexually transmitted diseases and the conceptus

Although many STDs can result in infertility among women, not every woman infected with an STD becomes infertile. When a woman becomes pregnant, we are concerned not only for her well-being, but also for that of the child she is carrying. An at-risk pregnant woman should be screened for STDs, because if she is found to have an STD that is treatable with antibiotics, she can be treated safely even though pregnant. She should also take precautions against further infection.

Acquired immunodeficiency syndrome

An HIV-infected woman can pass the HIV virus to her fetus in the womb or during birth (an estimated 1 in 6 babies is infected). Moreover, an HIV-infected mother who breast-feeds increases the risk of infection being passed to her baby. A baby delivered by cesarean section, however, has a reduced risk of infection.

A baby who is not infected with HIV may have a positive HIV test result because the test is for antibodies against the HIV virus and all babies born to HIV-infected mothers are born with the HIV antibodies. Babies who have not been infected with the virus lose the HIV antibodies within about 18 months after birth.

Congenital syphilis

During the late 1980s and early 1990s, the number of primary and secondary syphilis cases in young women rose dramatically. This was to a great extent due to illicit drug use and the exchange of drugs for sex. As a result, there was an increase in congenital syphilis. Thus, socioeconomic factors continue to constitute a major obstacle to the prevention of congenital syphilis.

Congenital syphilis is tragic. Up to 50% of infants born to mothers with primary or secondary syphilis are premature or stillborn or die in the neonatal period. Furthermore, children who survive are born with congenital disease that may not be apparent for years. Although the development of congenital syphilis is not completely understood, invasion of the placenta is presumed to be the major route.

There are two types of congenital syphilis: early (within the first 2 years of life) and late (diagnosed 2 or more years after birth). Therefore, infants born with congenital syphilis may appear healthy at birth, but problems (see symptoms below) begin to appear after 2 years.

Early symptoms Early symptoms of congenital syphilis are enlargement of the liver and spleen, inflammation of the liver (hepatitis), jaundice, mouth and skin lesions, rhinitis, inflammation of the long bones, adenopathy (large or swollen lymph nodes) and blood disturbances, and possibly low birth weight and failure to thrive.

Late symptoms Late symptoms of congenital syphilis are damaged developing tooth buds, resulting in abnormalities of the permanent teeth (e.g., notched edges of incisor teeth [Hutchinson's teeth]); photophobia (abnormal intolerance to light), pain, or blurred vision first in one eye and then in both eyes between the ages of 5 and 20 years; deafness (although uncommon) in the first decade of life in one or both ears; interstitial keratitis (inflammation of the cornea); facial abnormalities (e.g., protruding mandible [lower jaw] and a markedly depressed bridge [saddle nose]); snuffles (obstructed nasal respiration); mental retardation; seizure disorders; bone or joint disease (e.g., symmetric degeneration of joints [Clutton's joints]); and rhagades (fissures at mucocutaneous junctions).

Importance of treating pregnant women

Pregnant women with syphilis may be treated during pregnancy. However, treatment is not always successful, especially for women with secondary syphilis and those treated late in pregnancy. Altered movement of penicillin within the body in the later stages of pregnancy and inadequate fetal exposure time to the antibiotic may be causes of treatment failure in the third trimester of pregnancy. Although penicillin is the drug of choice for pregnant women, up to 5% to 10% of pregnant women report allergy to penicillin. In these instances, other antibiotics (e.g., doxycycline, erythromycin, and ceftriaxone) may be used.

Women with congenital herpes

Congenital herpes may be acquired by the baby during passage through the birth canal if the mother has an active infection. Infection of infants is very serious because the symptoms include blindness, brain infections, and potentially fatal disorders. Any woman who is having an active outbreak of herpes at the time of delivery is delivered by cesarean section. For women who are infected, but not experiencing an active outbreak at the time of delivery, the decision between cesarean section and vaginal delivery should be made on a case-by-case basis by both the mother and the physician.

Women with gonorrhea

When a mother with gonorrhea gives birth, her child can be afflicted with a variety of symptoms, such as blindness, meningitis, and septic arthritis. Because of the possibility of blindness, antibiotic drops are placed into the eyes of babies delivered in hospitals in the United States.

Chlamydia

Recall that chlamydia is often asymptomatic and therefore untreated. If a pregnant woman is not treated, her baby has a 50% chance of developing conjunctivitis (an eyesight-threatening inflammation of the eyes) and a 20% chance of getting pneumonia. Chlamydia can also lead to premature birth or low birth weight. These risks should reinforce the wisdom of testing for STDs or, more importantly, preventing them in the first place.

Preventing sexually transmitted diseases

Any sexual contact creates a risk of contracting an STD. Of course, not everyone contracts an STD. Some people can consider themselves merely lucky, but others have taken the precautions necessary to protect themselves.

The simplest and most obvious solution for preventing STDs is to refrain from all sexual activity (abstinence). However, for many people, this is a very unpopular solution. Short of abstinence, the safest course to follow is to be in a monogamous relationship in which both partners are uninfected. This sounds simple enough, but remember that many people with STDs are asymptomatic. Therefore, laboratory testing is the only safe route for determining whether you carry an STD. HIV testing is a very reasonable precaution for couples, because a person can harbor the virus for years with no symptoms and the consequences of contracting HIV can be ultimately fatal.

Condom use

In our discussion of birth control (Chapter 20), we mentioned that a condom can be used for disease prevention in addition to its use as a contraceptive device. For example, because the bacteria that cause gonorrhea are typically transmitted during intercourse when they enter the male urethra or female vagina, effective use of a condom prevents transmission of the bacteria by blocking contact between urethra and vagina. Similarly, the HIV virus is transmitted in semen, vaginal secretions, and blood: use of a condom by an infected man can prevent transmission of the virus in his semen, whereas use by an uninfected man can prevent blood or vaginal secretions from contacting his urethra. Although use of a condom can prevent transmission of herpes, note that the disease may be transmitted by or to areas not covered by the condom.

Prevention of herpes

Special consideration should be given to prevention of herpes. As mentioned, a condom might not prevent transmission of herpes, and consideration should also be given to whether the disease is in its active or latent phases. In the active stage of herpes (when lesions are present), sexual contact should be avoided, because this is when the virus is most easily transmitted. However, the virus can also be transmitted when the disease is in its latent phase, and condoms provide an additional amount of protection. It is worth a reminder that oral sex with a partner with a cold sore (HSV-1) can result in a genital infection. Therefore, such contact must be avoided.

Prevention of acquired immunodeficiency syndrome

Despite all the destruction it has caused, the HIV virus is actually very fragile and cannot live in the environment. Nonetheless, if it is passed in body fluids from an infected person to a healthy person, the disease will be transmitted. Consequently, behavior that encourages transfer of body fluids should be avoided, such as intercourse with an infected person—vaginal, oral, or anal; transfusion of infected blood; and sharing of syringe needles with infected persons. How do you determine whether the person, blood, or needle is infected? As Shakespeare's Hamlet said, "ay, there's the rub."

Hopefully, you have learned more about preventing AIDS than you have learned about Shakespeare. Therefore, you know that the only safe course is to assume that the person with whom you are going to have sexual intercourse is infected and act accordingly. Or, to assume that the blood you are to receive in a transfusion is tainted until proven otherwise. An increasing number of people are being tested with their partners for the HIV virus before they have unprotected sex (without a condom) and some people anticipating surgery have their own blood stored for use during the surgery. If you use intravenous drugs, you must never use anything but an unused, sterile, one-use, disposable needle, and then destroy it.

Study questions

1. Name the sexually transmitted diseases (STDs) caused by bacteria. Name the STDs caused by viruses.
2. What are the two chief routes of infection by the HIV virus?
3. What are the symptoms of AIDS?
4. What causes syphilis and in what two ways, both developmentally significant, can it be acquired?
5. What causes gonorrhea? What are developmentally relevant consequences of gonorrhea?
6. Why is genital herpes, unlike syphilis, gonorrhea, and chlamydia, not curable with antibiotics? What is a consequence of this?
7. What are complications of untreated chlamydia in women?
8. What is the significance of pelvic inflammatory disease (PID) for human development?
9. What is an ectopic pregnancy? What is the most common site for an ectopic pregnancy and how serious is this?
10. What is indicated for a woman who has an active outbreak of herpes at the time of delivery?
11. Why may an *un*infected baby have a positive HIV test?
12. Why are antibiotic drops placed into the eyes of babies delivered in U.S. hospitals?

Critical thinking

1. A number of sexually transmitted diseases (STDs) are easily cured with treatment. From a human development point of view, why should a young woman try to avoid being infected in the first place?
2. Syphilis was not treatable 100 years ago and AIDS is not effectively treatable now. What does this say, in a general way, about medicine's ability to deal with two different kinds of pathogens?

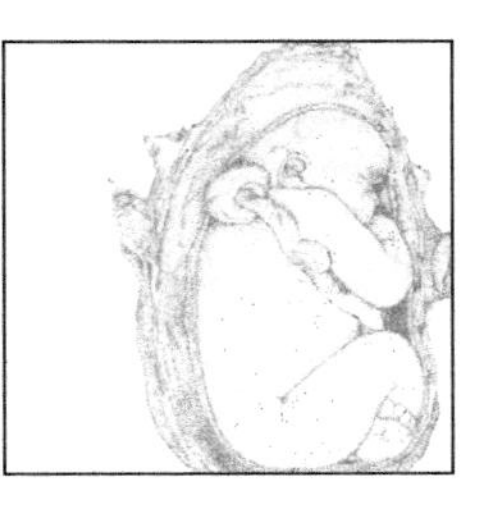

CHAPTER TWENTY-THREE

Cultural aspects of development

CHAPTER OBJECTIVES

After studying this chapter, you should be able to:

1. Understand various expressions referring to pregnancy, such as womb and molar pregnancy.
2. Understand various expressions referring to birth, such as "her water broke," labor, and afterbirth.
3. Explain the science underlying some common expressions used in the media, such as "blue baby" and "test tube baby."

You now have a good basic understanding about how the human body develops, starting with the parents' reproductive systems, through the creation and union of sperm and egg, and through the embryonic and fetal periods up to the time when the child is finally born. You have learned about the development of individual systems and about topics that affect development, such as birth control, reproductive technology, and sexually transmitted diseases. You have also seen what happens when development does not occur entirely according to plan. It is hoped that you have enjoyed what you have learned and that all this information will prove useful to you—in your role as a parent or perhaps as a health care professional.

It is also hoped that this book has broadened your knowledge of many aspects of development that you may have heard about for years but have not entirely understood—something that I like to call the "cultural aspects of development." I like to draw a distinction, for example, between the "cultural egg" and the "biological egg." When the average person hears the word "egg," the hen's egg usually comes to mind. Biologically speaking, a hen's "egg" is really just the yolk, for the biological egg is the single cell that is fertilized to give the zygote. (Yolk, in biological terms, is the stored food in the egg.)

Before bringing this book to a close, let's review some of the cultural aspects of development, looking at some terms and phrases that have been in the lay vocabulary for years. We'll look at some of the terms used to describe pregnancy, birth, and twins, and we'll also look at some of the terms that you have heard on the news.

Terms related to pregnancy

Pregnancy is an amazing process in which the male half of the population can share only by association. Pregnancy is something that fascinates the child whose mother is on her way to bringing a younger brother or sister into the family or, as we might say, carrying a child in her womb.

Another name for the uterus is **womb**, the female organ for protection and nourishment of the fetus. Ann Oakley, in *The Captured Womb: A History of the Medical Care of Pregnant Women*, discusses the gradual medicalization of motherhood.

The term **mole** has different meanings for the lay person, including a skin growth and a small animal. In the context of human development, you may have heard of a **molar pregnancy**. Here, the adjective "molar" is shorthand for **hydatidiform mole** and refers to abnormal development in which there is no embryo or fetus, only an abnormal placenta. In addition, the abnormal placenta looks like a cluster of grapes or large water droplets. The word "hydatidiform" means water droplet-shaped (Figure 23.1).

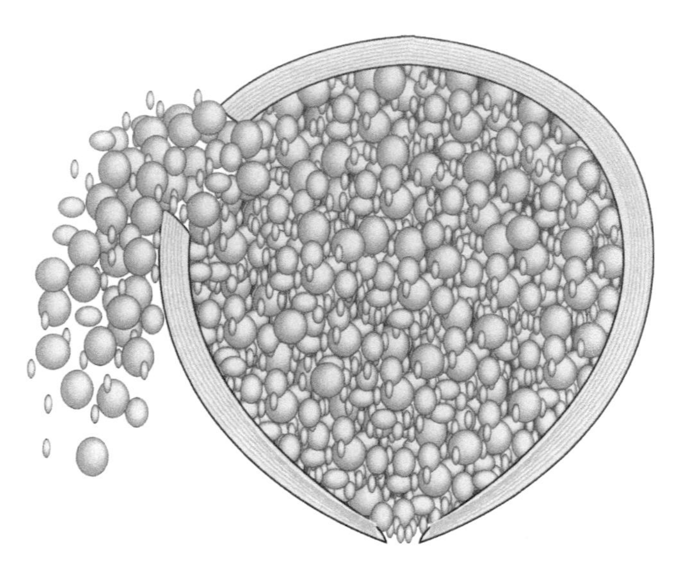

FIGURE 23.1 Drawing of a distended uterus containing a hydatidiform mole. Superficially, the abnormal placenta looks like a cluster of large water droplets. (From Eastman, N.J. and L.M. Hellman. *Williams Obstetrics*, 13th edition. Appleton & Lange, 1966: Figure 1, p. 571 with permission from The McGraw-Hill Companies.)

The term **miscarriage** has been commonly used to refer to a rather sad event during a pregnancy, the **spontaneous abortion** or termination of a pregnancy. As disheartening as this experience may be for the couple anticipating the birth of a child, it is made somewhat less painful to realize that often this is "nature's way" of eliminating an abnormal embryo or fetus. Alternatively, **abortion**, as generally used (including in this book), refers to an *induced* termination of a pregnancy as opposed to a spontaneous abortion or miscarriage.

Most people have heard that **Rh** has something to do with genetic considerations in having children. Rh refers to a particular gene that is found in the human population in two forms, Rh-positive (Rh+) and Rh-negative (Rh–). Because we inherit two copies of most of our genes, one copy from each parent, there are three possible combinations of the two forms of this gene: Rh+Rh+, Rh+Rh–, Rh–Rh–. The Rh+ form of the gene makes a protein that is not made by the Rh– form of the gene.

If a protein not made by our body is introduced into our circulatory system, our body's immune system rejects it as foreign. Thus, a problem arises when an Rh– mother (Rh–Rh–) is carrying an Rh+ conceptus (Rh+Rh–); that is, the conceptus inherited the Rh+ copy of the gene from its father and the Rh– copy from its mother. This means that the conceptus will make a protein perceived as foreign by the mother's immune system. In general, the fetal and maternal bloodstreams are kept separated, but it is common for some mixture of the blood to occur near the end of a pregnancy. When this happens, the mother's immune system mounts an attack against the foreign protein, but this is not important for this fetus because birth is about to occur. However, the mother's immune system has been sensitized and, if a subsequent pregnancy is also Rh incompatible, antibodies produced by the mother will enter the fetus's circulatory system and destroy its red blood cells (a condition called **erythroblastosis fetalis**), with possible fatal consequences for the fetus (Figure 23.2).

Terms related to birth

As with pregnancy, you may have heard discussions of childbirth that you did not fully understand. Moreover, childbirth is something that the mother goes through with a varying level of support and understanding of the father. In the common expression, "her water broke," the **water** referred to is the amniotic fluid released from the ruptured amniotic cavity at the beginning of labor. Usually, the membrane that makes up the wall of the amniotic cavity breaks by its self; if not, it is cut by the obstetrician, a procedure called **amniotomy**.

In the context of childbirth, **labor** is associated with a difficult time for the mother-to-be leading up to the delivery of the child. During this process called labor, contractions of the smooth muscle of the uterine walls dilate the uterine cervix and force the fetus out of the uterine cavity through the birth canal.

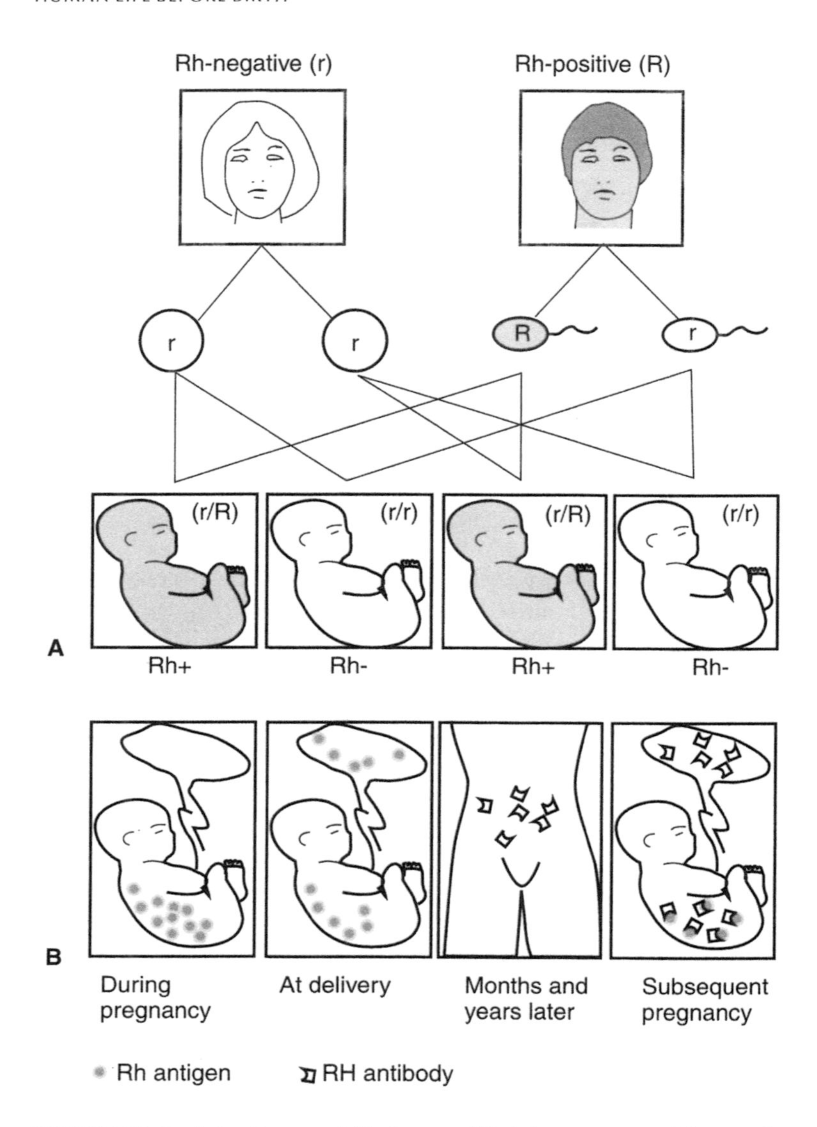

FIGURE 23.2 Inheritance of Rh factor. (A) A homozygous Rh– mother produces eggs, all of which are Rh–. An Rh+ father, who is heterozygous for Rh factor, produces sperm, half of which are Rh+ and half of which are Rh–. Therefore, statistically half of their children will be Rh– and half of their children will be heterozygous Rh+. (B) The first Rh+ fetus of an Rh– mother will sensitize her to the Rh+ factor, as a result of which the mother produces antibodies against the Rh+ antigen. During a subsequent pregnancy with an Rh+ fetus, the maternal antibodies may cause the destruction of the fetal red blood cells (erythroblastosis fetalis). R = Rh-positive; r = Rh-negative.

In the minds of many people, **afterbirth** is something to be put out of mind as a rather nebulous mass of material that is part of childbirth. Afterbirth is that which is born after the baby and consists of the placenta, membranes, and an attached part of the umbilical cord (Figure 23.3).

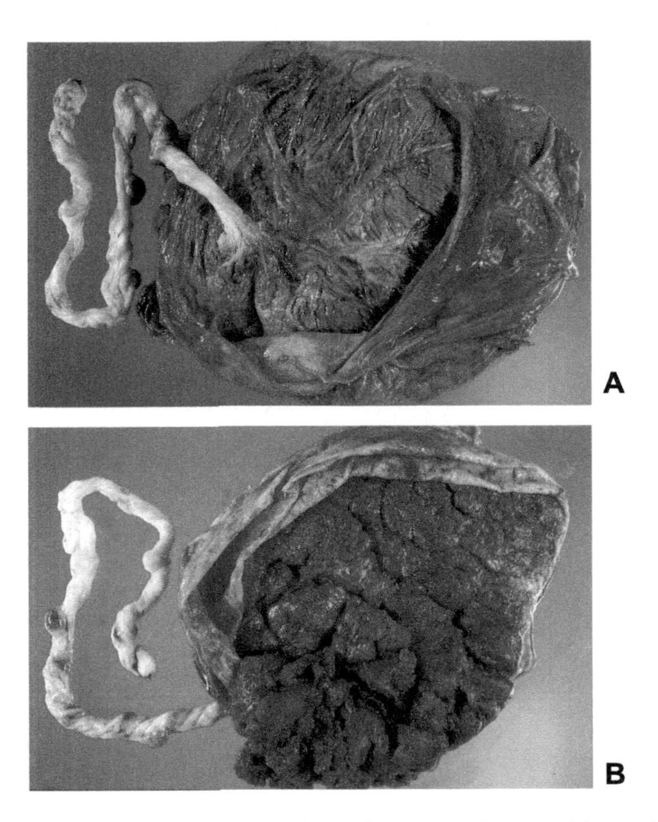

FIGURE 23.3 (A) The fetal side of the placenta and most of the umbilical cord, including its attachment to the placenta. (B) The maternal side of the same placenta. (Courtesy of Jack Wolk, MD, The Danbury Hospital.)

Often, a newborn's skin is covered by **vernix caseosa,** a cheesy, whitish, material. This is primarily composed of the secretions of the sebaceous glands of the skin and is believed to play a role in protecting the skin of the fetus from the amniotic fluid in which it bathes. If one is in attendance at childbirth and has never seen a newborn covered with vernix caseosa, it can be a somewhat startling experience.

Sometimes a baby is born with what appears to be a shroud or veil over its head. In earlier times, superstitious significance was attached to this. In fact, this caul is simply the torn fetal membranes that happen to drape over the head of the newborn. It is of no significance.

At the time of childbirth, the newborn may not have that pinkish color that we associate with a new baby, but rather a bluish cast to the skin. This has led to the common expression **blue baby**. Such coloration is referred to as **cyanosis**, and the baby is said to be **cyanotic**. This bluish color is due to a deficit of oxygen in the blood caused by a heart defect. Here's how it happens. At the time of birth, the **foramen ovale** between the two atria of the heart should functionally close and, by so doing, keep the deoxygenated blood, returning from the body and entering the right atrium of the heart, from mixing with the oxygenated blood, returning from the lungs and entering the left atrium of the heart. When this closure does not occur, some of the deoxygenated blood enters the left atrium, passes into the left ventricle, and is pumped out the aorta to the rest of the body. Partially deoxygenated blood going to the various parts of the body gives it a bluish cast and thus the cyanotic appearance.

Types of twins

Children encounter twins long before they learn anything about human development. And they are sharp enough to pick up on the fact that some twins are more alike than others. There are the twins who look (and often are dressed) alike, and there are the twins who share their birthday party, which is more unusual if one is a girl and the other is a boy.

Identical twins arise from a single conception and zygote. Because they are genetically identical, it is not surprising that they have many characteristics in common. The fact that they become less identical with age testifies to the influence of environment on human development. They are also called **monozygotic twins** and are always of the same sex.

Arising from two separate conceptions, **fraternal twins** are not genetically identical. Genetically, they have no more in common than other siblings except that they happen to occupy their mother's uterus at the same time. Also called **dizygotic twins**, they may or may not be of the same sex.

The expression "Siamese twins" evokes the image of two newborns anatomically joined together. More correctly, these are called **conjoined twins**. Siamese twins actually refers to conjoined twins, Chang and Eng, born in Siam (Thailand) during the nineteenth century. These brothers were part of a circus troupe and became a world-famous curiosity. They married sisters and each sired a large number of children. There are many kinds of conjoined twins, based on how intimately they are connected. This connection ranges from trivial joining easily separated by surgery to **fetus in fetu**, in which one twin is a tiny parasiticlike growth on the larger twin. Conjoined twins result when the embryos either do not completely separate or, if they do, are so close together that they fuse with each other (see Figure 11.3).

Terms in the news

Newspapers and other forms of media have a knack for coining sensational terms. Two such examples applied to human life before birth are **test tube babies**, a poorly used expression referring to babies resulting from in vitro fertilization, and **popsicle babies**, a sensational expression referring to frozen embryos or the babies resulting from them.

A plethora of letter words (i.e., abbreviations and acronyms) has permeated biology in recent years. Examples familiar to many laymen are DNA (deoxyribonucleic acid), PCR (polymerase chain reaction), and ATP (adenosine triphosphate). This is also true in the realm of human development. Common examples are **IVF** (in vitro fertilization) and **IVF-ET** (in vitro fertilization followed by embryo transfer).

A woman who carries a pregnancy but is not the biological mother of the conceptus she carries is called a **surrogate mother**. In other words, the egg from which the conceptus derives is another woman's egg.

A facility for the **cryogenic** (frozen) maintenance of semen specimens with the intention of future use of the specimens for artificial insemination is called a **sperm bank**. The introduction of semen into the female reproductive tract by a means other than a penis is called **artificial insemination**. More recently, **embryo banks** have also been established.

A **clone** is a population of cells derived from a single cell by mitoses. In the context of human reproduction and development and in the mind of the average person, a clone conjures up the image of genetically identical humans. Although, as far as we know, humans have not been cloned, there appears to be little general sentiment to do so, even though in the near future, it will be technically possible. In a rush to enact legislation to prevent human cloning, it would be unfortunate to cast too wide a net and restrict cloning research in general. Such research, not using human cloning, has great potential benefit for human application in both medicine and agriculture.

Fetal sentience and abortion: Bonnie Steinbock, in *Fetal Sentience and Women's Rights*, Hastings Center Report (The Hastings Center) 41(6) p c3 (2011), concludes the following regarding the onset of fetal sentience. All we can say now is that the fetus almost certainly cannot feel pain during the first trimester and almost certainly can during the third. The exact onset of sentience in the second trimester is not just unknown; it may even be unknowable.* Susan Tawia, in *When Is the Capacity for Sentience Acquired during Human Fetal Development*, Journal of Maternal-Fetal and Neonatal Medicine 1(3):153–165 (2009), maintains that the question of when the human fetus develops the capacity for sentience is central to many contentious issues, and that the answer could and should influence attitudes toward IVF and embryo experimentation, abortion, and fetal and neonatal surgery. Furthermore, she maintains that in order for a fetus to be described as sentient, the somatosensory pathways from the periphery to the primary somatosensory region of the cerebral cortex must be established and functional. Although, as she points out, it is concluded that the basic neuronal substrate required to transmit somatosensory information develops by midgestation (18–25 weeks), the functional capacity of the neural circuitry is limited by the immaturity of the system; thus, 18–25 weeks is considered to be the earliest stage at which the lower boundary of sentience could be placed. Furthermore, she maintains that at this stage of development, there is little evidence for the central processing of somatosensory information; before 30 weeks gestational age, EEG activity is extremely limited and somatosensory evoked potentials are immature, lacking components that correlate with information processing within the cerebral cortex. Thus, she concludes that 30 weeks is considered a more plausible stage of fetal development at which the lower boundary for sentience could be placed.†

* https://muse.jhu.edu/article/458046

† https://www.researchgate.net/publication/232061401_When_is_the_Capacity_for_Sentience_Acquired_During_Human_Fetal_Development

Study questions

1. What is a hydatidiform mole?
2. What is Rh incompatibility and what life-threatening condition may it cause?
3. What is meant by the expression "her water broke?"
4. The "caul" of superstition is actually what?
5. "Blue babies" are actually cyanotic babies resulting from what heart defect?
6. Distinguish among monozygotic, dizygotic, and conjoined twins.
7. Distinguish among Siamese, fraternal, and identical twins.
8. For what purpose are sperm banks used?
9. What is artificial insemination?

Critical thinking

1. Is it likely that human cloning will ever produce *identical* humans? Explain.
2. If the mother's body mounts an immune response against Rh factor protein, why doesn't it mount an immune response against other proteins for which the father has the dominant allele and the mother the recessive allele?

Answers to study questions

Part I. An overview of human development

Chapter 1. Before you begin

1. True.
2. False. Cephalic refers to the head end of the body.
3. False. Laterad is movement toward the side of the body.
4. True.
5. True.
6. False. The frontal plane separates the dorsal and ventral surfaces of the body.
7. False. The sagittal plane separates the left from the right side of the body.

Chapter 2. Cells

1. Nucleus, cytoplasm, and plasma membrane.
2. Within the nucleus.
3. When chromosomes are dispersed, they are called chromatin; when chromatin is condensed, it forms chromosomes.
4. Cytoplasm and nucleoplasm.
5. Between the plasma membrane and the nuclear membrane.
6. Plasma membrane.
7. Adenosine triphosphate (ATP).
8. To generate heat.
9. To package substances for export from the cell, to direct molecular traffic in the cell, to chemically modify glycoproteins.
10. Digestive enzymes; genetically programmed cell death and fertilization.
11. Lysosomal storage diseases.
12. Assisting in cell division; organizing part of the cytoskeleton.
13. The microtubular core of the sperm tail.
14. To bend and propel the spermatozoa.

Chapter 3. Cell division

1. Mitotic and meiotic.
2. The daughter cells have the same number of chromosomes as the original cell.
3. G_1, S, G_2, M; cell growth, DNA synthesis, preparation for mitosis, mitosis.
4. G_1, S, G_2.
5. Chromatin.
6. A metacentric chromosome has arms of equal length; an acrocentric chromosome has one arm shorter than the other.
7. Prophase, prometaphase, metaphase, anaphase, telophase.
8. G_1, one; S, one going to two; G_2, two; metaphase, two; anaphase, one.
9. Prophase—condensing; prometaphase—fully condensed in the cytoplasm; metaphase—aligned on equator of spindle; anaphase—moving toward spindle poles; telophase—dispersing.
10. Centrosomes.
11. One equator and two poles.
12. To reduce the chromosome number from diploid to haploid.
13. Two; meiosis I and meiosis II.
14. The species-specific number (diploid); half the species-specific number (haploid); homologous chromosomes undergo synapsis (pair up) during prophase I; bivalent, a pair of homologous chromosomes; tetrad, the four chromatids making up a bivalent.
15. Reciprocal exchange of homologous segments of homologous chromosomes; genetic diversity.
16. Epigenetic. Because the sequence of nucleotides is not changed.
17. In symmetric cell division, both daughter cells have the same fate, whereas in asymmetric cell division, the two daughter cells have different fates.

Chapter 4. Genetics

1. Mendel made two contributions. (1) He proposed that hereditary material has a particulate nature and (2) he proposed two “laws” for the distribution of these particles: the Law of Segregation and the Law of Independent Assortment; 1856–1864, Austria, and pea plants.
2. That in the formation of gametes, the members of the pair of alleles for each unit character separate.
3. Polygenic characters involve a number of genes; multifactorial inheritance occurs when environmental factors also cause variation in the character.
4. The intimate relationship between Mendelian genetics and cytology.

5. The theory that the alleles (genes) are carried on the chromosomes.
6. Down's syndrome; 1959.
7. Discovery of the structure of DNA.
8. Three billion; to determine the sequence of all 3 billion base pairs.
9. Role of transcription: when the information encoded in DNA is expressed, a sequence in DNA is used to dictate a sequence of ribonucleotides in a molecule of RNA; role of translation: the process involves the conversion of the information in a sequence of ribonucleotides in a messenger RNA molecule to information in a sequence of amino acids in a protein molecule.
10. Messenger RNA (mRNA), transfer RNA (tRNA), and ribosomal RNA (rRNA). mRNA encodes the genetic information transcribed from DNA, tRNA carries amino acids to the ribosomes, and rRNA is part of the structure of ribosomes.
11. Hundreds of individual chemical reactions that make up the biochemistry and metabolism of the cell (and, therefore, the embryo).
12. Expressed; different kinds of cells express different sets of genes.
13. Epigenesis is the concept that the developing organism gradually comes into being. Preformation is the concept that the organism exists from the beginning, in either the sperm or egg, and development basically involves the growth of the preformed organism.
14. Epigenesis is the concept that the developing organism gradually comes into being. Epigenetics refers to alteration of genes not by alteration of deoxyribonucleotide sequences, but rather by chemical modification (e.g., by methylation) of the deoxyribonucleotides.
15. By means of a DNA-cutting enzyme guided to a target gene (DNA sequence) by a guiding specific RNA molecule; the target gene may be destroyed (cut) or replaced by another gene.

Chapter 5. Reproduction

1. Penis and testes; the penis is the male organ of sexual intercourse, and the testes produce semen and testosterone.
2. Leydig cells.
3. Head (caput), body (corpus), and tail (cauda).
4. Cauda of the epididymis.
5. Seminal vesicles, prostate gland, bulbourethral glands (Cowper's glands), Littre's glands; they nourish and protect sperm.
6. Fructose.
7. One-third of a billion.
8. Vulva; mons veneris, clitoris, labia majora, labia minora, fourchette, and vestibule.
9. The membranous support of the ovary.
10. Cortex.

11. Release of the egg from the ovary about every 28 days; ciliary beating and smooth muscle contraction aid in egg movement.
12. Corpus (body) and cervix (neck). The corpus contains the uterine cavity wherein the baby develops and the cervix is a passageway for sperm and the fetus.
13. Menstrual flow—much of the endometrium is lost; proliferative stage—the endometrium is rebuilt; secretory stage—the endometrium is making and secreting nutrients; ischemic stage—the endometrium undergoes degenerative changes.
14. Skene's glands and Bartholin's glands; their secretions apparently play a role in lubricating the female genitalia in preparation for the penetration of intercourse.
15. Nerve and endocrine signaling are often used for long-distance signaling, while autocrine, juxtacrine, and paracrine signaling are used for short-distance signaling.
16. Polar signals attach to cell surface receptors, while nonpolar signals attach to cytoplasmic receptors.

Chapter 6. Gametogenesis

1. Somatic cells and germ cells; germ cells are directly involved in reproduction and somatic cells are not.
2. The subset of germ cells directly involved in fertilization; spermatozoa and ova; gametogenesis, spermatogenesis, and oogenesis.
3. On the yolk sac on the 21st day (after fertilization).
4. Spermatogonia; oogonia.
5. Primary spermatocytes; secondary spermatocytes; spermatids.
6. Four spermatozoa, haploid.
7. "No" (but see Question 16); in the fetus before birth.
8. Hundreds of thousands; tens of thousands; none.
9. Four; one ovum and three polar bodies, haploid.
10. Head carries the paternal genes; midpiece provides the energy for sperm motility; tail is the motility mechanism.
11. First, to reach the egg; second, to fuse with the egg and deliver the paternal genes; third, to activate the egg to begin development.
12. pH.
13. Size—the sperm is smaller than the egg; motility—the sperm is motile and the egg is not; nutrient storage—the egg stores food and the sperm does not.
14. A secondary oocyte, a first polar body, a surrounding zona pellucida, a number of follicle cells (collectively called the corona radiata).
15. This question is not yet settled (see Chapter 6).
16. Oogenesis. Spermatogenesis.

Chapter 7. Fertilization

1. 12 hours, 2 days.
2. Corona radiata and the zona pellucida.
3. Sperm plasma membrane and egg plasma membrane.
4. Nucleus and centriole.
5. First, the egg (secondary oocyte) completes meiosis II; second, the sperm nucleus forms the male pronucleus; third, the sperm's centriole begins to organize the spindle; fourth, the human diploid number of 46 chromosomes is reestablished; fifth, the egg is activated.
6. Sex chromosomes, autosomes; sex chromosomes.
7. Father's.
8. On the X chromosome.
9. Female.
10. Fertilization, fusion of myoblasts to form skeletal muscle fibers, and formation of the syncytiotroblast part of the placenta.

Chapter 8. The embryonic period

1. The genome or genes, maternal inheritance information, genomic imprinting information.
2. Cleavage.
3. Eggs with little yolk, uniformly distributed, undergo cleavage of the whole egg, producing equal-size blastomeres.
4. Inhibitory.
5. In the fallopian tube; zona pellucida.
6. From the fertilization of the egg until the end of the eighth week of development.
7. Blastocyst; blastocyst cavity.
8. Inner cell mass and trophoblast; gap junctions and tight junctions.
9. Embryologist—266 days from time of fertilization; obstetrician—280 days from beginning of last menstrual period.
10. Hatch from the zona pellucida.
11. The burrowing of the conceptus into the endometrium.
12. Syncytiotrophoblast; fluid-filled spaces that appear between strands of the syncytiotrophoblast, which give rise to intervillous spaces.
13. Germ cells and blood cells.
14. Ectoderm, mesoderm, and endoderm.
15. Gastrulation.
16. Primitive streak, primitive node.
17. Chordates develop a notochord; vertebrates develop a vertebral column.
18. Formation of the neural tube.

19. Anencephaly and spina bifida.
20. Somitic mesoderm, intermediate mesoderm, and lateral mesoderm.
21. Kidneys; muscles of the gut and the body wall.
22. Human eggs are not provided with sufficient yolk to support their entire development.
23. Cardiovascular system.
24. Forebrain protuberance, cardiac prominence, liver bulge, somite bulges, lens thickenings, branchial (gill) arches, limb buds, and tail.
25. By 21 days (after fertilization).
26. Invagination, epiboly, involution, convergence, extension, and ingression.
27. Embryonic stem cells, trophoblast stem cells, and a 3D scaffold of extracellular material.
28. The two types of cells began to talk to each other.

Chapter 9. The fetal period

1. Growth and differentiation of previously established structures.
2. By the dramatic morphogenetic changes that occur.
3. The eyes close at the beginning of the early fetal stage; they open by the end of the middle fetal stage.
4. By about 12 weeks.
5. Vernix caseosa is a whitish, cheesy coat on the skin of the fetus; lanugo is fine, downlike hair confined to the skin of the fetus.
6. The wide-set closed eyes have moved from the sides to the front of the head.
7. The midgut loop extends into the umbilical cord; the midgut herniation withdraws from the umbilical cord into the enlarged abdomen during the first half of the early fetal stage (9–13 weeks).
8. Suck on the thumb.
9. The fetus normally drinks, as well as urinates into, amniotic fluid.
10. From the upper neck onto the sides of the head.
11. Fetal movements felt by the mother during the second half of the early fetal stage (14–18 weeks).
12. Brown fat, which produces heat for the newborn infant.
13. During the middle fetal stage, 19–28 weeks.
14. The cerebral hemispheres develop gyri (convolutions) and sulci (furrows).
15. During the fetal period.
16. Organoids are small versions of organs. derived from stem cells, by 2D and 3D culture.
17. Derived from stem cells, by 2D and 3D culture.
18. Cells must communicate.

Chapter 10. The placenta and the umbilical cord

Placenta

1. The placenta; it is the only organ that results from the cooperation of two separate individuals—the mother and the fetus.
2. Villi; intervillous spaces.
3. Chorion laeve and chorion frondosum.
4. Decidua; decidua basalis, decidua capsularis, and decidua parietalis.
5. The placenta provides for embryonic and fetal nourishment (food in), respiration (oxygen in and carbon dioxide out), and excretion (waste out).
6. The placenta separates prematurely from the lining of the uterus.
7. The placenta forms across the opening into the cervical canal.
8. To prevent infants born prematurely from developing complications of organ immaturity, especially chronic lung disease, resulting from current neonatal intensive care.
9. Lamb development for 28 days.
10. Embryo or fetus and placenta.
11. The body stalk of the early conceptus. A layer of amnion; yolk stalk, two umbilical arteries, one umbilical vein, and the allantoic diverticulum.
12. Two umbilical arteries carry blood toward the placenta and carry away carbon dioxide and wastes; one umbilical vein, carries blood toward embryo/fetus, replenished with oxygen and nutrients.
13. Blood stem cells, which are able to produce new blood cells as well as maintain themselves as a stem cell population.
14. Bone marrow transplant.
15. Malignancies (a number of leukemias, lymphoma, and neuroblastoma), blood disorders (including aplastic anemia and sickle cell anemia), immunodeficiencies (including severe combined immunodeficiency disorders [SCID]), and a number of inborn errors of metabolism (including Gaucher's disease and Hurler's syndrome).

Chapter 11. The pregnant woman, childbirth, and multiple pregnancies

The pregnant woman

1. Hormones and the increasing size of her uterus.
2. "Morning sickness," constipation, and heartburn; unknown origin, hormone-induced decreased motility of the bowel and pressure on the bowels from the enlarging uterus, and relaxation of the cardiac sphincter muscle, respectively.

3. Hemorrhoids, frequent urination, and a waddling gait.

Childbirth

4. Labor; the dilatation stage, the expulsion stage, and the placental stage; these stages are shorter in the multigravida than they are in the primigravida.
5. Expulsion begins with full dilatation of the cervix and ends with the birth of the baby.
6. Attitude, lie, presentation, and position.
7. Engagement, descent, flexion, internal rotation, extension, external rotation, and expulsion.
8. Uterine contractions that functionally clamp off uterine blood vessels.

Multiple pregnancies

9. Monozygotic and dizygotic; identical and fraternal.
10. Conjoined (Siamese) twins.
11. They are like ordinary brothers and sisters, having come from different fertilized eggs and just having developed in the same uterus at the same time.
12. Heteropaternal superfecundation involves two different fathers, homopaternal superfecundation involves one father.

Part II. Development of the systems of the body

Chapter 12. The skin

1. Epidermis and dermis; by birth.
2. Cell-mediated immunity, the immune system.
3. Pigmentation; they arise from the neural crest.
4. Merkel cells.
5. Dermis; dermatome, somatic mesoderm, and neural crest.
6. Epidermis and dermis.
7. In the 12th week of development; fine, down-like hair, confined to the fetal period.
8. Melanin pigment, provided for the skin by the melanocytes.
9. Sebaceous glands, sweat glands, and mammary glands.
10. Early in the fetal period (at 10 weeks); fingernails.
11. Epithelium and mesenchyme.

Chapter 13. The nervous system

1. The fluid-filled lumen that runs through the neural tube and gives rise to the spinal canal of the spinal cord; cerebrospinal fluid.

2. Ependymal cells, which make up the inside lining of the neural tube; neuroblasts, which give rise to the various kinds of neurons (nerve cells) found in the brain and spinal cord (central nervous system or CNS); and glioblasts, from which two types of CNS glial cells arise.
3. The ventricular layer, containing the ependymal cells; the intermediate layer, containing the cell bodies of the neuroblasts, which contain the nuclei of the nerve cells; and the marginal layer, made up of the processes (neurites: axons and dendrites) of the developing neurons.
4. The intermediate layer (containing the cell bodies of the neuroblasts) has a gray appearance and is therefore called gray matter; the marginal layer (made up of the processes of the developing neurons) is similarly referred to as white matter because of its appearance.
5. The spinal nerves have both sensory and motor components.
6. The anterior region of the neural tube.
7. The cortex (outer portion) of our cerebrum.
8. The telencephalon and diencephalon arise from the forebrain, the midbrain persists as the mesencephalon, and the hindbrain gives rise to the metencephalon and the myelencephalon.
9. Two outgrowths of the telencephalon, the cerebral hemispheres, contain ventricles I and II; these ventricles communicate with ventricle III, making up the remainder of the telencephalon and diencephalon by means of the foramina of Monro; ventricle III in turn communicates—through the aqueduct of Sylvius (lumen of the mesencephalon)—with ventricle IV, which is the lumen of the metencephalon and myelencephalon.
10. The brain itself, the neural crest, and thickenings of the ectoderm called ectodermal placodes.
11. Nuclei.
12. Nerve tracts.
13. From neural crest cells.
14. A process in which most axons, which make up nerves in the PNS and nerve tracts in the CNS, become coated with a fatlike substance called myelin; myelination enables nerves and nerve tracts to become functional.
15. Developmental neurogenesis occurs before birth, adult neurogenesis occurs after birth.

Chapter 14. The skeleton and muscles

Skeleton

1. The axial skeleton is composed of the skull, vertebral column (backbone), sternum, and ribs; the appendicular skeleton is composed of our limbs, pelvic girdle, and pectoral girdle.
2. Osteogenesis is bone formation; intramembranous ossification and endochondral ossification.

3. Somites.
4. The nucleus pulposus, which occupies the core of each intervertebral disc.
5. Neurocranium and viscerocranium; both parts are formed partially from intramembranous ossification and partially from endochondral ossification.
6. Newborn babies have unclosed sutures (in which two bones come close together) and fontanelles (spaces in which more than two bones come together); these temporary discontinuities aid passage of the head through the birth canal at childbirth.
7. Epiphyseal plates provide for interstitial growth and periosteum provides for appositional growth.
8. Fusion of the epiphyseal plates.

Muscles

9. Mesoderm.
10. Dorsal (upper) somatic mesoderm and ventral (lower) splanchnic mesoderm; splanchnic mesoderm provides muscle tissue for the circulatory, gastrointestinal, and respiratory systems; somatic mesoderm gives rise to the appendicular skeleton.
11. The coelom; the peritoneal cavity, the pericardial cavity, and the pleural cavity.
12. Skeletal muscle powers arms and legs; smooth muscle enables our stomachs and bladders to contract; cardiac muscle of the heart pumps our blood.
13. Sarcomeres.
14. Satellite cells.
15. It is not endowed with a large amount of yolk, and the rapidly growing embryo must develop a functional circulatory system to obtain food and eliminate wastes.
16. Unique cell junctions.

Chapter 15. The circulatory system and hormones

1. The head fold.
2. A type of connective tissue between the outer epimyocardial tube and the inner endocardial tube.
3. A single pericardial cavity; the amniocardiac vesicles.
4. The dorsal mesocardium.
5. Sinus venosus, atrium, ventricle, and truncus arteriosus.
6. The interventricular sulcus is a depression over the surface of the apex of the ventricular loop formed by bending of the cardiac tube; the interventricular septum is opposite the interventricular sulcus. Strands of heart muscle coalesce to form the interventricular septum on the interior of the ventricular wall; this septum will contribute

to the partitioning of the ventricular region of the cardiac loop into right and left ventricles.

7. The right atrium and left atrium; the left atrium.
8. Between the pulmonary trunk, leaving the right ventricle, and the aorta, leaving the left ventricle; it provides the right ventricle with somewhere to pump its load of blood, which is necessary for its proper development.
9. Arteries, veins, and capillaries; mesoderm.
10. The two umbilical arteries carry blood that is deficient in oxygen and nutrients out to the placenta, and the single umbilical vein carries replenished blood back into the body of the fetus.
11. Clusters of mesenchyme cells form on the yolk sac, which give rise to endothelial linings of blood vessels and to primitive blood cells.
12. The formation of blood cells.
13. Hormones.
14. Pituitary gland (ectoderm), thyroid gland (endoderm), gonads (mesoderm).
15. Plays a role in the onset of puberty, triggers ovulation, rebuilds the lining of the uterus after menstruation, stimulates the sweeping of the end of the fallopian tube over the surface of the ovary at the time of ovulation (in addition to maintaining pregnancy), rescues the corpus luteum and saves the early pregnancy, stimulates the smooth muscle contraction initiating labor and childbirth, and prepares the breasts for postnatal lactation.
16. Fetal heart tissue.
17. The regenerative capacity of immature human heart tissue in response to injury.
18. Regenerate.

Chapter 16. The digestive system and the respiratory system

Digestive system

1. Endoderm.
2. Foregut and hindgut; midgut.
3. Membranes that suspend the digestive organs.
4. The pharynx, esophagus, stomach, liver, biliary apparatus, pancreas, and a portion of the small intestine.
5. The esophagus may become occluded (blocked) by an overproduction of epithelial cells.
6. Two; the ventral pancreatic bud and the dorsal pancreatic bud.
7. Much of the colon and small intestines, the cecum, and the appendix.
8. Space for development.
9. A large part of the colon and the rectum.

10. During the seventh week of development.
11. Meconium is composed of cells and hair from the amniotic fluid, cells from the lining of the digestive tube, and other matter.
12. It does not normally defecate into it; meconium-stained amniotic fluid found during amniocentesis is a sign of fetal stress.

Respiratory system

13. From the early digestive system; endoderm.
14. The distal portion of the trachea; growth and branching of the lung buds.
15. Pseudoglandular period, canalicular period, terminal sac period, and alveolar period. They overlap because of the cephalic-to-caudal progression of lung development.
16. Only from the beginning of the terminal sac period.
17. A lipid (fatty material) mixture that prevents the lungs from collapsing with each breath; during the terminal sac period.
18. Lung development.
19. Lung disease.

Chapter 17. Mouth and throat, face, and the five senses

Mouth and throat

1. An inpocketing of ectoderm on the ventral side of the embryo's head; the oropharyngeal membrane.
2. In the boundary between the oral cavity and the throat or pharynx; during the 24th day of development.
3. In the anteriormost portion of the foregut.
4. The mandibular arches; formation of much of the jaws.
5. The thymus gland and parathyroid glands.

The face

6. Nasal placodes; nasal pits.
7. The nasopharynx.
8. On the roof of each nasal cavity; sensory olfactory cells, olfactory nerves.
9. From the sides of the neck onto the sides of the head.
10. The premaxillary portion of the upper jaw.
11. The original oral cavity into an oral cavity and a nasal cavity; a pair of tissue folds grow backward (caudad) from the posterior margins of the hard palate and by their fusion give rise to the soft palate and uvula.
12. Cleft palate.

Five senses

13. The inner layer of the cup (the nervous layer) develops photoreceptor cells (called rods and cones), which will receive light energy and change it into nerve impulses. The outer layer of the cup (the pigmented layer) develops pigment cells and becomes increasingly pigmented as development proceeds so as to absorb light not intercepted by the nervous layer.
14. The cornea.
15. Ossified cartilage of the upper ends of the first two branchial arches; expanded distal portions of the first pharyngeal pouches.
16. The foramen cecum, the thyroid diverticulum.
17. Taste cells, supporting cells.
18. Free nerve terminations and encapsulated nerve endings.

Chapter 18. The urinary and reproductive systems and the external genitalia

Urinary and reproductive systems

1. Because of their close structural relationship in both time and space; intermediate mesoderm.
2. Pronephros, mesonephros, and metanephros; they all develop in pairs, grow progressively from cephalic to caudal locations, are bilaterally disposed across the midline, and have a tubular nature.
3. From the mesonephric ducts, cephalic to the level where the ducts enter the cloaca; ureters, calyces, and collecting tubules of the kidneys.
4. The cloaca.
5. From splanchnic mesoderm behind the peritoneum; the membranous support for a testis is the mesorchium and for an ovary is the mesovarium.
6. Female; the developmental pathway is shifted to male development.
7. The cephalic portion of the mesonephric duct and some of the mesonephric tubules become parts of the epididymides, whereas much of the rest of the mesonephric ducts become the vasa deferentia and ejaculatory ducts; their cephalic ends remain separate, giving rise to the paired fallopian tubes, whereas their caudal ends fuse, giving rise to the uterus and vagina.
8. Skene's glands are homologous to the male prostate glands; the glands of Bartholin are homologous to the male bulbourethral glands.

External genitalia

9. The external genitalia of the male are the penis and scrotum. The penis is the male sex organ and also delivers urine to the outside of the body; the scrotum contains the male gonads and maintains them

at a temperature lower than that of body temperature, apparently a necessary condition for normal spermatogenesis. The external genitalia of the female are the mons veneris, clitoris, labia majora, labia minora, fourchette, and vestibule. The clitoris is highly erogenous, and the vagina (which opens into the vestibule) carries the products of the female reproductive system and is also the female organ of intercourse.

10. By the 12th week of development.
11. In the male, the genital tubercle gives rise to the penis; fusion of the labioscrotal folds gives rise to the scrotum. As the penis forms (which involves growth of the genital tubercle and fusion of the urogenital folds), a portion of the urogenital sinus is incorporated into the penis as the penile urethra. In the female, growth of the genital tubercle gives rise to the clitoris. The labioscrotal folds grow into the labia majora; the vestibule is the persistent urogenital sinus.
12. Clitoris, labia majora, vestibule.

Part III. The impact of development on life

Chapter 19. Birth defects

1. The study of abnormal development; a distinct biological discipline resulting from the marriage of genetics (the study of heredity) and cytology (the study of cells).
2. Chemical (thalidomide, tetracycline), physical (x-rays), and biological (viruses).
3. Thalidomide.
4. Down's syndrome.
5. Sex chromosomes (XO, Turner's syndrome) and autosomes (trisomy 21, Down's syndrome).
6. Chromosome number (aneuploidy, triploidy, tetraploidy) and structure (*cri du chat*).
7. An individual who has more than one type of cell, as far as chromosome number is concerned, making up his or her body.
8. When organ systems are forming and, therefore, when teratogens have their most dramatic effects; from the beginning of the third week through the end of the eighth week.
9. Direct viewing of the fetus; sometimes greater detail than can be provided by ultrasound is needed, but it is much more invasive and risky.
10. Withdrawal of a sample of amniotic fluid; fluid and cells.
11. The ratio of lecithin to sphingomyelin.

Chapter 20. Birth control

1. Complete abstinence from sexual intercourse or any other sexual activity in which semen may enter into the female reproductive system.

2. Identify the fertile days of the month; rhythm (calendar) method, basal body temperature method (BBT), and cervical mucus method.
3. To interfere with the union of the sperm and egg.
4. Prevention of the sperm from getting from the cauda epididymis to the ampulla of the fallopian tube, where fertilization occurs. Vasectomy prevents sperm from entering the ejaculate, condoms prevent semen from entering the vagina, coitus interruptus (withdrawal) is supposed to prevent semen from entering the vagina.
5. Diaphragms and cervical caps physically prevent sperm from entering the cervical canal, tubal ligation removes the egg's pathway for union with sperm, spermicidal foams immobilize and kill sperm, "the pill" interferes with the production of eggs.
6. Pregnancy (or implantation).
7. RU 486, IUD (intrauterine device).
8. Emergency contraceptive drugs that contain a high dose of certain common birth control pills; they prevent pregnancy by rapidly elevating a woman's hormone levels, which prevents the fertilized egg from implanting into the endometrium.
9. Spontaneous abortion.
10. (a) Suction method, (b) dilation and evacuation, (c) injection of a hypertonic salt solution or a prostaglandin solution into the amniotic fluid to induce uterine contractions. Third-trimester induced abortions are rare.
11. (a) A hormonal gel, which is rubbed onto the upper arms and shoulders of men once a day; (b) a male pill; and (c) a nonhormonal method, wherein a polymer gel is injected into the vas deferens.

Chapter 21. Reproductive technology

1. Rendezvous of gametes; by bringing them together outside of the woman's body.
2. Ultrasound.
3. To develop embryos from zygotes; to transfer the embryos into the woman's uterine cavity.
4. Because, generally, several embryos are transferred into the uterus to increase the chances of a single pregnancy.
5. Gamete intrafallopian transfer (GIFT) begins with gathering of sperm and eggs, as if in preparation for IVF; next, however, first the eggs and then the sperm are transferred into the fallopian tubes. Fertilization occurs more naturally with GIFT than with IVF, and the embryo formed subsequently enters the uterus by the normal route.
6. Transfer takes place into the fallopian tube; zygotes (rather than gametes) are transferred.
7. Genetic mothers produce the egg, gestational mothers carry the pregnancy for the genetic father (usually their husbands), surrogate mothers carry the pregnancy of a couple unrelated to them.

8. By injecting a single spermatozoon into a single egg's cytoplasm.
9. NSA aspirates the spermatozoa directly from the testis through a thin needle; MESA aspirates epididymal sperm. NSA and MESA are used on men with no spermatozoa or only dead spermatozoa in their ejaculates.
10. A single blastomere; polymerase chain reaction (PCR).
11. The creation of organisms from preexisting organisms by asexual means. Cloning may be accomplished with animals either naturally (some animals reproduce asexually) or artificially (as by shaking apart a four-cell sea urchin embryo to produce four sea urchins).

Chapter 22. Sexually transmitted diseases

1. Syphilis, gonorrhea, chlamydia (bacterial); AIDS, genital herpes (viral).
2. Direct introduction into the circulatory system by intravenous drug use or blood transfusions and indirect introduction into the circulatory system by sexual intercourse, either vaginal or anal.
3. Unexplained tiredness; unexplained fever; shaking chills; soaking night sweats; long-lasting swollen glands (over several weeks); weight loss not due to dieting; white spots on the tongue or in the mouth; persistent diarrhea; unexplained dry cough; and pink or purplish flat or raised blotches that don't go away on or under the skin and inside the mouth, nose, eyelids, or anus.
4. A bacterium (spirochete), *Treponema pallidum*; it can be acquired (usually during sexual intercourse) or congenital (passed from mother to child *in utero*).
5. The gonococcal bacterium *Neisseria gonorrhoeae*. Gonorrhea can lead to pelvic inflammatory disease in women and sterility by scarring in the vas deferens in men, which inhibits the travel of sperm.
6. Herpes is caused by a virus and is therefore not curable with antibiotics. Once infected with herpes, you are infected for life.
7. Pelvic inflammatory disease, ectopic pregnancy, infertility, and dangerous complications during pregnancy and birth.
8. PID can lead to infertility and ectopic pregnancy.
9. A pregnancy outside the uterus; the fallopian tube. A tubal pregnancy is a medical and life-threatening emergency.
10. Delivery by cesarean section.
11. Because the test is for *antibodies* against the HIV virus and all babies born to HIV-infected mothers are born with the HIV antibodies.
12. The eyes of the newborn can be infected during birth from a mother with gonorrhea.

Chapter 23. Cultural aspects of development

1. Abnormal development in which there is no embryo or fetus, only an abnormal placenta, which somewhat resembles a cluster of grapes or large water droplets.
2. When an Rh– mother (Rh–Rh–) is carrying an Rh+ conceptus, the conceptus makes a protein perceived as foreign by the mother's immune system. When this happens, the mother's immune system mounts an attack against the foreign protein, resulting in erythroblastosis fetalis.
3. The amniotic fluid was released from the ruptured amniotic cavity at the beginning of labor.
4. The torn fetal membranes that happen to drape over the head of the newborn and have no significance.
5. A defect in which closure of the foramen ovale does not occur at the time of birth.
6. Monozygotic twins arise from one zygote, dizygotic twins arise from two zygotes, and conjoined twins are twins that are anatomically joined together.
7. Siamese twins are conjoined twins, fraternal twins are dizygotic twins, and identical twins are monozygotic twins.
8. Cryogenic (frozen) maintenance of sperm specimens in sperm bank facilities is intended for the future use of such specimens for artificial insemination.
9. Introduction of semen into the female reproductive tract by means other than a penis.

Appendix A: Examples of birth defects by terminology

Roots

amnios

Referring to the amnion and, in the context of human development, a part of terms describing abnormal amounts of amniotic fluid. Examples:

polyhydramnios an excessive volume of amniotic fluid.

oligohydramnios a deficiency of amniotic fluid.

cele

Referring to a rupture or hernia. A number of significant birth defects involve ruptures. Also see *hernia*. Examples:

meningocele a protrusion of the cerebral or spinal meninges (membranes) through a defect in the skull or vertebral column, forming a cyst filled with cerebrospinal fluid.

meningocephalocele a protrusion of part of the cerebellum and meninges through a defect in the skull.

meningomyelocele a protrusion of a portion of the spinal cord and membranes through a defect in the vertebral column.

omphalocele a result of a failure of the developing gut to return completely to the abdominal cavity from its embryonic location in the umbilical cord. It is a congenital umbilical hernia into which intestine and peritoneum protrude.

cephaly

Referring to the head. A number of birth defects involve the abnormal development of the head. Examples:

acrocephaly a condition in which the head is roughly conical in shape, caused by premature suture closure; also called *oxycephaly*.

macrocephaly an abnormal largeness of the head; also called *megalocephaly*.

microcephaly a congenitally smaller-than-normal head, often but not always associated with mental retardation. This condition may be due to congenital hypoplasia of the cerebrum.

plagiocephaly a type of strongly asymmetric cranial deformation due to a number of causes, such as a disordered sequence of suture closure.

scaphocephaly a long, narrow skull resulting from early closure of one of the sutures of the early skull.

crani

Referring to the skull, specifically the part that encloses the brain. Example:

acrania partial or complete absence of the cranium.

dactyly

Referring to the fingers or toes. A number of specific kinds of birth defects involve dactyly, depending on what happens to the digits during development. Recall that during development, the digits are "sculpted" out of a hand or foot plate by specific, genetically programmed cell death. Examples:

brachydactyly abnormal shortness of the fingers or toes.

polydactyly existence of extra (supernumerary) fingers or toes.

syndactyly adhesion of fingers or toes; webbed fingers or toes.

encephaly

Referring to the brain. In addition to the mental retardation associated with many chromosome abnormalities, structural malformations of the brain are represented in human development. Examples:

anencephaly without a brain. More specifically, it refers to the absence of the cerebrum, cerebellum, and flat bones of the skull.

meroanencephaly partial absence of the brain.

glossia

Referring to the tongue. Examples of birth defects involving the tongue:

ankyloglossia an abnormal attachment of the tongue (sometimes referred to as tongue-tie) to the floor of the oral cavity (mouth).

macroglossia a large tongue.

microglossia a small tongue.

hydro

Referring to water; in the context of development, an abnormal accumulation or deficit of water. Examples:

hydrocephalus a condition involving distention of the ventricular system (the ventricles are the cavities within the brain), which may be caused by overproduction or defective absorption of cerebrospinal fluid. In infants, it leads to increased head size and thinning of skull bones, with suture separation.

hydronephrosis a condition of gross dilation of the kidney pelvis and calices with urine, caused by an obstruction of the urinary tract.

Both of these conditions have been successfully treated with *in utero* surgery. For an example of an abnormal deficit in fluid accumulation, see **oligohydramnios**.

melia

Referring to the limbs. Examples:

amelia congenital absence of limbs.

meromelia deficits of limb development, in which parts of limbs are missing.

phocomelia birth defects in which the hands or feet or both are attached directly to the shoulders or hips, with the intervening parts being absent.

The drug thalidomide caused these types of birth defects.

plasia

Denotes formation, development, or growth. *plasia* is used with a number of prefixes to form the names of birth defects having to do with abnormal growth. Examples:

achondroplasia typical congenital dwarfism due to abnormal bone formation, specifically abnormal chondrification and ossification of the ends of the long bones; often inherited as a dominant trait.

adrenal hyperplasia abnormally excessive growth of the adrenal glands due to an inherited enzyme deficiency; leads to overproduction of male hormones, which causes masculinization in females and may cause early sexual development in males.

enamel hypoplasia underproduction of the enamel of the teeth, which may be caused by tetracycline (an antibiotic) use, as well as by other factors.

schisis

Referring to a cleft or split. A number of important birth defects include such an abnormality. Examples:

myeloschisis a complete or partial failure of the neural plate to form a neural tube, resulting in a cleft spinal cord.

rachischisis a condition characterized by fissure, incomplete closure, or absence of vertebrae of the spinal column; also called *spina bifida*.

stomia

Referring to the mouth. Examples:

macrostomia an abnormally large mouth.

microstomia an abnormally small mouth.

Types of birth defects

Agenesis

The absence of an organ or other structure from early development. The precise kind of agenesis refers to the missing organ. Examples:

anal agenesis literally, the lack of the anal opening (anus), but also used to refer to an abnormal opening as into the vulva in females or the urethra in males. The latter variety results from incomplete partition of the cloaca by the urorectal septum.

renal agenesis the lack of formation of one or both kidneys, the latter causing death soon after birth; caused by the failure of metanephric diverticulum development.

Atresia (imperforation)

An abnormal closure of a normal opening or canal. Examples:

imperforate anus an abnormal closure of a normal anal opening.

imperforate hymen an abnormal closure of a normal vaginal opening into the vestibule.

esophageal atresia an abnormal closure of a canal, in this case the esophagus. Because this condition prevents the fetus from swallowing amniotic fluid, polyhyramnios results.

Birth defects involving sexuality

Examples:

hermaphroditism the presence of recognizable ovarian *and* testicular tissue, in which sexual structures and secondary sex characteristics may show any combination of maleness and femaleness. About 50% of these individuals have an XX sex chromosome makeup, about 25% have XY makeup, and the remainder are XX/XY mosaics. In Greek mythology, Hermaphrodites, a son of the Greek gods Hermes and Aphrodite, coalesces with a nymph who is in love with him.

pseudohermaphroditism a genetic disorder in which the external genitalia resemble one sex whereas the gonads are those of the opposite sex.

testicular feminization a result of a sex-linked mutation in which tissues that normally respond to androgens (male hormones, primarily testosterone) do not respond. As a consequence, affected individuals, though carrying the XY sex chromosome makeup, are phenotypically females.

Exstrophy

The eversion or turning inside out of a part. Example:

exstrophy of the bladder the condition in which part of the abdominal wall and the anterior wall of the bladder are missing, and the posterior wall of the bladder bulges through the opening.

Fistula

An abnormal connection between two internal structures or from an internal structure to the external surface of the body. Examples:

brachial fistula a fistula between an internal structure and the external surface of the body, from the pharynx to an opening on the side of the neck.

rectovaginal fistula a fistula between two internal structures—the vagina and the rectum.

Hernia

An abnormal protrusion of an organ or a part through the containing wall of its cavity is called a hernia or rupture. Also see *cele*. Examples:

diaphragmatic hernia a hernia that passes through the diaphragm into the thoracic cavity; it may be congenital or acquired.

umbilical hernia a hernia occurring through the umbilical ring, which may be congenital (a birth defect) due to imperfect closure of the umbilical ring. The umbilical ring is a dense fibrous ring surrounding the umbilicus at birth. An umbilical hernia may also be acquired later in life.

Stenosis

A narrowing. A stenosis can have disastrous developmental effects, depending on what is being narrowed. Examples:

anal stenosis a narrowing of the anal canal, probably due to urorectal septum deviation in partitioning of the cloaca.

pulmonary stenosis a narrowing of the pulmonary trunk, coming from the right ventricle of the heart, due to unequal partitioning of the truncus arteriosus.

pyloric stenosis unlike the latter two examples of stenosis due to abnormal partitioning of a cavity, the narrowing of the pyloric region of the stomach is due to thickening (hyperplasia) of the wall of the pylorus.

Appendix B: Examples of birth defects by organ or system

Eyes

cataract a partial or complete opacity of the lens; although usually thought of as a result of aging, cataracts may be caused by rubella virus during the embryonic period.

glaucoma congenital glaucoma results from a defect of the eye that impairs the flow of aqueous humor, leading to increased intraocular pressure and eye tissue damage; also generally associated with aging. Glaucoma may result from rubella virus infection during the embryonic period.

Kidneys

ectopic kidney a congenital anomaly in which the kidney is held in an abnormal position.

horseshoe kidney a greater or lesser degree of congenital fusion of the two kidneys, usually at the lower poles.

pelvic kidney a kidney abnormally located in the pelvis.

Lungs

hyaline membrane disease associated with the initial ventilation of lungs that lack adequate surfactant stores. Lack of surfactant leads to uneven distribution of ventilation, inadequate gas exchange, and respiratory distress. The hyaline membrane is the layer of fibrin and cellular debris found adhering to the walls of alveoli, alveolar ducts, and respiratory bronchioles on postmortem examination. Also called *respiratory distress of the newborn*.

Mouth

cleft lip a congenital defect of the upper lip; sometimes referred to as *harelip*.

cleft palate a congenital defect due to failure of fusion of embryonic facial processes (the lateral palatine processes), resulting in a fissure through the palate.

Skin and derivatives

albinism hereditary absence of pigment from the skin, hair, and eyes.

alopecia congenital loss of hair.

hypertrichosis excessive growth of hair.

ichthyosis congenita a severe form of a skin condition characterized by dry, harsh skin with adherent scales. The affected fetus may be stillborn or live a few days.

polymastia presence of more than two breasts.

supernumerary nipples presence of more than two nipples.

Circulatory system

atrial septal defect a common congenital defect characterized by an abnormal opening in the septum between the right and left atria; also referred to as *ASD*.

ventricular septal defect a congenital anomaly characterized by an abnormal opening in the ventricular septum between the right and left ventricles; also called *VSD*.

patent ductus arteriosus a ductus arteriosus that remains open after birth.

hemolytic disease of the newborn a hemolytic anemia seen in the newborn, caused by transfer of maternal antibodies across the placenta in response to Rh incompatibility; also called *erythroblastosis fetalis*.

Limbs

clubfoot a congenital malformation in which the forefoot is inverted and rotated.

congenital dislocation of the hip a potentially crippling abnormality, commonly involving one or both hip joints and, though present at birth, often discovered only after the child starts to walk.

lobster claw deformity congenital absence of all fingers or toes except the first and fifth; also called *bidactyly*.

Skeleton

lumbar rib a common type of accessory rib formed by abnormal development of lumbar vertebrae.

scoliosis abnormal lateral curvature of the spine.

Glossary

2-dimensional culture: *in vitro* cultures where the cells or tissues grow flat on their substratum (glass, most often plastic, or extracellular matrix).

3-dimensional culture: *in vitro* cultures where the organs or embryos grow in three dimensions rather than flat on their substratum.

abdominal cavity or peritoneal cavity: derived from the coelom and containing much of the viscera (internal organs).

ABO blood typing system: a system of typing human blood for the purpose of blood transfusion.

abortion: the process of terminating a pregnancy with the demise of the conceptus; induced rather than spontaneous.

abortus: that which results from an abortion; an aborted embryo or fetus.

abruptio placentae: premature separation of the placenta from the lining of the uterus.

accessory sex glands: in the male, seminal vesicles, prostate gland, Cowper's glands, and the glands of Littre; in the female, Bartholin's glands and Skene's glands.

acrocentric chromosome: a chromosome in which the centromere is not equidistant from the two ends of the chromosome. Therefore, the chromosome's two arms are of unequal length.

acrosome: an organelle, found in the head of the spermatozoon, containing hydrolytic enzymes that play a role in the sperm reaching the plasma membrane of the egg.

actin: a rather ubiquitous protein in eukaryotic cells, occurring in especially abundant and organized form in skeletal muscle (where it is organized into thin filaments).

adenosine triphosphate (ATP): a so-called high-energy molecule, which is the almost ubiquitous source of immediate chemical energy for biochemical reactions.

adrenal glands: a pair of endocrine glands found attached to the cephalic poles of the kidneys.

adrenaline: see **epinephrine**.

adult neurogenesis: that which, recently discovered, occurs in adult mammals.

afterbirth: that which is born after the baby, namely the placenta, membranes, and part of the umbilical cord.

AIDS: acquired immune deficiency syndrome, a viral disease that may be transmitted from mother to offspring.

albino: an organism lacking pigment.

alcohol: a chemical teratogen.

alkaline phosphatase: an enzyme involved in the deposition of calcium phosphate crystals during ossification.

allantoic diverticulum: an evagination or diverticulum of the hindgut, composed of endoderm and mesoderm. In human development, the endoderm of the allantoic diverticulum is rudimentary, but the mesoderm gives rise to important blood vessels of the umbilical cord.

allantois: one of the four extraembryonic membranes; arises from the hindgut and is not conspicuous in human development.

allele: most genes are present as two copies; each copy is called an allele.

alveolar period: a period during lung development when the alveoli characteristic of the lung form.

alveoli, primitive: alveoli (terminal air sacs of the lungs) before their linings have become as thin as they will later in lung development.

ameloblasts: cells of the enamel organ, derived from oral epithelium, of the developing tooth that give rise to the enamel of the developing tooth.

amino acids: the building blocks of protein molecules.

amniocentesis: a prenatal diagnostic procedure that is initiated by obtaining a sample of amniotic fluid from the amniotic cavity. This procedure provides fetal cells as well as fluid for analysis.

amniocyte: cell found in the amniotic fluid.

amnion: one of the four extraembryonic membranes formed during the development of higher vertebrates, including humans; it makes up the wall of the amniotic cavity.

amniotic cavity: a fluid-filled cavity that begins to form on the eighth day of human development. During their development, both the embryo and fetus float in this, their private aquarium.

amniotic fluid: the fluid that fills the amniotic cavity and in which the embryo and later the fetus float during their development. Amniocentesis is a prenatal diagnostic procedure that begins with procuring a sample of this fluid.

amphibians: members of the class of vertebrates (Amphibia) that includes frogs and salamanders.

ampulla: the portion of the fallopian tube, closest to the ovary, in which fertilization normally occurs.

anal membrane: the membrane formed of proctodeal ectoderm and hindgut endoderm. Its normal breakdown creates the opening called the *anus*.

analgesics: drugs that relieve pain.

anaphase: the stage of mitosis or meiosis when the daughter chromosomes are moving toward opposite poles of the spindle.

anatomic position: standing erect with the arms at the sides and the palms forward.

anencephaly: a type of birth defect in which most of the brain is missing.

aneuploid: a number of chromosomes in a cell that is not an exact multiple of the haploid number of chromosomes for the species; see **euploid**.

anovulation: the lack of ovulation.

anterior neuropore: the temporary opening at the cephalic end of the early neural tube.

anticodon: the triplet of nucleotides in a transfer RNA molecule that hydrogen bonds to a codon of messenger RNA.

antral vacuoles: tiny fluid-filled spaces found among the follicle cells of a developing ovarian follicle.

anus: the caudal opening of the gut to the outside of the body.

aorta: the largest arterial blood vessel in the body, which carries blood away from the heart to various parts of the body.

aortic arches: a series of six arterial blood vessels that arise in early development in conjunction with the pharyngeal arches. Although all six aortic arches do not persist, several give rise to important prenatal and adult arterial blood vessels.

appendicular skeleton: the skeleton of the arms and legs and the associated pectoral and pelvic girdles, respectively.

appendix: a small, blind pouch attached to the cecum.

appositional growth: the growth of an object by the addition of material to its surface, as in the growth of the long bones.

aqueduct of Sylvius: the cerebrospinal fluid-filled cavity of the brain, which connects ventricles III and IV.

arrector pili muscle: the smooth muscle fiber, derived from the dermis, associated with each hair of the skin.

arteries: blood vessels that carry blood away from the heart. Although this blood is generally rich in food and oxygen, the paired umbilical arteries carry depleted blood from the fetus out to the placenta.

artificial insemination: introduction of sperm into the vagina by a means other than a penis.

asymmetric cell division: cell division where the two daughter cells have different fates; e.g., division of a primary oocyte, during oogenesis, to produce a larger secondary oocyte and a smaller first polar body.

astrocyte: a type of glial cell found in the central nervous system; derived from the neuroepithelium.

ATP: see **adenosine triphosphate**.

atrioventricular canal: the passageway between the atrial and ventricular regions of the heart, which is subsequently divided by endocardial cushion tissue into right and left channels.

atrium: one of two kinds of chambers that make up the four-chambered heart. The atria pump blood through valves and into ventricles.

attitude: one of four descriptions of the fetus's alignment in the uterus. Attitude is the posture that the fetus assumes in the uterus near the end of pregnancy.

auditory nerves: the eighth pair of cranial nerves. These nerves connect the inner ear with the brain and provide the sensory input for hearing and balance.

auditory vesicles: see **otic vesicles**.

autocrine signaling: a mode of cell–cell communication in which signaling molecules (autocrine factors) attach to receptors on the same cell that produced them, e.g., the explosive proliferation of placental cytotrophoblast cells in response to platelet-derived growth factor (PDGF), which these cells themselves produce.

autonomic ganglia: ganglia of the autonomic nervous system.

autonomic nerves: nerves of the autonomic nervous system.

autonomic nervous system: the portion of the nervous system involved with control of involuntary activity, such as that of the internal organs.

autosomes: all the chromosomes other than the sex chromosomes.

axoneme: the core of the flagella of sperm, made up of microtubules, the bending of which propels the sperm.

axial skeleton: the portion of the skeleton composed of the skull, vertebral column, sternum, and ribs.

axons: nerve cell processes, generally less numerous and longer than dendrites, which carry nerve impulses away from the cell body of the nerve cell.

azoospermia: the absence of sperm in semen.

AZT: azidothymidine (also zidovudine); a drug used to slow the replication of the HIV virus in patients with AIDS.

B cells or B lymphocytes: a subset of white blood cells that produce antibodies.

backbone: a term in common usage referring to the bony vertebral column of vertebrates, including humans.

balloon catheterization: a procedure used to open a blocked or partially blocked body tube by the expansion of a balloonlike device in the blocked portion of the tube.

banding techniques: chemical treatments of chromosomes that cause them to have a consistent pattern of bands (stripes).

Bartholin's glands: female auxiliary sex glands, the ducts of which open into the vestibule of the vulva. Their secretions provide lubrication at the time of sexual intercourse.

basal body temperature (BBT) method: a fertility awareness method based on an increase in the BBT at the time of ovulation.

basal layer: the inner layer of the two-layered skin of the embryo at the end of the embryonic period; see **periderm**.

base pair: a pair of nitrogenous bases (each a part of a nucleotide in a nucleic acid molecule) hydrogen-bonded to each other.

base-pairing rules: adenine hydrogen bonds to thymine in DNA, adenine hydrogen bonds to uracil in RNA, and cytosine hydrogen bonds to guanine in DNA and RNA.

bifurcate: split into two parts; branch.

bilateral symmetry: a type of symmetry exhibited by some organisms in which only one plane, the midsagittal, divides the organism into mirror image halves.

binocular vision: using both eyes synchronously to see one visual image, as opposed to double vision.

biochemical cell differentiation: cell differentiation involving an elaboration of biochemicals rather than an elaboration of structural features.

birth: the process by which an infant normally leaves the uterus (womb) through the birth canal.

birth canal: the conduit, partially the uterine cervix and partially the vagina, through which the baby passes during birth.

birth defect: a defect in development that arises before birth and is usually, *but not always*, apparent at birth.

bivalent: a pair of homologous chromosomes in synapsis during prophase I of meiosis.

bladder: see **urinary bladder**.

blastocoele: the cavity of the blastocyst.

blastocyst: an early embryonic form of mammals, containing two types of cells (trophoblast cells and inner cell mass cells) and a fluid-filled cavity (blastocoel cavity). This embryonic form is the equivalent of the blastula in other animals. However, an important difference is that all the cells of a blastula contribute to the developing organism, but only the inner cell mass, not the trophoblast, contributes to the developing mammal. The trophoblast contributes to the development of extraembryonic membranes.

blastocyst cavity: the blastocoele or cavity of the blastocyst.

blastomeres: the cells of an embryo undergoing cleavage.

blastula: an early embryonic form of animals generally consisting of a layer of cells (blastomeres) surrounding a fluid-filled cavity, the blastocoele.

blood islands: aggregations of mesenchyme cells, derived from the splanchnic mesoderm of the yolk sac, which give rise to the yolk sac blood vessels and the embryo's first blood cells.

blood plasma: the liquid portion of the blood, that is, not including the blood cells or so-called formed elements.

body cavity: the coelom is the original, single, body cavity of the embryo. During development, it is subdivided into pericardial, peritoneal, and pleural cavities, which contain the viscera.

body folds: folds in the germ layers of the developing embryo, which initially mark off embryonic regions from extraembryonic regions, and, subsequently, form the sides and ventral surface of the developing embryo.

body stalk: see **connecting stalk**.

bone marrow: hematopoietic (blood-forming) tissue found within certain bones.

Bowman's capsule: the infolded, blind end of the uriniferous tubule in which the glomerulus (tuft of capillaries) is found.

brain vesicles: subdivisions of the early developing brain; consisting first of three vesicles (prosencephalon, mesencephalon, and rhombencephalon) and then of five vesicles (telencephalon, diencephalon, mesencephalon, metencephalon, and myelencephalon).

branchial (gill) arches: the successive bulges in the lateral walls of the pharynx found between successive branchial grooves or branchial pouches or both.

branchial grooves: the ectodermal invaginations of the lateral walls of the pharynx.

breech presentation: the relation of the part(s) of the fetus to the birth canal at the beginning of labor where the buttocks or feet or both are closest to the birth canal.

bronchioles: small subdivisions of the bronchi.

brown fat cells: cells found in the late fetus and early newborn, containing mitochondria that convert food energy into heat rather than into ATP. Brown fat cells are an important source of warmth for the newborn.

bulbourethral glands: male auxiliary sex glands that empty their secretions into the vestibule of the vulva; also called Cowper's glands.

bushy chorion: see **chorion frondosum**.

calcium phosphate: a salt, crystals of which are deposited during bone formation.

calvarium: the flat bones of the skull collectively, which make up the cranial vault containing the brain.

calyces (calyx): subdivisions of the pelvis of the kidney.

canalicular period: a period during lung development when the lumens of the bronchi and terminal bronchioles enlarge.

capillaries: the smallest blood vessels. Through the thin walls of these vessels, exchange of materials occurs between the blood and tissue fluids.

caput epididymis: the head (cephalic) end of the epididymis.

carbohydrate: a class of organic molecules that carries out various functions in the cell. An especially important function is to serve as an energy source for cellular activities.

cardiac cells: heart cells.

cardiac jelly: the connective tissue found between the endocardium and the epimyocardium of the early developing heart.

cardiac muscle: the muscle of the heart; one of three general kinds of muscle tissue in the body.

cardiac myoblasts: myoblasts derived from splanchnic mesoderm that are the specific precursors of cardiac muscle cells.

cardiogenic mesoderm: that portion of the mesoderm that gives rise to the heart.

cardiac prominence: the bulge of the chest region found on the surface of the embryo during the middle of the embryonic period.

carotid arteries: the pair of major arteries found in the neck, which supplies the head with blood.

carrier: a person who carries a single recessive gene (allele) for a given characteristic and therefore does not exhibit the recessive form of the characteristic in his/her phenotype, although the gene (allele) may be passed onto the person's offspring.

cartilage: a type of connective tissue; somewhat rigid, but less rigid than bone.

catalyst: a substance that increases the rate of a chemical reaction without itself becoming permanently changed; see **enzyme**.

catalyze: increase the rate of a chemical reaction.

catheterization: see **balloon catheterization**.

cauda epididymis: the terminal portion of the epididymis, where it joins the vas deferens; the major storehouse of sperm in the male.

caudad: toward the tail (cauda) end of the body.

caudal: a term referring to the tail or tail end of an organism or structure.

caul: the membrane that sometimes covers the face of a newborn; consists of the amnion and chorion.

cecum: the blind pouch of the intestine where the large intestine begins.

cell cycle: the series of stages a cell passes through between successive cell divisions.

cell death: genetically programmed cell death is a normal part of development in many animals. Examples range from the resorption of the tail of the frog tadpole during metamorphosis to the sculpting of human fingers out of the original hand plate.

cell differentiation: the process by which a cell becomes specialized or differentiated; also called *cytodifferentiation*.

cell hypertrophy: increase in cell size.

cell junctions: specialized regions of contact between cells; generally falling into three categories: tight junctions, adhering junctions, and gap junctions.

cell-mediated immunity: that portion of immunity that directly involves cells (T lymphocytes), as opposed to humoral immunity, which involves antibodies produced by cells (B lymphocytes).

cell proliferation: an increase in cell number by cell division.

central nervous system (CNS): that portion of the nervous system derived from the neural tube; that is, the brain and spinal cord.

centrioles: organelles found in animal cells that are frequently associated with microtubule-organizing centers, such as at the poles of mitotic and meiotic spindles. Centrioles themselves are made up of microtubules.

centromere: the part of a chromosome by which it attaches to the spindle, which is necessary for chromosome movement during cell division.

centrosome: the cell organelle that is found at the so-called cell center or at the poles of a cell division spindle. It apparently acts as a microtubule-organizing center.

cephalad: toward the head end of the body.

cephalic: a term referring to the head or head end of an organism or structure.

cephalic presentation: the relation of the part(s) of the fetus to the birth canal at the beginning of labor where the head is closest to the birth canal.

cerebellum: that part of the brain that derives from the metencephalon and controls balance, posture, and movement.

cerebral hemispheres: the two major subdivisions of the cerebrum, derived from the telencephalon. The cerebrum is responsible for the "higher" brain functions, such as intellect and memory.

cerebral vesicles: the early evaginations (outpocketings) of the telencephalon, which give rise to the cerebral hemispheres.

cerebrospinal fluid (CSF): the fluid, derived from the choroid plexuses, that circulates through the ventricles of the brain and the spinal canal of the spinal cord.

cerebrum: that portion of the brain derived from the telencephalon and that is responsible for the "higher" brain functions, such as intellect and memory.

cervical canal: the lumen of the cervix of the uterus.

cervical cap: a contraceptive membrane that fits over the cervix of the uterus where it projects into the blind end of the vagina.

cervical flexure: the bend in the brain in the neck region of the embryo.

cervical mucus: the fluid found in the cervical canal, the consistency of which varies with the stages of the menstrual (uterine) cycle. The consistency of cervical mucus can be used as a fertility awareness method, because the mucus becomes thinner and wetter at the time of ovulation.

cervical plug: highly viscous cervical mucus that fills the cervical canal during pregnancy; believed to prevent infections from ascending out of the vagina and into the uterus.

cervix: the narrow lower portion of the uterus that projects into the vagina.

cesarean section: the surgical removal of the fetus from the womb, as opposed to vaginal childbirth; usually carried out because vaginal childbirth is contraindicated.

chancre: the initial lesion (ulcer) of syphilis.

chemotherapeutic drugs: drugs used for the treatment of cancer.

childbirth: the vaginal delivery of a baby.

chlamydia: a sexually transmitted disease caused by the bacterium *Chlamydia trachomatis.*

chloasma: the mask of pregnancy; patchy hyperpigmentation located chiefly on the forehead, temples, and cheeks.

chondrocytes: cartilage cells, those cells, derived from mesodermal mesenchyme, that lay down the connective tissue, cartilage.

Chordata: a major phylum (classification division) of animals. It includes the vertebrates (including humans) and the protochordates.

chordate: an animal placed into the phylum Chordata because it possesses a notochord sometime during its development. Vertebrates are a subgroup of the chordates possessing a vertebral column.

choriocarcinoma: a form of cancer originating in the trophoblast.

chorion: one of the four extraembryonic membranes formed during the development of higher vertebrates, including humans. It provides the fetal contribution to the formation of the placenta.

chorion frondosum or bushy chorion: the portion of the chorion that retains its villi and becomes the placenta fetalis or fetal contribution to the placenta.

chorion laeve or smooth chorion: the portion of the chorion that does not retain its villi and does not contribute to the placenta.

chorionic sac: saclike portion of the conceptus, which contains the embryo, amniotic cavity, and amnion.

chorionic villi: featherlike extensions of the chorion. Those that make up part of the placenta are the sites where actual exchanges of materials occur between fetus and mother.

chorionic villus sampling (CVS): a very early prenatal diagnostic procedure made possible because the chorionic villi contain dividing cells derived from the conceptus.

choroid fissure: the break in the continuity of the optic cup, resulting from the eccentric invagination of the optic vesicle that gives rise to the optic cup.

choroid plexus: one of several thin, highly vascular membranes that hang down into ventricles of the brain; sites of cerebrospinal fluid formation and of the blood-brain barrier.

chromatid: from anaphase to S-phase of the cell cycle, a chromatid is a chromosome. From G_2 of the cell cycle to metaphase, a chromatid is a longitudinal half of a chromosome. During S-phase of the cell cycle, chromatids are replicated.

chromatin: a dispersed state of the chromosomes found during the interphase phase of the cell cycle.

Chromosomal Theory of Heredity: the concept that the genes are found on the chromosomes.

chromosomes: a condensed state of the chromatin (DNA and protein) of the cell nucleus, these "colored bodies" are visible during the mitotic or division stage of the cell cycle.

cilia: hairlike organelles that project from the surfaces of some cells; may move cells through a liquid medium or may move a liquid medium over cells.

cleavage: a process undergone universally by mammalian embryos after formation of the zygote at fertilization; characterized by mitotic divisions without intervening growth. During cleavage, the size of blastomeres decreases even as their number increases.

cleavage patterns: embryos of different animal species, because of differences in amount and distribution of yolk in their eggs and genetic makeup, undergo different modes of cleavage called cleavage patterns. For example, human embryos undergo holoblastic, equal, rotational cleavage.

cleft palate: a fissure through the palate.

cloaca: the common chamber at the caudal end of the embryo/fetus into which the digestive, urinary, and reproductive systems open.

cloacal membrane: the boundary in the embryo between the rectum and proctodeum.

cloning: as applied to organisms, the process (natural or artificial) of making several genetically identical, independent organisms from a single initial organism.

CNS: central nervous system.

cochlea: the auditory part of the inner ear; a cone-shaped tube.

codominance: the form of inheritance wherein both alleles of genes are expressed.

codons: the triplets of nucleotides in messenger RNA molecules that specify (through the intermediary of transfer RNA with its anticodons) the position of a specific amino acid.

coelom: the original body cavity that appears between layers of somatic and splanchnic mesoderm.

coitus interruptus: a form of birth control that entails the withdrawal of the penis from the vagina before ejaculation occurs.

collecting tubules: the tubules of the kidney that collect the urine from the uriniferous tubules and transfer it to the calyces of the kidney.

colliculi: mounds, as in mounds of tissue.

compaction: early during cleavage in mammalian development, the blastomeres become so intimately associated that their individual boundaries become obscured; this is compaction.

comparative embryology: comparison of developmental processes in different species, providing greater insight into developmental processes.

conception: fertilization.

conceptus: that which results from conception (fertilization), namely, the embryo or fetus and its associated membranes.

condensation: as it pertains to chromosomes, to become compact and visible (as opposed to dispersed chromatin).

congenital syphilis: see **syphilis**.

conjoined twins: twins that have, during their development, fused (or failed to separate) to a greater or lesser degree; also called *Siamese twins.*

connecting stalk: the connection between the trophoblast and the inner cell mass as it develops into the embryo. It gives rise to part of the umbilical cord; also called *body stalk.*

contraception: prevention of conception (fertilization).

contractile proteins: proteins, such as actin and myosin, that are directly involved in muscle contraction.

contraction: as applied to muscles, the shortening of the length of muscle fibers.

contragestion: prevention of the establishment of a pregnancy, that is, implantation.

convergence: the movement of cells toward each other.

copulation: sexual intercourse.

cornea: the portion of the eye in front of the lens, which is responsible for most of the refraction (bending) of light that focuses the image on the retina.

corona radiata: collectively, those follicle cells that remain adherent to the outer surface of the zona pellucida after ovulation.

corpora quadrigemina: four eminences of tissue, the paired superior colliculi, concerned with vision, and the paired inferior colliculi, concerned with hearing; derived from the roof of the mesencephalon.
corpus: the major portion of the uterus, which contains the uterine cavity.
corpus epididymis: the body (corpus) or main portion of the epididymis, found between the caput epididymis and cauda epididymis.
corpus luteum: formed from the mature (graafian) follicle in the ovary during the second half of the menstrual (uterine) cycle. It is an important source of the sex steroids, estrogen and progesterone.
corpuscle of Meissner: ovoid corpuscles attached to nerve fibers found at the tips of fingers and toes. A corpuscle is a small round body.
cortex: the outer portion of something, as in ovarian cortex or cortex of the adrenal gland.
covalent bond: the type of chemical bond resulting from a sharing of a pair of electrons.
Cowper's glands: see **bulbourethral glands**.
cranial ganglia: ganglia associated with some of the cranial nerves.
cranial nerves: 12 pairs of nerves associated with the human brain, each pair of which is one of three types of nerves: sensory, motor, or mixed.
cranial neural crest: that portion of the neural crest found in the head.
cranial vault: see **calvarium**.
cri du chat: cry of the cat; name of a syndrome resulting in early death of the child, caused from a deletion of the short (p) arm of chromosome 5.
CRISPR-Cas9: a gene editing technique.
crossing over: reciprocal exchange of homologous segments of homologous chromosomes during prophase I of meiosis; increases genetic diversity of gametes and, therefore, of the species.
cryogenic storage: storage at low temperature, as with sperm and embryos.
cryptorchidism: hidden testis. This refers to an undescended testis that has not appeared in the scrotum.
crystallins: proteins made by the cells of the lens of the eye; gives the lens its crystal-clear property.
CSF: cerebrospinal fluid.
cytodifferentiation: see **cell differentiation**.
cytogenetics: the marriage of the disciplines of cytology (the study of cells) and genetics (the study of heredity).
cytokines: growth factors that regulate blood cells.
cytokinesis: division of the cytoplasm as a part of mitosis or meiosis.
cytomegalovirus: example of a teratogenic virus.
cytoplasm: that portion of the cell between the plasma membrane and the nuclear membrane.
cytoplasmic matrix: that portion of the cytoplasm found between the organelles and containing the cytoskeleton of the cell; also called *cytosol*.

cytoplasmic organelles: subcellular structures specialized for specific functions; examples: mitochondria, lysosomes, and Golgi apparatus.

cytoskeletal proteins: proteins such as tubulin, actin, and keratin, which make up the microtubules, microfilaments, and intermediate filaments of the cytoskeleton.

cytoskeleton: that portion of the cytoplasm of the cell made up of fibrillar protein components—namely microtubules, microfilaments, and intermediate filaments—that is largely responsible for cell shape and cell movement.

cytotrophoblast: the inner cellular component of the trophoblast.

daughter cells: those cells resulting from the division (mitotic or meiotic) of preexisting cells.

decidua: the endometrium during pregnancy.

decidua basalis: that portion of the decidua underlying the conceptus; gives rise to the placenta materna or maternal contribution to the placenta.

decidua capsularis: that portion of the decidua overlying the conceptus.

decidua parietalis: that portion of the decidua that neither underlies nor overlies the conceptus.

deletion: as applied to chromosomes, a missing portion of a chromosome.

dendrites: nerve cell processes, generally more numerous and shorter than axons, which carry nerve impulses toward the cell body of the nerve cell.

deoxyribonucleic acid: see **DNA**.

deoxyribonucleotides: monomers (building blocks) of DNA.

dermatome: the portion of a somite that gives rise to the dermis of the skin of the back.

dermis: the inner layer of the skin, derived from mesoderm; see **epidermis**.

dermomyotome: that portion of the somite, not including sclerotome, which gives rise to dermis of the back and muscles.

descent: as applied to the gonads, the normal movement of the testes from their site of origin into the scrotum and the normal movement of the ovaries from their site of origin to their final location in the pelvic cavity.

desiccation: drying out; loss of water.

developmental anatomy: the study of the dynamic (changing) anatomy of the developing human.

developmental geneticist: a geneticist who specializes in the genetics of development; one who is especially concerned with the relationship between gene expression and development.

developmental neurogenesis: that which occurs in developing organisms; i.e., prenatal or perinatal neurogenesis.

diaphragm: the sheet of muscle and tendon found between the thoracic and abdominal cavities, the contraction of which is essential for breathing.

diaphysis: shaft of a long bone.

diencephalon: the second division, from the rostral end, of the five-vesicle early brain; derives from the forebrain (prosencephalon) and gives rise to the optic vesicles, pineal gland, and neurohypophyseal portion of the pituitary gland.

digestive proteins: those proteins whose function (purpose) is the digestion (hydrolysis; breakdown) of large molecules.

digestive tube: the tubular portion of the digestive system found between the mouth and anus.

digit: a finger or toe.

dilatation and evacuation (D & E): a method of abortion used early in pregnancy.

dilatation stage: the dilatation stage begins with the first true contractions of labor; contractions that begin to cause dilatation (enlargement) of the cervix.

diploid: twice the haploid number of chromosomes; the normal and characteristic number of chromosomes for a given species, for example, for humans, the diploid number is 46.

diplotene stage: a stage of prophase I of meiosis. Human oocytes may remain in this stage for decades before continuing on to produce "eggs" capable of fertilization.

distal: an anatomic term referring to a structure that is distant from an anatomic reference point; see also **proximal**.

dividing egg: an expression sometimes used for the early embryo during the cleavage stage of development.

dizygotic twins: see **fraternal twins**.

DNA: deoxyribonucleic acid, the chemical substance that actually makes up the genes or hereditary material of almost all organisms, with the exception of some RNA viruses.

DNA replication: process by which a DNA molecule produces two identical daughter DNA molecules.

DNA synthesis: the formation of DNA.

dominant: descriptive of one allele of a nonidentical pair of alleles that is expressed.

dominant-recessive inheritance: the form of inheritance wherein dominant (expressed) and recessive (not expressed) forms (alleles) of genes are involved.

dorsal: an adjective referring to the dorsum or back of an animal.

dorsal mesocardium: a membrane, derived from a double layer of splanchnic mesoderm, which for a time suspends the developing heart in the pericardial region of the coelom.

dorsal mesoderm: see somitic mesoderm

dorsum: the back of an animal.

Down's syndrome: a syndrome, including mental retardation, resulting from an extra (trisomy) chromosome 21.

ductus arteriosus: a blood vessel shunt between the pulmonary trunk and arch of the aorta, which shunts blood past the undeveloped lungs into the systemic circulation so that the right ventricle may get the exercise required for its normal development.

duodenum: the first (most cephalic) portion of the small intestine.

duplication: as applied to chromosomes, a duplicated portion of a chromosome.

dynamic anatomy: dramatically changing anatomy, as in the embryo or fetus, as opposed to the adult.

dynein: a type of protein motor, found in various cells including sperm ells, that converts ATP energy into the kinetic energy of microtubule bending which, in turn, propels sperm.

ectoderm: the outer of the three primary germ layers; literally, outer skin.

ectodermal placodes: circumscribed thickenings of the ectoderm of the head that contribute to sensory structures and cranial ganglia.

ectopic pregnancy: a pregnancy that becomes established in an abnormal location, such as the fallopian tube or abdominal cavity, rather than in the normal uterine location.

edema: swelling of part(s) of the body due to retention of water.

effacement: loss of form; as of the uterine cervix during labor.

efferent ductules: as applied to the epididymis, the tiny tubules emerging from the testis and joining the duct of the epididymis.

egg: or ovum, the female gamete.

egg coats: coverings surrounding an egg when it is ovulated from the mammalian ovary. Around human eggs, two such egg coats are found: the outer, cellular corona radiata, and the inner, noncellular zona pellucida.

ejaculate: a portion of semen ejected from the penis during one ejaculation event.

ejaculation: the process of discharging one portion of semen from the penis.

ejaculatory ducts: the pair of highly muscularized ducts that pierce the substance of the prostate gland, thereby providing a passageway between the vasa deferentia and the urethra.

electron microscope: a type of microscope that uses a beam of electrons rather than light; provides great detail in images.

embryo: the early developing animal before it begins to look like the adult of the species.

embryo bank: a collection of embryos kept in cryogenic storage.

embryoblast: the inner cell mass; that portion of the blastocyst that actually gives rise to the embryo.

embryogenesis: the formation of the embryo.

embryologist: a biologist who specializes in the study of embryology or developmental biology.

embryonic: an adjective referring to the embryo.

embryonic disc: the early embryo when it consists of only the two or three germ layers.

embryonic ectoderm: the outermost germ layer of the embryo.

embryonic endoderm: the innermost germ layer of the embryo.

embryonic induction: a developmental phenomenon wherein one part of the embryo determines the fate of another part of the embryo.

embryonic mesoderm: the germ layer of the embryo found between the outer ectoderm and the inner endoderm.

embryonic period: the first 8 weeks of human development during which the developing organism does not resemble a human.

embryonic pole: that portion of the blastocyst where the inner cell mass is located.

embryo transfer (ET): the artificial movement of an embryo into a uterine cavity.

emulsify: to bring a lipid (e.g., oil) into aqueous (watery) suspension by means of molecules that are partially soluble in oil and partially soluble in water.

encapsulated nerve endings: nerve terminations that are wrapped in connective tissue.

endocardial cushion tissue: a type of embryonic connective tissue that plays an important role in heart development. It divides the atrioventricular canal into two channels, it contributes to the formation of the interventricular septum, and it provides the valves of the heart.

endocardial tube: the first cardiac tube formed. It arises from mesenchyme cells, derived from the medial faces of splanchnic mesoderm in the region of the amniocardiac vesicles, which first organize themselves into a pair of tubes, which then fuse into the single endocardial tube that becomes the lining of the heart.

endocardium: the lining of the heart.

endochondral ossification: ossification that occurs within cartilage.

endocrine cells: cells that produce hormones.

endocrine signaling: a mode of cell–cell communication in which signaling molecules (endocrine factors) are released into the circulatory system that may affect cells (target cells) that are some distance from the signaling cell, e.g., the effect of the anterior lobe of the pituitary gland hormones on the gonads.

endoderm: the inner of the three primary germ layers; literally, inner skin.

endolymphatic duct: the dorsal evagination of the otocyst in the embryo, which gives rise to the endolymphatic sac.

endometriosis: the disease resulting from the migration of portions of the endometrium into abnormal locations, such as the pelvic cavity or the fallopian tubes.

endometrium: lining of the uterus.

endoskeleton: the skeleton found inside the body, as in humans; as opposed to on the surface of the body as in insects (exoskeleton).

endothelial cells: the cells lining the heart, blood vessels, and lymph vessels.

engagement: one of the seven cardinal movements of labor; the entrance of the presenting part of the fetus into the upper pelvic passage.

enzymes: protein molecules that increase the rates of biochemical reactions.

ependyma: the cellular membrane that lines the ventricles of the brain and the spinal canal of the spinal cord.

ependymal cells: the cells derived from the lining of the neural tube, which make up the ependyma.

epiblast: the early human embryo is composed of two layers; the upper layer, the epiblast, gives rise to the three germ layers.

epiboly: the spreading of cells upon a surface

epicardium: the covering of the heart; the visceral pericardium.

epidermal cells: cells of the epidermis, which are derived from the ectoderm.

epidermal ectoderm: the portion of the ectoderm that gives rise to the epidermis; see **neuroectoderm**.

epidermis: the outer layer of the skin, derived from ectoderm and itself composed of multiple cell layers; see **dermis**.

epididymis: the proximal portion of the male reproductive duct consisting of several efferent ductules of the epididymis and the epididymal duct. The epididymal duct is highly contorted and consists of three parts: caput (head), corpus (body), and cauda (tail).

epigenesis: the concept that the developing human is not preformed in either the sperm or the egg, but rather gradually comes into existence, as opposed to the concept of preformation.

epigenetics: the alteration of a genetic message not by the alteration of its DNA nucleotide sequence, but rather by chemical modification of the nucleotides in the sequence or by chemical modification of the histone proteins associated with the unaltered nucleotide sequence.

epimyocardial tube: derived from the coming together of two layers of splanchnic mesoderm around the endocardial tube; gives rise to the muscular wall of the heart, myocardium, and the covering of the heart, epicardium.

epinephrine: the hormone produced by the adrenal glands, which is associated with the "fight or flight" reaction of animals.

epiphyseal fusion: fusion of an epiphysis with the diaphysis of a long bone, accompanied by ossification of the cartilage of the epiphysis and bringing an end to growth in length of the bone.

epiphyseal plates: cartilaginous regions of long bone growth in the epiphyses of long bones.

epiphyses: the ends of a long bone.

epithelial–mesenchymal interactions: the development of many organs results from an inductive interaction between two types of tissues, epithelium and mesenchyme; e.g., embryonic tooth development results from interaction between oral epithelium, giving rise to the enamel organ, and mesenchyme, giving rise to the dentine of the tooth.

epithelium: a layer of cells that lines a cavity or covers a surface.

equator: as applied to the spindle of a dividing cell, the imaginary plane located midway between the poles of the spindle. When the centromeres of all the chromosomes of the cell are aligned on the equator the cell is said to be in metaphase of mitosis or meiosis.

erythroblastosis fetalis: destruction of red blood cells by maternal antibodies produced in response to Rh incompatibility.

esophagus: the portion of the digestive tube between the pharynx and the stomach.

estrogen: female sex hormone; actually a small family of closely related female sex steroids. It is produced by ovarian follicles, as well as by other cells, and is responsible for the development and maintenance of female secondary sex characteristics.

euploid: an exact multiple of the haploid number of chromosomes.

eustachian tubes: the pair of tubes that extends between the pharynx and the cavity of the middle ears.

evagination: a morphogenetic movement involving an out-pocketing of a layer of cells, as in the formation of the optic vesicles from the lateral walls of the embryonic brain.

evolution: the change in the gene frequency of a population over time.

excretion: the discharge of waste.

exocoelom: the extension of the coelom (body cavity) between the extraembryonic regions of the mesoderm; see **extraembryonic coelom**.

exoskeleton: the skeleton found on the surface of the body, as in insects, as opposed to inside the body (endoskeleton), as in humans.

expulsion: one of the seven cardinal movements of labor; the forcing of the fetus out of the birth canal.

extension: one of the seven cardinal movements of labor; the straightening out of the fetal head.

extension: the elongation of a structure as a result of cell convergence.

external auditory meatus: external ear canal.

external os: the opening of the cervical canal of the uterus into the vagina.

external rotation: one of the seven cardinal movements of labor; rotation of the fetal head after it has emerged from the birth canal.

extracellular matrix: the biochemically complex material outside of and secreted by cells.

extraembryonic: those portions of the conceptus not incorporated into the developing embryo.

extraembryonic coelom: the extension of the coelom (body cavity) between the extraembryonic regions of the mesoderm; see **exocoelom**.

extraembryonic membranes: the four membranes, amnion, chorion, yolk sac, and allantois, developed from the conceptus, that will not be incorporated into the embryo. Although only temporary structures, these membranes are essential for mammalian development.

extraembryonic mesoderm: derived from the trophoblast, this mesoderm does not become part of the embryo.

extrauterine fetus: a premature baby may be thought of as an extrauterine fetus and, in terms of brain development, human babies have been considered, in general, to be extrauterine fetuses.

fallopian tubes: these tubes convey eggs or embryos from the site of ovulation at the ovary into the uterine cavity; also called *oviducts* or *uterine tubes*.

fecund: means the ability to produce offspring.

fertility awareness methods: methods that identify the fertile days of the month. These methods are the rhythm (calendar) method, the basal body temperature method (BBT), and the cervical mucus method.

fertilization: the interaction between gametes culminating in the formation of the zygote; generally refers to the fusion of a spermatozoon with an ovum; see **in vitro fertilization**.

fetal period: in human development, from the end of the eighth week until birth.

fetoscopy: a prenatal diagnostic technique allowing direct, visual observation of the fetus in the womb.

fetus: when the developing mammal begins to resemble the adult, it is called a *fetus*. Until then, it is called an *embryo*.

fetus in fetu: a variation on conjoined twins in which one fetus is small and "parasitic" on or in the other relatively normal fetus.

fimbria: literally, a fringe. The free end (near the surface of the ovary) of each fallopian tube is fimbriated.

flagellum: also referred to as a tail, the organelle of the spermatozoon that provides it with the function of motility.

flexion: the bending of the longitudinal axis of the body around a transverse axis. When you try to touch your toes, you undergo flexion.

flexures: bends in the body axis resulting from flexion, such as cervical flexure.

foam: see **spermicide**.

follicle: a structure formed in ovaries, usually consisting of one germ cell and multiple follicle cells, in which germ cells are maintained and differentiate, culminating either in ovulation or death of the contained oocyte.

follicle cells: somatic cells found in ovaries in association with germ cells as part of a structure referred to as a follicle; also called *granulosa cells*.

fontanelles: the openings between the bony plates of the skull, most conspicuous at about the time of birth.

foramen cecum: the opening from the pharynx into the thyroglossal duct of the thyroid diverticulum.

foramen ovale: an oval window in the septum secundum. It becomes part of the important valvular mechanism between the atria in the prenatal heart. The foramen ovale allows the left atrium to receive blood so that it may have a load to pump and, therefore, get the exercise it needs for its normal development.

foramen primum: the first opening in septum primum, between the two atria of the developing heart, which subsequently closes.

foramen secundum: the second opening in septum primum, which persists during prenatal development, the edges of which provide the flaps for the valve at foramen ovale.

foramina of Monro: openings between the lateral ventricles of the cerebral hemispheres and the third ventricle of the brain through which cerebrospinal fluid circulates.

forebrain: the most rostral of the early brain vesicles of the three-vesicle brain.

forebrain protuberance: the bulge found on the surface of the head of the embryo during the middle of the embryonic period, reflecting the dramatic development of the forebrain beneath the surface of the head.

foregut: the cephalic-most portion of the early embryonic gut; see **midgut** and **hindgut**.

fraternal twins: twins derived from two different zygotes. Also called **dizygotic twins**.

free nerve termination: a nerve termination that simply moves in among the cells of epithelial or connective tissues and develops branches.

frontal plane: the plane that, when passed through the body, separates the dorsal part from the ventral part of the body.

frontonasal prominence: the surface of the developing face of the embryo between and above the two nasomedial processes.

fructose: a simple sugar (monosaccharide) containing six carbons (hexose), preferentially utilized as an energy source by spermatozoa, whereas other cells of the body preferentially use glucose.

fundus: the upper (superior) part of the uterus, between the attachment sites of the two fallopian tubes. The height of the ascent of the fundus out of the pelvic cavity is used by obstetricians to judge the stage of pregnancy (gestation).

gallbladder: a saclike organ, attached to the undersurface of the liver, which stores and releases bile.

gamete intrafallopian transfer (GIFT): artificial transfer of gametes (sperm and egg) into the normal site of fertilization in the ampulla of the fallopian tube.

gametes: a subset of germ cells actually involved directly in fertilization; also called *sex cells, sperm* and *eggs*.

gametogenesis: formation of the gametes; see **oogenesis** and **spermatogenesis**.

ganglion: (plural, ganglia) an aggregate of nerve cell bodies outside of the central nervous system.

gap junctions: specialized contacts formed between cells that establish cytoplasmic continuity between the cells. Cells with these junctions rapidly communicate with each other.

gastrulation: an early morphogenetic process, following cleavage, which results in the establishment of the three primary germ layers.

gene: unit of hereditary material, made up of stretches of DNA (RNA in some viruses) nucleotide sequences.

genetic code: the relationship between the codons of messenger RNA and the amino acids of the corresponding protein.

genetic diversity: the degree of variety of genes in a population.

geneticist: a person engaged in the study of heredity (genetics).

genetic sex: the normal pattern of sex; in humans, two X chromosomes result in a female, and one X and one Y chromosome result in a male.

genital herpes: a viral, highly contagious, sexually transmitted disease.

genitalia: reproductive organs.

genital tubercle: the embryonic precursor of the penis in the male or the clitoris in the female.

genome: the genetic makeup of an individual, found in a haploid set of chromosomes.

genomic imprinting: the nonidentity of maternal and paternal chromosomes received by a zygote, because of differential chemical modification (methylation) of the DNA by the two parents.

genotype: the fundamental hereditary constitution of an individual.

germ cells: those cells, as opposed to somatic cells, that have the responsibility for continuity of the species. That is, they provide the cellular basis for reproduction.

germ layers: cellular layers resulting from gastrulation, which make up the developing body of the early embryo.

germplasm: germ cells, collectively.

gestation: pregnancy.

gestational mother: a woman who carries a pregnancy, regardless of the origin of the conceptus.

gill arches: the thickened masses of tissue that make up the walls of the pharynx in the early developing embryo; see **pharyngeal arches**.

gill clefts: see **hyomandibular furrows**

glands of Littre: multiple male auxiliary sex glands found along and emptying into the penile urethra.

glial cells: one of three kinds of cells derived from the neuroepithelium of the neural tube. Originally thought to be a sort of connective tissue in the central nervous system, glial cells have been found to play a more dynamic role in the physiology of the nervous system. Also, mesenchyme cells give rise to glial cells (microglial cells) outside of the central nervous system.

glioblasts: cells derived from the neuroepithelium of the neural tube, which give rise to the glial cells of the central nervous system.

glomerular filtrate: the liquid that passes from the glomerulus into Bowman's capsule of the uriniferous tubule.

glomerulus: the tuft of capillaries that projects into Bowman's capsule.

glucose: an example of a simple type of sugar called a hexose. Glucose is the sugar most commonly used by the cells of the body.

glycoprotein: a molecule that is partially carbohydrate and partially protein; extremely important molecules within cells, on the surface of cells, and in the extracellular matrix.

Golgi apparatus: a cellular organelle concerned with glycosylation of proteins and macromolecular traffic in the cell.

gonads: the organs responsible for the production of gametes; ovaries producing eggs in females and testes producing sperm in males.

gonadotropins: hormones, released by the anterior lobe of the pituitary gland, that act to stimulate the gonads; e.g., FSH (follicle-stimulating hormone) and LH (luteinizing hormone).

gonadotropin-releasing hormones (GnRH): peptide hormones released from the hypothalamus that stimulate the pituitary to release the gonadotropins, follicle-stimulating hormone, and luteinzing hormone, which are required for mammalian gametogenesis and steroidogenesis.

gonorrhea: a bacterial, highly contagious, sexually transmitted disease.

graafian follicle: a mature ovarian follicle, consisting of a single oocyte and numerous follicle cells (and their products).

granulosa cells: see **follicle cells**

gray matter: those portions of the central nervous system that have a large concentration of nerve cell bodies and therefore appear gray in the fresh state.

gut: a term commonly used to refer to the digestive tube.

gyri: the ridges found on the surface of the cerebrum, separated from each other by the sulci (furrows).

haploid: half the diploid number of chromosomes. For humans, the normal haploid number is 23.

hard palate: the anterior part of the palate, containing bone.

hatching: escape of an embryo from the confines of the original egg coats.

hCG: human chorionic gonadotropin.

head fold: one of four body folds; the head fold sculpts the head of the embryo out of the cephalic region of the embryonic disc.

heart murmur: a roaring sound coming from the heart; detectable with a stethoscope.

Hellin's Law: a mathematical prediction of the occurrence of multiple births; for example, twins once in 89^1 pregnancies, triplets once in 89^2 pregnancies, and so on.

hematopoiesis: formation of blood cells.

hemizygous: genes that are found on the male Y chromosome and not on the female X chromosome, its pairing partner. Genes usually come in pairs because they are found on paired chromosomes. Since such genes are only present in a single dose, they are said to be hemizygous (half in the zygote).

hemoglobin: the oxygen- and carbon dioxide–carrying protein found in blood. Packaged in red blood cells, this protein is a cell differentiation product of red blood cells.

hemorrhage: abnormal loss of blood.

hepatic diverticulum: an evagination of the embryonic foregut, which gives rise to the liver, gallbladder, and their ducts.

herpes simplex virus (HSV): a virus that may be passed from mother to baby during birth as the baby passes through the birth canal, with the resulting infection and possible death of the baby.

heteropaternal superfecundation: the fertilization of two separate eggs by two different fathers during the same uterine cycle.

heterozygous: descriptive of an organism in which its two alleles for a given characteristic are not identical (different in the zygote).

hindbrain: the most-caudal of the early brain vesicles of the three-vesicle brain.

hindgut: the caudalmost portion of the early embryonic gut; see **foregut** and **midgut**.

histogenesis: the formation of tissues.

histones: small, basic proteins that combine with DNA to form chromatin.

HIV: human immunodeficiency virus; the causative agent of AIDS (acquired immunodeficiency syndrome).

holoblastic: a type of cleavage pattern wherein the entire egg undergoes cleavage. Human eggs undergo holoblastic cleavage; see **meroblastic**.

homolecithal egg: a type of egg that has its yolk uniformly distributed throughout its volume. Human eggs are homolecithal; see **telolecithal**.

homologous chromosomes: the members of a pair of chromosomes that undergo synapsis (pairing) during prophase I of meiosis.

homopaternal superfecundation: the fertilization of two separate eggs by the same father during the same uterine cycle, resulting in *fraternal twins*.

homozygous: descriptive of an organism in which its two alleles for a given characteristic are identical (same in the zygote).

hormones: a chemically diverse group of messenger molecules secreted by endocrine cells and designed to act on cells at some distance from the hormone-secreting endocrine cells.

hydrocephaly: a birth defect arising from an abnormal accumulation of cerebrospinal fluid in the ventricles of the developing brain.

hydronephrosis: a birth defect arising from an abnormal accumulation of urine in the developing kidneys.

hypoblast: the early human embryo is composed of two layers; the lower layer is the hypoblast.

Human Genome Project: an international, multiyear project with the objective of discovering the sequence of all 3 billion base pairs that make up the human genome.

humoral immunity: the portion of immunity that involves antibodies produced by cells (B lymphocytes), as opposed to cell-mediated immunity, which directly involves cells (T lymphocytes).

hydatidiform mole: a birth defect in which the fetus is absent and only a very abnormal placenta is present.

hydrogen bonds: weak chemical bonds resulting from proton (hydrogen ion) sharing by oxygen or nitrogen atoms or both.

hydrolytic: to split (break a chemical bond) by means of water. Digestion is hydrolytic.

hyoid arches: the second pair of branchial arches.

hyomandibular furrows: the pair of furrows located between the first (mandibular) and second (hyoid) pairs of branchial arches.

hypobranchial eminence: a small bulge posterior to the foramen cecum, which gives rise to the posterior third (pharyngeal part) of the tongue.

ICM: see **inner cell mass**.

ICSI: see **intracytoplasmic sperm injection**.

identical twins: twins derived from the same zygote; also called *monozygotic twins*.

implantation: the attachment of the conceptus to and its incorporation into the lining of the uterus (endometrium).

imprinting: see **genomic imprinting**.

ingression: the movement of single cells out of a cell layer into a preformed cavity, e.g., ingression of epiblast cells along the primitive streak to give rise to endoderm and mesoderm, or formation of the notochord by cells that ingress through the primitive (Hensen's) node.

***in situ*:** growing in the organism in the normal location, as opposed to growing in the organism, but not in the normal location in the organism.

***in utero*:** in the uterus.

***in vitro*:** literally, in glass; refers to growing cells, tissues, organs, or embryos in culture; however, the culture vessels currently used are plastic rather than glass.

in vitro fertilization (IVF): human fertilization occurring under artificial conditions outside the body.

***in vivo*:** growing in the organism, as opposed to growing in culture.

induced labor: the initiation of labor by artificial means, such as by injection of the hormone, oxytocin.

induced pluripotent stem cells (iPSC): pluripotent stem cells derived from differentiated cell types.

inferior: below; as the chin is inferior to the forehead.

inferior vena cava: a major venous (vein) blood vessel returning blood from the body (inferior to the heart) to the right atrium of the heart.

infundibulum: the funnel-shaped opening at the ovarian end of the fallopian tube.

inner cell mass: that portion of the blastocyst that gives rise to the embryo; also called *embryoblast*.

innervate: to supply with and control with nerves.

intercalated discs: the specialized cell junctions between the contractile cells of the heart.

intercourse: the insertion of the male penis into the female vagina and subsequent ejaculation of semen into the vagina.

intermediate filaments: one of the three fibrous, proteinaceous components of the cytoskeleton; made up of fibrous proteins such as keratin.

intermediate inheritance: descriptive of an organism in which two alleles for a given characteristic are not identical and both are expressed in the phenotype.

intermediate layer: (1) skin: a layer of skin produced during the early fetal period between the outer periderm and the inner basal layer. (2) spinal cord: the portion of the wall of the spinal cord made up of the cell bodies of nerve cells between the inner ventricular layer and outer mantle layer, that is, gray matter.

intermediate mesoderm: that portion of the mesoderm found bilaterally between the somitic mesoderm and the lateral mesoderm.

internal rotation: one of the seven cardinal movements of labor; the rotation of the fetal head necessary for it to fit through the pelvic outlet.

interphase: that portion of the cell cycle in which the cell is not dividing.

interstitial cells of Leydig: testosterone-secreting cells found between the seminiferous tubules of the testes.

interstitial growth: the growth of an object by the addition of material throughout its volume, as in the growth of the liver.

interstitial tissue: tissue found between the seminiferous tubules of the testes.

interventricular septum: the partition between the two ventricles of the heart.

interventricular sulcus: the groove on the surface of the ventricular region of the developing heart opposite the internal, forming interventricular septum.

intervertebral discs: discs of soft tissue found between successive vertebrae in the vertebral column, which originate from the somites and notochord.

intervillous spaces: placental spaces, filled with maternal blood, found between the chorionic villi of the placenta.

intracellular signal transduction: to convey information carried by a signal from the cell surface to the interior of the cell (e.g., the nucleus) where the information is acted upon, the cell makes use of cascades of chemical reactions that make up intracellular signal transduction pathways.

intracytoplasmic sperm injection (ICSI): the injection of a sperm, incapable of fertilizing an egg in the normal way, directly into the cytoplasm of an egg.

intramembranous ossification: ossification occurring within condensed mesenchyme, usually involved in the formation of flat bones, such as in the calvarium.

intrauterine device: see **IUD**.

invagination: the movement of a sheet of cells into a preformed cavity; e.g., the formation of the retinas from the optic cups.

inversion: turned inside out.

involution: the turning in of cells over a rim.

iris: the pigmented membrane located between the cornea and lens of the eye, surrounding the pupil and capable of contraction.

islets of Langerhans: scattered accumulations of insulin-secreting cells found in the pancreas.

isolecithal: see **homolecithal**.

IUD: intrauterine device; a birth control device inserted into the uterus that prevents the embryo from implanting into the endometrium; see **contragestion**.

IVF: see **in vitro fertilization**.

juxtacrine signaling: mode of cell–cell communication in which signaling molecules are retained on the surface of the signaling cell and interact with receptor proteins on adjacent cell surfaces.

keratin: a fibrous protein produced by cells of the epidermis.

Klinefelter's syndrome: a syndrome that includes a male phenotype, tall stature, and sterility; resulting from an XXY sex chromosome constitution.

Kupffer's cells: cells found in the liver; immobile macrophages.

labiogingival lamina: a fold of tissue between the forming lips and gums.

labioscrotal folds: the pair of bulges that makes up part of the indifferent stage of the external genitalia and that give rise to the labia majora in the female and the scrotum in the male.

labor: the process during which contractions of the smooth muscle of the uterine walls dilate the uterine cervix and force the fetus out of the uterine cavity through the birth canal.

labyrinth: a system of intercommunicating canals and cavities that make up the inner ear; it has membranous and bony components.

lactation: the production of milk.

lactiferous ducts: the ducts that carry milk from the mammary glands to the nipples.

lacunae: fluid-filled spaces found within the syncytiotrophoblast during implantation, which become the intervillous spaces of the placenta.

lanugo: fine, down-type hair found on the skin of the fetus.

Langerhans cells: cells of the immune system found in the skin; see **islets of Langerhans**.

laparoscopy: a technique for examining the interior of the peritoneal (abdominal) cavity by means of a peritoneoscope.

laparotomy: the making of an incision through the abdominal wall.

laryngotracheal groove: an evagination of the floor of the embryonic pharynx that gives rise to the lining of the respiratory system.

laterad: to move in a lateral direction.

lateral: side.

lateral body folds: two of four body folds; the lateral folds sculpt the sides of the embryonic body out of the embryonic disc.

lateral lingual swellings: two eminences found anterior to the foramen cecum, which give rise to the anterior two thirds (oral part) of the tongue.

lateral mesoderm: that part of the mesoderm, found bilaterally, lateral to the intermediate mesoderm.

lateral palatine processes: bilateral processes that grow out from the inner medial borders of the maxillary processes and fuse with each other and the median palatine process to give rise to the hard palate.

Law of Independent Assortment: one of Gregor Mendel's profound generalizations about heredity. The way in which members of one pair of alleles segregate (separate) in the formation of the gametes has no influence on the way in which members of another pair of alleles segregate. This is not always true, such as when pairs of alleles for two different genes are on the same chromosome.

Law of Segregation: one of Gregor Mendel's profound generalizations about heredity. Members of pairs of alleles segregate (separate) in the formation of the gametes. Consequently, a gamete has a single allele for each gene.

LD_{50}: the dose of a chemical (e.g., toxin) that is lethal for 50% of a population of organisms.

lens thickenings: a pair of bilateral thickenings of the ectoderm of the head of the early embryo. Induced by the optic vesicles, they give rise to the lenses of the eyes.

Leydig's cells: see **interstitial cells of Leydig**.

levonorgestrel: one form of a variety of progestogen hormones found in birth control pills. It has activity similar to the natural hormone progesterone.

lie: the relationship of the long axis of the fetus to that of the mother; one of four descriptions of the fetus's alignment in the uterus.

ligation, tubal: see **tubal ligation**.

limb buds: the first indications of the developing limbs; found bilaterally on the flanks of the early 4- to 5-week human embryo.

limbs: arms and legs.

linea nigra: black line; a line of pigmentation, running from the pubic region to the umbilicus (navel) in pregnant women.

lipid: a class of organic molecules that carry out various functions in the cell. An especially important function is to serve as the basic structure of cell membranes for both the plasma membrane and the membranes of various organelles.

litter: a collective term for the multiple progeny of some animals, such as cats and dogs.

Littre's gland: see **glands of Littre**.

liver bulge: the bulge of the abdominal region found on the surface of the embryo during the middle of the embryonic period.

locus: (plural loci) the site (location) along the length of a chromosome where an allele of a specific gene is located.

lysosomes: cellular organelles, containing a wide variety of hydrolytic (digestive) enzymes, which help the cell remove waste materials.

lysosomal storage diseases: diseases resulting from the lack of specific enzymes in the lysosomes of a cell. As a consequence, the lysosome is not able to digest (hydrolyze) specific kinds of molecules, resulting in an accumulation of these molecules with often fatal consequences for the cell and, ultimately, the organism.

macromolecules: large molecules. Three classes are found in cells and organisms: polysaccharides (some carbohydrates), polypeptides (proteins), and polynucleotides (nucleic acids).

macula: a spot. The macula acustica is the acoustic nerve termination in both the sacculus and utriculus.

mammary glands: glands found in the breasts of female mammals that secrete milk for the nourishment of young offspring.

mammary ridges: approximately parallel ridges of ectodermal tissue extending from the armpits to the groins of the embryo, along which mammary glands may develop at any point; normally, a single pair of mammary glands form on the chest of a human female.

mandible: the lower jaw.

mandibular arches: the first pair of branchial (pharyngeal) arches, which give rise to the upper and lower jaws.

mantle layer: see **intermediate layer**.

marginal layer: the outer layer of the wall of the spinal cord, made up of the processes of the nerve cells, that is, white matter.

marrow: see **bone marrow**.

maternal inheritance: the inheritance of genetic information from one's mother through the cytoplasm of the egg (rather than through the typical route, the nuclear genes). For example, mitochondrial genes are inherited only from one's mother, not from one's father.

maxilla: the upper jaw.

maxillary processes: a pair of processes arising from the mandibular arches, which give rise to the maxilla (upper jaw).

mechanoreceptors: sensory receptors responding to the sensations of sound and pressure.

meconium: the waste material that accumulates in the gut of the fetus and normally is not excreted until after birth.

mediad: movement toward the median plane of the body.

median palatine process: grows out from the innermost part of the premaxilla and fuses with the lateral palatine processes to give rise to the anterior, medial part of the hard palate.

median plane: an imaginary plane that separates the right and left sides of the body from each other.

mesolecithal eggs: eggs that have an intermediate amount of yolk; more than microlecithal and less than macrolecithal eggs. Frogs have mesolecithal eggs.

medulla: (1) general: the inner portion of something, as in the ovarian medulla or medulla of the adrenal gland. (2) brain: the caudal-most portion of the brain, derived from the myelencephalon of the early five-vesicle brain.

meiosis: one of two general types of cell divisions undergone by eukaryotic cells; an integral part of sexual reproduction.

melanin: a dark pigment produced by melanocytes of the skin.

melanocytes: pigment cells derived from the neural crest, which produce the pigment melanin.

membranous labyrinth: see **labyrinth**.

menarche: the beginning of the menstrual cycle with the onset of puberty in a female child.

Mendel, Gregor: the ninteenth-century Austrian monk considered the father of genetics. He proposed that hereditary material has a particulate nature (which we call *genes*), and he proposed two laws for their distribution (Law of Segregation and Law of Independent Assortment).

mendelian: referring to Gregor Mendel.

menopause: the end of menstrual cycles in a woman at the end of her child-bearing years.

menses: the menstrual flow or discharge that occurs during the menstruation portion of the menstrual or uterine cycle.

menstrual cycle: the cyclic changes, especially in the lining of the uterus, undergone by a woman during her potential child-bearing years; also called *uterine cycle.*

Merkel cells: cells derived from the neural crest, which give rise to mechanoreceptors in the skin.

Merkel's tactile disc: a type of free nerve termination, each composed of a Merkel cell and the leaflike expansion of a nerve terminal. Merkel cells are thought to play a role in sensation as skin mechanoreceptors, sensitizing the skin to touch.

meroblastic: a type of cleavage pattern in which only part of an egg undergoes cleavage. Chicken eggs undergo meroblastic cleavage; see **holoblastic**.

MESA: see **microsurgical epididymal sperm aspiration**.

mesencephalon: the midbrain.

mesenchymal cell: a cell that makes up part of mesenchyme.

mesenchyme: a loosely arranged type of embryonic tissue, with substantial amounts of extracellular matrix; can arise from any germ layer and is not necessarily of mesodermal origin.

mesenteries: membranes that suspend various parts of the digestive system in the abdominal (peritoneal) cavity.

mesoderm: the middle of the three primary germ layers; literally, middle skin.

mesogastrium: the part of embryonic mesentery that suspends the stomach.

mesonephric ducts: ducts of the mesonephros that give rise to the vasa deferentia (sperm ducts) in males.

mesonephric tubules: the tubular components of the mesonephros.

mesonephros: the chronologically intermediate, embryonic, nephroi (kidneys) of human development; between the earlier pronephros and the later metanephros.

mesorchium: the membranous support for the testis.

mesovarium: the membranous support of the ovary in the abdominal cavity.

messenger RNA (mRNA) molecules: a subclass of nucleic acids that carry the genetic message from DNA to the ribosomes, where protein is made.

metacentric chromosome: a chromosome in which the centromere is equidistant from the two ends of the chromosome. Therefore, the chromosome's two arms are of equal length.

metafemales: human females with more than the normal number (two) of X chromosomes; they exhibit varying degrees of mental retardation.

metanephric diverticulum: an evagination from the caudal end of each mesonephric duct, which gives rise to a ureter, pelvis, calyces, and collecting tubules of the metanephric kidney.

metanephrogenous mesoderm: the portion of intermediate mesoderm that is induced by the metanephric diverticula to give rise to the functional units (nephrons) of the metanephric kidneys.

metanephros: the definitive human kidney; chronologically, the final nephros (kidney) of human development, following sequentially the pronephros and the mesonephros.

metaphase: the stage of mitosis or meiosis when the centromeres of the chromosomes are aligned on the equator of the spindle.

metencephalon: the fourth vesicle from the rostral end of the early five-vesicle brain; gives rise to the cerebellum and pons varolii.

methylation: the chemical addition of a methyl group to a molecule, such as to DNA.

microfilament: that component of the cytoskeleton of the cell made up of long, solid filaments of actin protin.

microlecithal eggs: eggs that have a small amount of yolk, less than mesolecithal and macrolecithal eggs. Humans have microlecithal eggs.

microsurgical epididymal sperm aspiration (MESA): the withdrawal (by aspiration; suction) of sperm directly from any part of the epididymis by means of a tiny cannula (tube).

microtrabecular lattice: a controversial (as to its existence) part of the cytoplasm, revealed by high-voltage electron microscopes. It may be the physical basis of the classic cytosol or cytoplasmic matrix; it is regarded as an artifact by some cell biologists.

microtubule-organizing centers (MTOCs): organelles in the cell that facilitate the polymerization (incorporation) of tubulin protein into microtubules, such as centrosomes.

microtubules: a type of proteinaceous, fibrillar structure found in the cytoplasm of eukaryotic cells; making up part of structures as diverse as the cytoskeleton, spindle, centrioles, cilia, and flagella.

micturition: the frequent urination of a small amount of urine by a pregnant woman because of pressure on the urinary bladder.

midbrain: the middle of the early brain vesicles of the three-vesicle brain.

midgut: the middle portion of the early embryonic gut; see **foregut** and **hindgut**

midgut loop: the part of the elongating midgut of the embryo that extends into the umbilical cord.

migration: the movement of single cells, e.g., the movement of neural crest cells throughout the developing embryo to give rise to a wide diversity of structures.

miscarriage: see **spontaneous abortion**

mitochondria: cellular organelles with several functions, one of the most important of which is providing the cell with readily available biological energy in the form of ATP.

mitosis: one of two general types of cell divisions undergone by eukaryotic cells. It is the more ubiquitous of the two types, playing a prominent role in the growth and maintenance of organisms.

mitral valve: the valve between the left atrium and left ventricle of the heart, formed from endocardial cushion tissue.

mixed cranial nerves: cranial nerves composed of both sensory and motor components.

mixed nerves: nerves composed of both sensory and motor components.

molar pregnancy: see **hydatidiform mole**.

monomer: the single unit of a polymer. For example, amino acids are monomers of polymeric protein molecules.

monosomy: one body; as it pertains to chromosomes, a single copy of normally paired chromosomes.

monozygotic twins: see **identical twins**.

morning sickness: daily episodes of nausea, often in the morning, during early pregnancy.

morphogenesis: literally, the origin of form; refers to those processes during development that give rise to the form or shape of the organism or a structure.

morphogenetic movements: cellular movements carried out by groups of cells, which result in the origin of or change in form of the developing organism or part of it.

morula: an early stage in the development of animals, including humans, when cleavage is going on and the embryo consists of a solid ball of cells (blastomeres).

mosaic: as it pertains to development, an organism made up of two distinct cells or cell populations. The expression "mosaic development" pertains to those kinds of embryos not able to compensate for a missing part of the embryo; see **regulative development**.

motor cranial nerves: cranial nerves that innervate muscles, but lack sensory components.

motor proteins: proteins that carry cargo (e.g., mitochondria) along tracks (e.g., microtubules) within cells. Dynein, kinesin, and myosin are motor proteins.

müllerian ducts: a pair of ducts that arises from intermediate mesoderm and gives rise to the fallopian tubes, the uterus, and at least part of the vagina in females.

multicellular: consisting of more than a single cell and, generally, of a large number of cells.

multicellularity: exhibiting a multicellular composition, which, during animal development, is initially provided by cleavage.

multifactorial inheritance: the inheritance of a characteristic (e.g., intelligence) that depends on more than one gene and on environmental factors as well.

multigravida: a woman who has been pregnant more than one time.

multinucleated: possessing more than one nucleus (generally, many) within the same mass of cytoplasm. For example, syncytiotrophoblast and skeletal muscle fibers are multinucleated.

multiple allelism: the existence of more than two different allele forms for a given gene in a given species. In humans, the ABO blood typing system exhibits three different allelic forms, I^A, I^B, and i.

multiple pregnancy: in humans, a pregnancy in which more than one conceptus is carried at a time.

multipotent stem cells: stem cells limited to giving rise to specific populations of cells; e.g., umbilical cord blood and tissue stem cells.

muscle spindles: mechanoreceptors in skeletal muscle; also called *neuromuscular spindles.*

mutation: a genetic change.

myelencephalon: the caudal-most vesicle of the early five-vesicle brain; gives rise to the medulla.

myelin: a lipid substance that ensheathes some nerves, which are then referred to as *myelinated nerves.*

myelination: the process of ensheathing nerves in a covering of myelin carried out by glial cells; by oligodendrocytes in the central nervous system and by Schwann cells in the peripheral nervous system.

myoblasts: cells that are the precursors of muscle cells or muscle fibers.

myocardium: the muscular wall of the heart.

myoepithelial cells: smooth muscle cells derived from ectoderm and found in lacrimal, mammary, salivary, and sweat glands.

myofibrils: bundles of myofilaments (elongated aggregates of contractile protein) found in muscle fibers.

myosin: a rather ubiquitous protein in eukaryotic cells, occurring in especially abundant and organized form in skeletal muscle (where it is organized into thick filaments).

myotome: the portion of each somite that gives rise to muscle.

myotubes: syncytia (multinucleated cells), derived from the fusion of unicellular myoblasts, which give rise to skeletal muscle fibers.

nares: nostrils.

nasal passages: communications between the nares and the nasopharynx.

nasal pits: paired depressions on the head of the early embryo that develop into parts of the nasal cavities.

nasal placodes: paired ectodermal thickenings on the head of the early embryo, which will sink beneath the surface of the head to form the nasal pits.

nasal septum: the septum (partition) between the two nasal cavities.

nasolateral processes: paired thickenings on the head of the embryo, lateral to and partially surrounding the nasal pits.

nasopharynx: the portion of the pharynx above the soft palate.

natural selection: the sum total of those factors in the environment that favor the reproduction of some organisms and work against the reproduction of others; the driving force behind organic evolution.

nerve: a group of nerve cell processes contained within a single sheath.

nerve tracts: bundles of nerve cell processes within the central nervous system; equivalent to nerves in the peripheral nervous system.

neural canal: the lumen of the neural tube; also called *neurocoel.*

neural crest: a group of cells derived from the edges of the neural plate during neurulation, which initially occupies a position dorsal or dorsolateral to the neural tube. Neural crest cells migrate throughout the developing embryo and give rise to such an array of derivatives that the neural crest is sometimes regarded as a fourth germ layer.

neural ectoderm: neuroectoderm.

neural folds: the elevated margins of the neural plate, which arise early during the process of neurulation.

neural groove: the medial depression of the neural plate during neurulation, which brings together the neural folds in anticipation of neural tube formation.

neural plate: the thickened neuroectoderm on the dorsal side of the developing embryo, the formation of which initiates the process of neurulation.

neural tube: the tubular precursor of the central nervous system of chordates, the formation of which completes the process of neurulation.

neurites: nerve cell processes; that is, axons and dendrites.

neuroblasts: cellular precursors of neurons (nerve cells), derived from the neural tube or neural crest.

neurocoele: the lumen of the neural tube; also called *neural canal.*

neurocranium: that portion of the cranium that houses the brain.

neuroectoderm: that portion of the ectoderm of the early embryo that normally gives rise to the neural tube and neural crest; see **epidermal ectoderm**.

neuroepithelium: the epithelium derived from the neuroectoderm.

neurogenesis: origin of the nervous system; also, the process by which neurons are generated from neural stem cells is called neurogenesis. See **developmental neurogenesis** and **adult neurogenesis**.

neurohypophysis: the portion (posterior lobe) of the pituitary gland derived from the infundibulum of the brain.

neurons: nerve cells.

neuropores: openings at both ends of the newly formed neural tube, which in human embryos places the neurocoele in continuity with the amniotic cavity until the neuropores close.

neurulation: the embryonic process, following or overlapping gastrulation, beginning with the appearance of the neural plate and ending with the completion of the neural tube.

niche: in the context of stem cells, the environment in which stem cells are found; e.g., spermatogonia are found in the basal compartments of seminiferous tubules.

nidation: implantation.

nitrogenous waste: waste containing nitrogen-containing chemical compounds, such as urea.

nonsister chromatids: the chromatids of different chromosomes in a pair of homologous chromosomes. Sister chromatids are of the same chromosome in a pair of homologous chromosomes.
nonsurgical sperm aspiration: the withdrawal (by aspiration; suction) of sperm directly from the testis by means of a tiny cannula (tube).
norgestrel: one form of a variety of progestogen hormones found in birth control pills; has activity similar to the natural hormone progesterone.
nostrils: the openings of the nasal cavities to the outside of the body.
notochord: a dorsal rod of mesoderm, formed during gastrulation in all chordates, which gives some structural support to the early embryo. In human development, the notochord is replaced by the vertebral column and only gives rise to the nuclei pulposi of the intervertebral discs of the adult.
NSA: see **nonsurgical sperm aspiration**.
nuclear envelope: the nuclear membrane when viewed with an electron microscope, consisting of two membranes separated by a space (perinuclear cisterna) and pierced by numerous nuclear pores. Intimately associated with the endoplasmic reticulum, the outer membrane may have attached ribosomes on its cytoplasmic side.
nuclear membrane: the membrane found at the periphery of the cell nucleus, separating the nuclear contents from the cytoplasm; not found in prokaryotic cells and breaks down during cell division in most eukaryotic cells, with some protists being important exceptions; see **nuclear envelope**.
nuclear pores: passageways through the nuclear envelope between the cytoplasm and the interior of the nucleus.
nucleus: (1) cells: a cellular organelle found in eukaryotic cells. In eukaryotic cells, it has a limiting membrane, the nuclear membrane. In prokaryotic cells, no such membrane exists, and the term *nucleoid* refers to the comparable DNA-containing region. (2) nervous system: an aggregate of nerve cell bodies in the central nervous system; the equivalent of a ganglion in the peripheral nervous system.
nucleus pulposus: the center of intervertebral discs derived from the notochord.
obstetrician: a physician who specializes in the care of pregnant women.
odontoblasts: cells of the dental papilla, derived from neural crest cells, of the developing tooth, which give rise to the dentin of the developing tooth.
olfactory cells: cells that act as receptors for the sense of smell.
olfactory epithelium: a sheet of olfactory cells found in a nasal cavity.
olfactory nerves: the first pair of cranial nerves, going from the olfactory epithelium in the nose to the olfactory lobes of the brain; concerned with the sense of smell.
oligodendrocytes: glial cells found in the central nervous system, where they myelinate nerve tracts.

omental bursa: the space dorsal to the stomach and lesser omentum, lined with peritoneum and communicating with the general abdominal cavity.

ontogeny: the development of the individual organism.

oocyte: a female germ cell after it has developed from an oogonium and begins its cell differentiation toward becoming an ovum (egg).

oogenesis: the type of gametogenesis that occurs in females and results in the production of ova (eggs).

oogonium: (plural, oogonia) a type of female germ cell that functions as a stem cell until it begins to undergo cell differentiation and becomes an oocyte.

ootid: a female germ cell; specifically the egg (ovum).

optic nerves: the second pair of cranial nerves; they go from the retinas of the eyes to the superior colliculi of the midbrain.

optic stalk: the constricted portion of the original optic vesicle; found between the optic cup and the lateral wall of the diencephalon.

optic vesicles: paired, lateral evaginations of the wall of the forebrain that give rise to the optic cups and optic stalks.

organelles: subcellular components specialized for specific functions, such as mitochondria and lysosomes.

organogenesis: the formation of organs.

organoids: small versions of organs derived from stem cells or from tissue cells that have been converted into stem cells called induced pluripotent stem cells (iPSCs).

oropharyngeal membrane: the membrane formed of stomodeal ectoderm and foregut endoderm. Its normal breakdown creates the opening called the *mouth*.

os: opening.

ossicles: small bones, such as the incus, malleus, and stapes of the middle ear cavity.

ossification: the formation of bone.

ossification center: a region from which bone formation spreads outward.

osteoblasts: cells derived from mesodermal mesenchyme cells, which are the precursors of osteocytes and are responsible for bone deposition.

osteogenic cells: the only bone cells that divide; they replenish osteoblasts and osteocytes, cells that do not divide.

osteocytes: bone cells derived from osteoblasts, which reside in already formed bone.

osteogenesis: bone formation.

ostium: an opening, such as the opening of the fallopian tube near the ovary.

otic vesicles: the bilateral pair of vesicles that arise by invagination from the surface of the head at the level of the hindbrain and give rise to the inner ears.

ovarian follicle: a follicle found in the ovary; consisting of a single oocyte and numerous follicle cells (and their products).

ovaries: the female gonads.

ovulation: release of an egg(s) from the surface of the ovary.

ovum: a mature female germ cell, the female gamete.

oxytocin: a hormone produced by the posterior lobe of the pituitary gland, which stimulates contraction of the smooth muscles of the uterus at the time of labor and is used clinically to induce labor.

pacemaker: the sinoatrial node, which functions to stimulate the heart and control the rate of heartbeat.

palate: the roof of the mouth; partially hard (with bone) and partially soft (without bone).

palatine shelf: see **lateral palatine processes** and **median palatine process**.

palmar erythema: redness of the palms of the hands often experienced by pregnant women.

pancreas: an organ of the digestive system found in the abdominal cavity. The pancreas produces both digestive enzymes (e.g., trypsin) and hormones (insulin).

paracrine signaling: mode of cell–cell communication in which signaling molecules (paracrine factors) act as local mediators and only affect cells in the immediate environment of the signaling cell; e.g., stem cell factor (SCF) is a paracrine protein that promotes cell division in numerous stem cell populations.

parathyroid glands: two pairs of glands that arise as evaginations of the third and fourth pharyngeal pouches. They play a role in the regulation of calcium levels in the blood.

paraxial mesoderm: that portion of the mesoderm that bilaterally flanks the embryonic axis and gives rise to somites; also called *somitic* or *segmental* or *dorsal mesoderm*.

parenchymal cells: the cells of an organ that make up the specialized parts of the organ, as opposed to the organ's connective tissue.

parturition: childbirth.

passive immunity: immunity passively acquired, as in the form of antibodies coming through the placenta or in the form of antibodies received through an injection.

pattern of cleavage: the manner of cleavage; the way in which the embryos of a given species undergo cleavage. Humans undergo a holoblastic, equal, rotational cleavage pattern.

pelvic cavity: the lowermost portion of the abdominal cavity. In females it contains the nongravid, internal reproductive organs—specifically the uterus, fallopian tubes, and ovaries.

pelvic inflammatory disease (PID): infection of any female reproductive organ.

penis: the male organ of intercourse, which, together with the scrotum, make up the male external genitalia.

pericardial cavity: the cavity, derived from the paired amniocardiac vesicles, that contains the heart and is lined by the pericardium.

perichondrium: The connective tissue covering of cartilage.

periderm: the outer layer of the two-layered skin of the embryo at the end of the embryonic period; see also **basal layer**.

periosteal buds: ingrowths of the periosteum into the cartilage of growing bone.

periosteum: The connective tissue covering of bones.

peripheral: at the periphery; not at the center.

peripheral nervous system (PNS): that portion of the nervous system outside the central nervous system (brain and spinal cord).

peristalsis: the spontaneous contraction of the smooth muscle of the wall of the gut, which moves food along the digestive tract.

peritoneal cavity: derived from the coelom and contains most of the viscera; its lining is the peritoneum; also called *abdominal cavity.*

peritoneum: lining of the peritoneal (abdominal) cavity and covering of the abdominal internal organs.

perivitelline space: the fluid-filled space between the plasma membrane of the egg and the zona pellucida.

pH: a numerical scale for expressing hydrogen ion concentration; the negative log of the hydrogen ion activity.

pharyngeal arches: see **branchial arches**.

pharynx: the cavity between the oral (mouth) cavity and the esophagus.

phenotype: the outward, frequently visible, expression of the genotype.

phenotypic sex: the apparent sex of an individual; having the characteristics of a male or a female.

philtrum: the groove on the medial surface of the upper lip, directly beneath the nose.

photoreceptor cells: sensory cells found in the retinas of the eyes, which are light receptors for the sense of vision.

phylogeny: the evolutionary history of a group of organisms.

PiED: see **preimplantation embryo diagnosis**.

pigmentation: having pigment.

pigment cells: cells specialized for the production of pigment; for example, melanocytes are pigment cells specialized for the production of the pigment melanin.

pineal gland: an endocrine gland derived from the roof of the diencephalon; secretes melatonin, which may play some role in the onset of puberty.

pinnae: the external flaps of the ears found on the sides of the human head.

pituitary gland: an endocrine gland that has a dual origin. Part of it (neurohypophysis) is derived from the floor of the diencephalon (infundibulum), and part (adenohypophysis) is derived from the roof of the stomodaeum. Although under the control of the hypothalamus of the brain, the pituitary gland secretes hormones that play central roles in reproductive physiology; also called *hypophysis.*

placenta: an organ, partially fetal and partially maternal in origin, through which the mother and fetus exchange materials such as oxygen, food, and waste products.

placenta fetalis: the chorion frondosum or bushy chorion, which is the fetal contribution to the placenta.

placental circulation: that portion of prenatal circulation that carries blood from the fetus to the placenta and back by way of the umbilical cord; that is, that portion of the circulation that leaves the fetus by way of the umbilical arteries and returns by way of the umbilical vein.

placental stage: that stage of labor extending from the birth of the baby to the birth of the placenta.

placenta materna: the decidua basalis, which is the maternal contribution to the placenta.

placenta previa: an abnormal position of the placenta in which it blocks the internal os, that is, the opening from the uterine cavity into the cervical canal that makes up part of the birth canal.

placodes: thickenings of head ectoderm that contribute to the formation of some of the sense organs (e.g., olfactory placodes, lens placodes) and some of the cranial ganglia.

planes: see names of individual planes; e.g., frontal plane.

plasma: see **blood plasma**.

plasma membrane: the part of every cell found at its periphery and separating the contents of the cell from the cell's environment. Its function is vital because it determines what does and does not enter and leave the cell. It makes the interior of the cell different from its surroundings, which is necessary for life.

pleural cavities: contain the lungs and are lined by pleurae; derived from the coelom.

pluripotent stem cells: stem cells capable of forming most, but not all, tissues in an organism; e.g., embryonic germ cells.

PNS: peripheral nervous system.

polar body: a cell formed during oogenesis, from the division of an oocyte, that normally plays no role in development.

poles: the two ends of the spindle found in dividing cells.

polygenic: involving many genes.

polymer: a large molecule made of small, repeating, subunits (monomers).

polymerase chain reaction (PCR): a highly efficient enzymatic method of cloning (making many identical copies of) specific pieces of DNA.

polyploidy: having many multiples, i.e., more than two, of the haploid number of chromosomes.

polyspermy: the fertilization of a single egg by more than one sperm.

pons: a part of the brain derived from the metencephalon, which has fibers connecting the cerebral cortex with the cerebellum.

pontine: relating to the pons of the brain.

position: one of four descriptions of the fetus's alignment in the uterus. Position of the fetus relates a chosen part of the fetus to the right or left side of the mother, such as face left or face right.

posterior fornix: the deepest and most posterior of the four spaces of the vaginal vault, subdivided by the uterine cervix that projects into it.

posterior neuropore: the temporary opening at the caudal end of the early neural tube.

postpartum: after birth.

preformationism: the doctrine that the new individual is already preformed in the spermatozoon or egg and that development involves essentially the growth of the preformed organism.

pregnancy: the condition of a woman who is carrying a conceptus (that is, the product of conception or fertilization).

pregnancy reduction: selective abortion of some of the fetuses of a multiple pregnancy.

preimplantation embryo diagnosis (PiED): removal of a cell from an early embryo for very early prenatal diagnosis. The embryo is the result of IVF and the procedure is carried out in a dish.

premature: in the context of human development, the birth of a child before completion of the normal, full, term of a pregnancy.

premaxilla: incisive bone; in humans is fused with the maxilla.

prenatal: before birth.

prenatal diagnosis: the evaluation of the condition of the fetus before birth.

presentation: one of four descriptions of the fetus's alignment in the uterus. Presentation of the fetus refers to the fetal part leading the way down the birth canal.

primary follicle: an early ovarian follicle consisting of a single oocyte and a single *layer* of follicle cells.

primary oocyte: an oocyte that has begun meiosis, but has not yet completed meiosis I.

primary ossification center: locations in the forming fetal skull where bone formation begins.

primary vesicles: the three early subdivisions of the embryonic brain.

primigravida: a woman who is pregnant for the first time.

primitive blood cells: the first blood cells to form; appear first in the blood islands of the yolk sac.

primitive node: the mound of cells at the cephalic end of the primitive streak. Cells passing through the primitive node during gastrulation give rise to the notochord. Also called *Hensen's node.*

primitive streak: a thickened streak of cells found in the upper layer of the two-layered human embryonic disc. Cells pass through it during gastrulation, giving rise to the germ layers. Other higher vertebrates also develop primitive streaks early in their development.

primordial germ cells: migratory germ cells first observed on the yolk sac of mammalian embryos. Once they enter the developing gonad, they are known as *oogonia* or *spermatogonia.*

progesterone: the steroid hormone responsible for the maintenance of pregnancy by maintaining the endometrial lining of the uterus.

programmed cell death: cell death due to a genetic program for it and not due to a traumatic occurrence; also called apoptosis.

prometaphase: the stage of mitosis or meiosis when the chromosomes have been released from the nuclear membrane, but are not yet aligned on the equator of the spindle.

pronephric ducts: ducts of the pronephros derived from intermediate mesoderm.

pronephric tubules: the tubular components of the pronephros.

pronephros: the most primitive type of kidney that arises during human embryonic development.

pronucleus: when gametes are formed, the number of chromosomes is reduced to the haploid number. If fertilization occurs, the egg and sperm nuclei will have only the haploid number of chromosomes; such nuclei are called pronuclei.

protein motors: category of proteins with moving parts. They use ATP as a source of energy to move: substances within cells, cells, or things past cells. The three categories of motor proteins are dyneins, actins, and kinesins.

prophase: the stage of mitosis or meiosis when the chromosomes are becoming visible but are still contained within the nuclear membrane.

prosencephalon: the embryonic forebrain.

prostaglandins: chemical signals that have numerous activities in the body, including stimulating contraction of smooth muscle of the female reproductive systems.

prostate gland: one of four types of male auxiliary sex glands, the prostate gland makes a major contribution to the liquid portion (seminal plasma) of semen.

proteins: the class of organic molecules that plays *the* important role in almost any cellular activity.

proto-oncogene: normal genes that regulate cell proliferation and differentiation. Mutations of these genes often result in what are called *oncogenes* (cancer genes), which play a role in the development of cancer.

protozoa: animal organisms made up of a single cell.

proximal: an anatomic term referring to a structure that is close to an anatomic reference point; see also **distal**.

pseudoglandular period: a period during lung development when the lungs resemble an endocrine gland.

puberty: the time period during which changes occur that transform a juvenile into a sexually mature individual.

pulmonary circulation: the portion of the circulatory system that carries blood from the heart to the lungs and back again; that is, the portion of the circulation from the pulmonary trunk leaving the right ventricle to the pulmonary veins returning to the left ventricle.

pulmonary trunk: the main arterial blood vessel of pulmonary circulation that leaves the right ventricle of the heart.

Purkinje fibers: modified heart muscle fibers that form the terminal part of the heart's conducting system.

quadruplets: a single pregnancy simultaneously carrying four conceptuses or the approximately simultaneous birth of four babies by a single woman.

radial symmetry: the form of symmetry in which many planes passing through a given body axis divide the organism into two mirror-image halves; see **bilateral symmetry**.

radiation: energy emitted and propagated through space or matter, such as x-rays and ultrasound.

Rathke's pocket: an evagination from the dorsal surface of the stomodeum, which gives rise to the anterior lobe (adenohypophysis) of the pituitary gland.

receptor proteins: proteins that bind other molecules or ions and, as a result of the binding, influence cellular activity; e.g., insulin receptors in the cell surface and progesterone receptors in the cytoplasm.

recessive: descriptive of one allele of a nonidentical pair of alleles for the form of a characteristic that is not expressed.

recombination: the appearance in offspring of gene combinations not found in the parents.

recumbent: leaning back.

red blood cells: cells in the circulatory system that lack nuclei, contain hemoglobin, and are responsible for oxygenation of the tissues of the body.

regulative development: that type of development in which the early embryo is able to compensate for the loss of part of the embryo. Consequences of this type of development include the existence of identical twins and the ability to do preimplantation embryonic diagnosis.

replication: the making of copies of something, such as DNA.

reproduction: the complex process by which organisms give rise to offspring.

respiratory distress syndrome: symptoms, possibly including death, resulting from incomplete lung development at the time of delivery.

retina: the portion of the eye derived from the inner layer of the optic cup containing the photoreceptor cells and from which the optic nerve originates.

retrovirus: a virus with RNA (rather than DNA) as its genetic material, such as HIV virus.

Rh+, Rh–: symbols for the dominant and recessive alleles, respectively, of the Rh factor gene.

Rh incompatibility: if a pregnant woman who is Rh– is carrying a fetus who is Rh+, an Rh incompatibility exists and the mother's immune system may mount an immune attack against the fetus's red blood cells; see **erythroblastosis fetalis**.

ribonucleic acid (RNA): a type of nucleic acid; three general types of which are found in cells: messenger RNA, ribosomal RNA, and transfer RNA. RNA molecules are responsible for the translation of the genetic message into specific protein molecules.

ribosomal RNA (rRNA): a specific family of RNA molecules that, together with protein molecules, make up ribosomes, the protein-making organelles of cells.

rhombencephalon: the embryonic hindbrain.

rostral: means "beak" and is an anatomic term often used to refer to structures in the head in which the term "cephalic" would not be precise.

rubella virus: or German measles virus, is a teratogenic virus.

sacculus: the lower, saclike chamber of the membranous labyrinth of the ear.

sagittal plane: that plane of the body that separates the left side of the body from the right side.

sarcomeres: contractile units of skeletal muscle fibers made up of highly ordered arrays (reflected in their striated appearance) of contractile proteins.

satellite cells: small cells, associated with skeletal muscle fibers, that retain the capacity for cell division and may contribute to the substance of muscle fibers.

Schwann cells: glial cells outside the central nervous system, which originate from the neural crest and myelinate nerves outside the peripheral nervous system.

sclera: the tough outer coat of the eye.

sclerotome: the portion of the somites that gives rise to the vertebrae of the vertebral column.

scrotum: the pouch containing the testes, which, with the penis, make up the male external genitalia.

sebaceous gland: gland of the skin that makes and secretes sebum.

sebum: the secretion of sebaceous glands.

second polar body: a tiny cell produced by the secondary oocyte when it undergoes meiosis II.

secondary follicle: an ovarian follicle with more than one layer of follicle cells, but not yet containing antral vacuoles.

segregate: separate.

semen: the complex mixture of cells (spermatozoa) and liquid (seminal plasma) that makes up the product of the male reproductive system.

semicircular canals: those three portions of each inner ear, which allow us to orient ourselves in space and thereby maintain equilibrium.

semiconservative replication: a method of copying something in which each new copy consists of a half of a previous copy, which has acted as a template. For example, when DNA replicates, it undergoes semiconservative replication.

seminal plasma: the fluid portion of semen.

seminal vesicles: one of four kinds of male auxiliary sex glands. These paired glands make a substantial contribution to semen.

seminiferous tubules: found in the testes, these tubules are the sites of spermatogenesis.

sensory cranial nerves: those cranial nerves, such as optic nerves, that carry only sensory nerve impulses into the brain.

septum primum: the first partition of the atrial region of the early developing heart.

septum secundum: the second partition of the atrial region of the early developing heart.

Sertoli cells: somatic cells found in seminiferous tubules, which assist spermatids undergoing cytodifferentiation (spermiogenesis) into spermatozoa (gametes).

seven cardinal movements of labor: those movements undergone by the fetus during the process of childbirth.

sex cells: gametes; those cells that actually undergo fertilization.

sex chromosomes: those chromosomes, other than the autosomes, that play a direct role in sex determination: the X and Y chromosomes.

sex determination: the process by which genetic sex is specified. Human females have an XX chromosome makeup and males an XY chromosome makeup. At the time of fertilization, sex is determined. If an X-bearing sperm fertilizes the egg, an XX chromosome (female) zygote is produced. If a Y-bearing sperm fertilizes the egg, an XY chromosome (male) zygote is produced.

sex glands: glands that make up parts of the reproductive systems and whose secretions play a direct role in reproduction. In the female: Bartholin's glands and Skene's glands. In the male: seminal vesicles, prostate gland, bulbourethral glands (Cowper's glands), and the glands of Littre.

sex hormones or sex steroids: hormones that play dramatic and essential roles in reproduction: estrogens, progesterone, and testosterone.

sex-linked genes: genes (alleles) found on the X chromosome.

sex-linked inheritance: inheritance of characteristics whose genes are located on the X chromosome.

Siamese twins: see **conjoined twins**.

sinus venosus: the caudal-most portion of the early cardiac tube.

skeletal muscle: one of three general types of muscles found in the body and, generally, muscle under conscious control; also called *voluntary* or *striated muscle*.

Skene's glands: a type of female auxiliary sex gland that releases its secretions into the vestibule of the vulva.

skin: the largest organ of the body, which provides a protective covering for the organism. It is made up of two layers: the outer epidermis derived from ectoderm and the inner dermis derived from mesoderm.

skull: the skeleton of the head.

smooth chorion: see **chorion laeve**.

smooth muscle: one of three general types of muscles found in the body; involuntary muscle, that is, not under conscious control.

soft palate: see **palate**.

somatic cells: all the cells of the body except the germ cells; not directly involved in reproduction.

somatic mesoderm: the dorsal layer of mesoderm derived from the lateral mesoderm.

somatoplasm: a term that refers to somatic cells collectively.

somites: paired aggregates of mesodermal cells derived from the somitic mesoderm; they appear in a cephalocaudal direction with such regularity that their appearance is used to time early development in many species of animals.

somite bulges: visible, surface bulges on the flanks of the trunk and tail of the embryo during the middle of the embryonic period, which are indicative of the underlying developing somites.

somitic mesoderm: the portion of the mesoderm that becomes segmented into pairs of somites; also called *dorsal, segmental*, or *paraxial mesoderm.*

sonogram: the picture resulting from the prenatal diagnostic procedure of ultrasonography (use of high-frequency [ultrasound] sound waves).

sperm: a shorthand term for *spermatozoa.*

spermatic cord: a composite structure including blood vessels, lymphatic vessels, nerves, and the vas deferens or sperm duct proper.

spermatocytes: cells that arise from spermatogonia and undergo meiosis and cell differentiation to become spermatozoa.

spermatogenesis: the type of gametogenesis undergone by males and resulting in the production of sperm.

spermatogonia: (singular, spermatogonium) male germ cells that act as stem cells and that may undergo cell differentiation to become spermatozoa (gametes).

spermatozoon: (plural, spermatozoa) the male gamete.

sperm bank: a collection of sperm kept in cryogenic storage.

spermicide: a chemical designed to intentionally kill sperm; a birth control substance.

spermiogenesis: the cell differentiation by which spermatids are transformed into spermatozoa.

spina bifida: a type of birth defect resulting from incomplete fusion of the neural folds in the formation of a defective neural tube.

spinal canal: the fluid-filled lumen of the spinal cord.

spinal cord: the portion of the nervous system derived from the caudal portion of the neural tube, which, with the brain, makes up the central nervous system.

spinal ganglia: ganglia, derived from the neural crest, that flank the spinal cord in pairs and are connected to the spinal cord by means of the dorsal roots of the spinal nerves.

spinal nerves: mixed (sensory and motor components) nerves of the peripheral nervous system. Their sensory components originate in spinal ganglia and their motor components originate in the ventral portion of the spinal cord.

spindle: a transient organelle found in dividing eukaryotic cells, the function of which is the equitable distribution of chromosomes to daughter cells.

spindle fibers: fibers made up of microtubules, which radiate from each pole of the spindle of a dividing cell and attach to the chromosomes or interdigitate with spindle fibers from the opposite pole. They provide for the elongation of the spindle and the movement of the chromosomes toward the poles of the spindle.

splanchnic mesoderm: the ventral layer of mesoderm derived from the lateral mesoderm.

spontaneous abortion: the spontaneous expulsion of the conceptus from the womb; also called *miscarriage.*

stasis: lack of movement; as in the cessation of blood circulation.

STD: sexually transmitted disease.

stem cell: a cell that undergoes mitotic cell division to give rise to the same type of cell. At some point, stem cells leave the pool of mitotically dividing cells to begin a process of cell differentiation.

steroidogenesis: the formation of steroid hormones; e.g., testosterone, estrogens, and progesterone.

stratum corneum: the outermost layer of the epidermis, composed of keratinized cells.

stratum germinativum: the innermost layer of the epidermis, which contains mitotic stem cells required for the continual replenishment of the epidermis.

stratum granulosum: an intermediate layer of the epidermis consisting of cells containing many granules.

stratum lucidum: a translucent layer of the epidermis consisting of transparent cells.

stratum spinosum: a layer of the epidermis consisting of cells with conspicuous intercellular bridges.

striated: striped.

suction method: a method of abortion used early in pregnancy, in which the conceptus is literally aspirated (sucked) out of the uterus.

superfecundation: the fertilization of two or more *eggs* during the same uterine cycle by *sperm* from separate acts of *sexual intercourse*.

superior vena cava: a major venous (vein) blood vessel returning blood from the head and neck (parts of the body superior to the heart) to the right atrium of the heart.

supernumerary breast: extra breast; breast in excess of the normal two for a human; may arise anywhere along the mammary ridges.

superovulate: to ovulate more than the normal number of eggs under the influence of hormonal stimulation of the ovary.

supine: lying on one's back.

supporting cells: cells that aid or protect the cells carrying out a specific function. (e.g., supporting cells in taste buds, the Sertoli cells of the seminiferous tubules, and the follicle cells of the ovarian follicles.)

surfactant: lipid molecules (lecithin and sphingomyelin) found in the mature lungs that allow terminal alveoli to expand with each inhalation.

surrogate mother: a woman carrying a pregnancy when the conceptus is not hers.

sutures: the narrow spaces between the bony plates of the skull, which are most conspicuous at about the time of birth.

sweat glands: skin glands derived from the epidermis of the skin, with pores that open onto the surface of the skin. The evaporation of their secretions, sweat, is an important part of dissipation of heat by the body.

Sylvian aqueduct: or aqueduct of Sylvius, is the portion of the ventricular system of the brain that connects the third and fourth ventricles and through which cerebrospinal fluid circulates.

symmetric cell division: cell division where the two daughter cells have the same fates; e.g., division of a primary spermatocyte, during spermatogenesis, to produce two secondary spermatocytes.

synapse: the junction between two nerve cells or between a nerve cell and a muscle fiber.

synapsis: the pairing of homologous chromosomes during prophase I of meiosis.

syncytiotrophoblast: the outer, noncellular, highly invasive component of the trophoblast.

syncytium: a multinucleated mass of cytoplasm; loosely, a multinucleated cell.

syndactyly: a type of birth defect involving fusion of the digits.

syphilis: a bacterial sexually transmitted disease.

systemic circulation: the general blood circulation of the body; exclusive of the pulmonary and placental circulations.

T cells or T lymphocytes: a subset of lymphocytes (which are a subset of white blood cells) that carry out cell-mediated immunity.

tail fold: one of four body folds. The tail fold sculpts the tail end of the embryo out of the caudal region of the embryonic disc.

taste cells: the cells found in a taste bud, which are receptors for the sense of taste.

telencephalon: the cephalic-most vesicle of the early five-vesicle brain; gives rise to the cerebral hemispheres.

telolecithal egg: a type of egg that has a concentration gradient of yolk distributed along its animal-vegetal axis (e.g., frog eggs); see **homolecithal**.

telophase: the stage of mitosis or meiosis when the chromosomes reach the poles of the spindle.

template: guide to the form of a piece being made; see **semiconservative replication**.

teratogen: that which causes a birth defect; examples: some viruses, some kinds of radiation, and some chemicals.

teratology: the study of birth defects.

terminal sac period: period during lung development when the alveolar ducts give rise to terminal air sacs.

tertiary follicle: an ovarian follicle with antral vacuoles or a forming antrum.

test tube baby: a loosely used expression for babies resulting from IVF.

testes: (singular, testis) the male gonads.

testosterone: a male sex hormone (androgen) responsible for the development of male secondary sex characteristics and necessary for spermatogenesis.

tetrad: the four chromatids found in a synapsed pair of homologous chromosomes during prophase I of meiosis.

tetraploid number: four times the haploid number of chromosomes.

thalidomide: a chemical teratogen that is especially detrimental to limb formation.

thymus: an endocrine gland that plays a role in the development of the immune system, especially in the development of T lymphocytes.

thyroid gland: an endocrine gland that produces the hormones thyroxin (which regulates metabolism and growth) and calcitonin (which helps to regulate calcium levels in the blood).

tight cell junctions: specialized contacts formed between cells that establish partitions between isolated compartments of the body.

tonsillar region (tonsils): two masses of lymphoid tissue, each between folds of tissue at the back of the mouth (tonsillar region).

totipotent stem cells: stem cells capable of forming an entire organism; e.g., embryonic stem cells.

toxoplasmosis: a protozoan parasitic disease that can be teratogenic to a fetus being carried by an infected mother.

trabeculae carneae: muscular extensions of the interior of the ventricles of the heart that coalesce to form part of the interventricular septum and the papillary muscles and tendinous cords that operate the valves of the heart.

trachea: the windpipe portion of the respiratory system.

transcription: the synthesis of RNA using DNA as a template.

transfer RNA (tRNA): a class of RNA molecules that carries amino acids to their correct positions (as specified by the genetic code) along mRNA on the ribosome.

translation: the synthesis of proteins using mRNA as a template.

transverse plane: the plane of the body that separates the anterior end of the body from the posterior end.

tricuspid valve: the valve between the right atrium and the right ventricle of the heart, formed from endocardial cushion tissue.

trimester: one-third of the duration of pregnancy.

triplets: a single pregnancy simultaneously carrying three conceptuses or the approximately simultaneous birth of three babies by a single woman.

triploid: three times the haploid number of chromosomes.

trisomy 21: the presence of three (rather than the normal two) copies of chromosome 21; results in offspring with Down's syndrome.

trizygotic: of three zygotes; as with fraternal triplets.

trophectoderm: trophoblast.

trophoblast: the outer of the two parts of the blastocyst, which does not contribute to the embryo, but only to extraembryonic structures.

truncoventricular region: the region of the early developing heart in which the ventricular region narrows down to the truncus arteriosus.

truncus arteriosus: the arterial trunk that arises from the heart of the embryo; gives rise to the aorta and the pulmonary trunk.

tubal ligation: a birth control method involving the cutting and tying of the fallopian tubes.

tubal pregnancy: see **ectopic pregnancy**.

tubercles: small nodules.

tubulin: the protein that forms microtubules of the cytoskeleton and spindle.

tunica albuginea: the tough connective tissue covering of the testis.

Turner's syndrome: a syndrome that includes a female phenotype, short stature, and possibly mental retardation; resulting from an XO sex chromosome constitution.

twins: two siblings that happen to occupy the same uterus (womb) at the same time; see **identical twins, fraternal twins**, and **conjoined twins**.

tympanic cavities: the cavities of the middle ears.

ultrasonography, ultrasound: a noninvasive prenatal diagnostic procedure that uses high-frequency sound waves to create images of the embryo or fetus.

umbilical arteries: the pair of blood vessels that carries blood from the fetus through the umbilical cord out to the placenta. The blood carried is deficient in food and oxygen.

umbilical cord: the cordlike structure that attaches the belly of the fetus to the placenta and through which materials exchanged between the fetus and mother flow.

umbilical vein: the blood vessel that carries blood to the fetus, through the umbilical cord, from the placenta. The blood carried is rich in food and oxygen.

unicellular: single-cell. The only stage in the human life cycle that is unicellular is the fertilized egg (zygote).

ureter: the pair of tubes that carry urine from the kidneys to the urinary bladder.

urethra: the duct running from the urinary bladder to the outside of the body. In females, it is a purely urinary duct leading to the urethral opening into the vestibule; in males, it is a urogenital duct carrying both urine and semen to the urogenital opening on the surface of the glans penis.

urinary bladder: the saclike organ that receives and temporarily stores urine until the time of urination.

urine: liquid nitrogenous waste formed from the circulatory system and excreted by the kidneys.

uriniferous tubules: tubules found in the kidneys, which participate in the formation of urine.

urogenital membrane: the membrane found between the labioscrotal swellings of the early external genitalia; its rupture opens the urethral groove.

urogenital sinus: the part (anterior) of the cloaca into which the embryonic urogenital ducts open.

uterine cavity: the cavity or lumen of the body (corpus) of the uterus.

uterus: the female reproductive organ concerned with the reception, retention, nurturing, and expulsion of the conceptus.

utriculus: the part of the membranous labyrinth into which the semicircular canals open.

uvula: the small piece of flesh that hangs down from the soft palate at the back of the throat.

vagina: the female organ of intercourse, which also serves as part of the birth canal during childbirth.

varicose veins: veins that have become abnormally dilated and tortuous.

vas deferens: (plural, vasa deferentia) the sperm duct proper in males.

vascular spiders: minute reddened elevations of the skin from which there is branching of radicles (little roots).

vasectomy: a surgical contraceptive procedure that entails cutting the vasa deferentia and tying (ligating) the cut ends.

vasovasostomy: the surgical reversal of a vasectomy.

vault: an arched structure.

veins: blood vessels that carry blood to the heart. Although this blood is generally deficient in food and oxygen, the umbilical vein carries replenished blood from the placenta into the fetus.

venae cavae: (singular, vena cava) the major venous blood vessels returning blood to the heart (right atrium).

venter: the belly.

ventral: referring to the venter or belly of an animal.

ventral mesocardium: a membrane, derived from a double layer of splanchnic mesoderm, that forms beneath the developing heart in the pericardial region of the coelom. This membrane disappears almost as soon as it forms.

ventricles: cavities. (1) heart: two of four chambers of the heart, they pump blood out of the heart. (2) brain: the fluid-filled spaces within the brain through which cerebrospinal fluid circulates.

ventricular layer: the inner layer of the spinal cord, made up of ependymal cells and adjoining the cerebrospinal fluid-filled spinal canal.

vernix caseosa: a whitish, cheesy coat on the skin of the fetus made up of secretions of sebaceous glands (sebum), periderm cells, and lanugo.

vertebrae: (singular, vertebra) the elements of the vertebral column or "backbone."

vertebral column: the "backbone" of the vertebrates; that portion of the axial skeleton derived from the sclerotomes of somites and responsible for the bony protection of the spinal cord.

vertebrates: animals placed into the subphylum Vertebrata, of the phylum Chordata, because they possess, in addition to a notochord, a vertebral column.

vesicle: a small sac; usually containing fluid.

vestibule: the almond-shaped opening that makes up part of the vulva or external genitalia of the female.

viability: ability to live.

viscerocranium: that portion of the cranium that surrounds the oral cavity, pharynx, and upper respiratory tract.

vitelline duct: the passageway between the yolk sac and the midgut of the embryo.

vulva: the female external genitalia.

Wharton's jelly: the connective tissue matrix of the umbilical cord.

white matter: those portions of the central nervous system that have a large concentration of nerve fibers and therefore appear white in the fresh state.

womb: the uterus.

X chromosome: the chromosome responsible for the human female phenotype when present as a pair, XX.

Y chromosome: the chromosome responsible for the human male phenotype when present in a pair with the X chromosome, XY.

yolk: nutritive material deposited in the developing egg (oocyte) as it differentiates in the ovary of the mother. It provides nutrition for the early developing embryo.

yolk duct: see **vitelline duct**.

yolk sac: one of the four extraembryonic membranes formed during the development of higher vertebrates, including humans. It contains yolk in most cases, but generally not in mammals; in humans, it is the site of the first appearance of the primordial germ cells and of the first blood cells.

yolk stalk: the cordlike structure that attaches the belly of the fetus to the yolk sac.

zygote intrafallopian transfer (ZIFT): transfer of zygotes into the fallopian tubes on the day after fertilization occurs in a culture dish, instead of waiting for the zygotes to develop into embryos.

zona pellucida: a noncellular egg coat produced in the ovary during oogenesis and found around the ovulated egg.

zygote: the fertilized egg.

References

Alberts, B. et al. *Molecular Biology of the Cell*, 3rd edition. Garland Publishing, New York, 1994, p. 12. "It is thought that all organisms living now on earth derive from a single primordial cell born more than 3 billion years ago."

Asimov, I. At stake: 500,000,000 years of life. *National Wildlife*. April–May, 1972. Warned that humankind's increasing numbers and burgeoning technology have cast a shadow across the future. Perhaps with the human population exploding, birth control is becoming less of a personal matter and more of a public concern.

Cassill, K.T. *Nature's Amazing Mystery*. Atheneum, New York, 1982. In her engaging book, *Twins: Nature's Amazing Mystery*, states that Hellin's Law still provides a generally accepted prediction about the frequency of all multiple births.

Eastman, N.J. and L.M. Hellman. *Williams Obstetrics*, 13th edition, Appleton-Century-Crofts, Englewood Cliffs, NJ, 1966. According to Eastman and Hellman, a mathematical relationship (referred to as Hellin's Law).

Grobstein, C. *Science and the Unborn*. Basic Books, New York, 1988. In his book, Science and the Unborn, uses the term "preembryo" to refer to the conceptus during the first 2 weeks.

Tables created from Hildreth, C. Definitive Guide to Perinatal Stem Cells. August 23, 2017.

Jaynes, J. *The Origin of Consciousness in the Breakdown of the Bicameral Mind*. Houghton Mifflin, New York, 1990. We appear to occupy a space just behind the eyes.

Lewis, R. *Beginnings of Life*. Wm. C. Brown, Dubuque, IA, 1992. Birth control tables.

Oakley, A. *The Captured Womb: A History of the Medical Care of Pregnant Women*. Oxford, New York, 1984. Discusses the gradual medicalization of motherhood.

Rorvik, D.B. *New Baby: Promise & Peril of the Biological Revolution*, Simon & Schuster, New York, 1979. Some have made unsubstantiated claims that human cloning has occurred.

Saunders, J.W., Jr. *Developmental Biology*. Macmillan, New York, 1980. Classical experiments by Saunders on chick skin development.

Wilkins, A.S. *Making Faces: The Evolutionary Origins of the Human Face*. Harvard University Press, Cambridge, MA, 2017.

Wilson, E.B. *The Cell in Development and Heredity*, 3rd edition. Macmillan, New York, 1925, p. 1. "Long ago it became evident that the key to every biological problem must finally be sought in the cell; for every living organism is, or at some time has been, a cell."

Wolpert, L. *The Triumph of the Embryo.* Oxford University Press, New York, 1991, p. 12. "It is not birth, marriage, or death, but gastrulation which is truly 'the important event' in your life."

Websites: In chapter order

https://www.ncbi.nlm.nih.gov/pmc/articles/PMC4376261/—**Chapter 6**

https://www.bioinformant.com/perinatal-stem-cells/—**Chapter 6**

http://www.cam.ac.uk/research/news/scientists-create-artificial-mouse-embryo-from-stem-cells-for-first-time—**Chapter 8**

https://www.viacord.com/references/—**Chapter 10**

http://blogs.discovermagazine.com/d-brief/2017/04/25/artificial-placenta-lamb/#.WuB6QMgpBgo—**Chapter 10**

https://courses.lumenlearning.com/boundless-biology/chapter/bone/—**Chapter 14**

https://www.nature.com/articles/ncb3510#auth-14—**Chapter 16**

https://www.cdc.gov/hiv/pdf/library/reports/surveillance/cdc-hiv-info-sheet-diagnoses-of-HIV-infection-2016.pdf—**Chapter 22**

https://www.cdc.gov/std/stats16/infographic.htm?s_cid=tw_SR_17003#STDreport—**Chapter 22**

https://muse.jhu.edu/article/458046—**Chapter 23**

https://www.researchgate.net/publication/232061401_When_is_the_Capacity_for_Sentience_Acquired_During_Human_Fetal_Development—**Chapter 23**

Index